Occult and scientific mentalities
in the Renaissance

Occult and scientific mentalities in the Renaissance

Edited by
Brian Vickers
Centre for Renaissance Studies
ETH Zürich

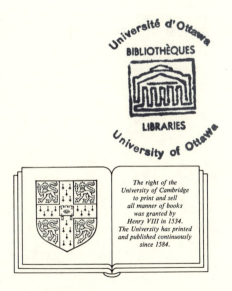

The right of the
University of Cambridge
to print and sell
all manner of books
was granted by
Henry VIII in 1534.
The University has printed
and published continuously
since 1584.

Cambridge University Press
Cambridge
London   New York
New Rochelle   Melbourne   Sydney

105558

Published by the Press Syndicate of the University of Cambridge
The Pitt Building, Trumpington Street, Cambridge CB2 1RP
32 East 57th Street, New York, NY 10022, USA
296 Beaconsfield Parade, Middle Park, Melbourne 3206, Australia

First published 1984

Printed in the United States of America

Library of Congress Cataloging in Publication
Main entry under title:
Occult and scientific mentalities in the Renaissance.
Includes index.
1. Occult sciences – History – Addresses, essays,
lectures. 2. Science, Renaissance – Addresses, essays,
lectures. I. Vickers, Brian.
BF1429.O26 1984 133'.094 83–15116
ISBN 0 521 25879 0

BF
1429
.O26
1984

For
Joseph Needham

CONTENTS

Contents   viii

CONTRIBUTORS

Robin Briggs has held fellowships at Balliol College and (since 1964) at All Souls College, Oxford, where he is a senior fellow and university lecturer in modern history. He has published *Early Modern France, 1560–1715* (Oxford, 1977) and is currently writing a book about the witches of Lorraine.

Stuart Clark took his first degree at the University of Wales and his doctorate at Cambridge. Since 1967 he has been a lecturer in history at the University of Wales, Swansea. He has published essays on Bacon's *Henry VII* in *History and Theory* (1974), on King James's *Daemonologie* in *The Damned Art: Essays in the Literature of Witchcraft*, ed. S. Anglo (London, 1977), and, in *Past and Present*, on witchcraft (no. 87) and on popular culture (no. 100). He is completing a book on learned witchcraft beliefs in early modern Europe.

Nicholas H. Clulee studied at the University of Chicago and since 1971 has been teaching at Frostburg State College, Maryland, where he is associate professor of history. He has published articles on John Dee's mathematics in *Ambix* (1971) and on his optics in *Renaissance Quarterly* (1977).

Mordechai Feingold took his first degrees at the Hebrew University of Jerusalem, and his doctorate at Oxford; he is now junior fellow at the Harvard Society of Fellows. His publications include *The Mathematicians' Apprenticeship: Science, Universities and Society in England, 1560–1640* (Cambridge, 1984) and *An Elizabethan Intellectual: John Rainolds, the Man and His Library* (forthcoming). He is currently working on intellectual patronage in England and the professionalization of scientific investigation, especially as it involves science and religion.

Judith V. Field read mathematics, astronomy, and history of science at the universities of Cambridge, Sussex, and London (Imperial College), and is now employed at the Science Museum, London. She has published articles on Kepler's star polyhedra in *Vistas of Astronomy*, 23 (1979); on his cosmological theories in *Quarterly Journal of the Royal Astronomical Society*, 23 (1982); and on cosmology in Kepler and Galileo in *Novità celesti e crisi del sapere* (*Convegno Internazionale di Studi Galileiani*), ed. P. Galluzzi and W. R. Shea (in press). She is contributing to the translation of Kepler's *Harmonice mundi* with Eric Aiton and Alastair Duncan.

William L. Hine's university degrees include the bachelor of science and doctorate in the history of science from the University of Oklahoma, and a degree (juris doctor) in law. He has held research posts at the Sorbonne and at Mansfield College, Oxford, and now teaches at York University, Downsview, Ontario. He has published essays on Mersenne in *Isis* (1973, 1976) and *Renaissance Quarterly* (1976), and has an essay, "Kircher and Magnetism," forthcoming in *Athanasius Kircher*, ed. J. Fletcher (Wolfenbüttel, 1984).

Ian Maclean took his first degree and doctorate at Oxford, where he is now university lecturer in French and fellow of The Queen's College. He has published *Woman Triumphant: Feminism in French Literature 1610–1652* (Oxford, 1977) and *The Renaissance Notion of Woman: A Study in the Fortunes of Scholasticism and Medical Science in European Intellectual Life* (Cambridge, 1980). He has edited and contributed to *Montaigne: Essays in Memory of R. A. Sayce* (Oxford, 1982).

Lotte Mulligan studied at the universities of Melbourne, Adelaide, and London (Westfield College), and has been reader in history at La Trobe University, Melbourne, since 1972. Her many articles on seventeenth-century politics and science include "Civil War Politics, Religion and the Royal Society," *Past and Present*, no. 59 (1973); "Anglicanism, Latitudinarianism and Science in Mid-Seventeenth-Century England," *Annals of Science*, 30 (1973); "Puritans and Mid-Seventeenth-Century Science: A Critique of Webster's Thesis," *Isis*, 71

(1980); and (with Glenn Mulligan) "Reconstructing Restoration Science: Styles of Leadership and Social Composition in the Early Royal Society," *Social Studies of Science*, 11 (1981).

Graham Rees took his doctorate at the Shakespeare Institute of the University of Birmingham and since 1974 has been senior lecturer in the history of science at the Polytechnic, Wolverhampton. His articles on Francis Bacon's natural philosophy have appeared in *Ambix*, 22 (1975) and 24 (1977), and in *Annals of Science*, 37 (1980) and 38 (1981). An essay on Bacon's concept of *spiritus vitalis* will appear in the *Atti del IV colloquio internazionale del Lessico Intelletuale Europeo*, ed. M. Fattori.

Edward Rosen received his doctoral degree from Columbia University in 1939. After teaching for more than half a century in the City University of New York, he was designated distinguished professor emeritus of the history of science. He has taught also at Indiana University and the Massachusetts Institute of Technology. His publications include *Three Copernican Treatises*, 3rd ed. (New York, 1971); *Kepler's Somnium* (Madison, 1967; Pfizer medal, History of Science Society); *Kepler's Conversation with Galileo's Sidereal Messenger* (New York, 1965); *Nicholas Copernicus Complete Works*, I (London/New York, 1972), II (Baltimore, 1978), III (in press); and *The Naming of the Telescope* (New York, 1947). He was the recipient of a Festschrift, edited by Erna Hilfstein and others: *Science and History: Studies in Honor of Edward Rosen* (Warsaw and New York, 1978; *Studia Copernicana*, XVI).

Brian Vickers was educated at Trinity College, Cambridge, and taught in the English faculty for ten years before moving to Zürich in 1972, where he has held chairs at both the university and the ETH. His publications include *Francis Bacon and Renaissance Prose* (Cambridge, 1968); *Classical Rhetoric in English Poetry* (London, 1970); and *Towards Greek Tragedy* (London, 1973). He has edited *Shakespeare: the Critical Heritage, 1623–1801*, 6 vols. (London, 1974–1981) and *Rhetoric Revalued* (Binghamton, N.Y., 1982). He is writing a book on the rejection of the occult sciences between 1580 and 1680.

Richard S. Westfall took his undergraduate and doctoral degrees at Yale University and taught history at the California Institute of Technology, the State University of Iowa, and Grinnell College, Iowa, before moving to Indiana University in 1963, where in 1976 he was designated distinguished professor of history and of history and philosophy of science. His publications include *Science and Religion in Seventeenth Century England* (New Haven, 1958); *Force in Newton's Physics* (London, 1971); *The Construction of Modern Science* (New York, 1971; Cambridge, 1977); and *Never at Rest: A Biography of Isaac Newton* (Cambridge, 1980).

Robert S. Westman took his undergraduate and graduate degrees in history at the University of Michigan and in 1971 moved to UCLA, where he is now professor of history. He edited *The Copernican Achievement* (Berkeley and Los Angeles, 1975) and is editing, with David C. Lindberg, *Reappraisals of the Scientific Revolution*. Forthcoming from Chicago University Press is a book-length study, *The Copernicans: Universities, Courts, and Interdisciplinary Conflict, 1543–1700*. His articles include "Kepler's Theory of Hypothesis and the 'Realist Dilemma,'" *Studies in History and Philosophy of Science*, 3 (1972); "The Melanchthon Circle, Rheticus and the Wittenberg Interpretation of the Copernican Theory," *Isis*, 66 (1975); and "Magical Reform and Astronomical Reform: The Yates Thesis Reconsidered," in *Hermeticism and the Scientific Revolution* (Los Angeles, 1977).

EDITOR'S PREFACE

The essays collected in this volume were originally given at a symposium that I organized in June 1982 at the Centre for Renaissance Studies of the ETH, Zürich. The Eidgenössische Technische Hochschule was founded in 1855 on the model of Napoleon's Ecole Polytechnique at Paris and in the wake of similar foundations at Berlin, Vienna, Munich, and Stuttgart, all of which were designed to supplement the arts curriculum of the older universities with teaching and research in science, technology, and architecture. Because the ETH has had a department of the humanities from its foundation (the first professor of art history was Jacob Burckhardt; Francesco de Sanctis held the first chair of Italian), and because its modern luminaries include both Einstein and Jung, it may be thought a not inappropriate setting for a conference on the relations between science and the occult. Whether or not the genius of the place exerted an influence on the proceedings is a question that had better be left open. At all events, the discussions were extremely lively and were marked by frequent challenging references to the texts (delegates seemingly happening to have with them copies of Thomas Aquinas, Newton, Cornelius Agrippa, and others). Among those who took a valuable part in the discussion, but who are not represented in this book, I should like to thank J. E. McGuire (University of Pittsburgh), Richard Gordon (University of East Anglia), G. A. J. Rogers (Keele University), and Keith Hutchison (University of Melbourne).

Contributors were chosen with an eye to balancing distinguished historians of science with less well-known scholars in a variety of subjects: mathematics, chemistry, astronomy, philosophy, history, English and French literature, and the history of universities. In the Introduction I have summarized the arguments of the various chapters and tried to show how they relate to the general historical issue under dispute. The contributors share a unity of purpose, namely, to define the issue more accurately than hitherto, but otherwise represent individual viewpoints. At the end of the Introduction I have developed an original suggestion of how the occult sciences can be better understood by an approach through anthropology, but this is to be taken as a personal opinion not necessarily shared by the others.

It remains only to thank the president of the ETH, Professor Heinrich Ursprung, and the director of research projects, Dr. Eduard Freitag, for the interest and encouragement they have shown. All who took part will want to thank my indefatigable assistant, Mrs. Ilse New-Fannenböck, for taking care of everyone and everything.

The volume is dedicated to a historian of science for whom we all feel the greatest admiration and gratitude.

B. V.

Introduction

BRIAN VICKERS

The scholars who took part in this symposium addressed themselves to a topic that has been much discussed in the history of science in the past twenty years. The extent to which the two great realms of "magic" and "science" – to give them their traditional names – influenced each other during the Renaissance is a fascinating and exciting question. One can distinguish, perhaps, three main stages in its elaboration so far. In the first the history of science was seen as a narrative of progress through inventions and discoveries, an ever-improving movement toward positive knowledge. In this history of scientific triumphs, magic and the occult could be simply dismissed as entertaining but irrelevant. Even as late as 1957 Herbert Butterfield, in *The Origins of Modern Science*, felt no qualms about dismissing the occult tradition and its historiographers in the most sweeping terms. Van Helmont, we are told,

> made one or two significant discoveries, but these are buried in so much fancifulness – including the view that all bodies can ultimately be resolved into water – that even twentieth-century commentators on Van Helmont are fabulous creatures themselves, and the strangest things in Bacon seem rationalistic and modern in comparison. Concerning alchemy it is more difficult to discover the actual state of things, in that the historians who specialise in this field seem sometimes to be under the wrath of God themselves; for, like those who write on the Bacon–Shakespeare controversy or on Spanish politics, they seem to become tinctured with the kind of lunacy they set out to describe.[1]

Such comments may have raised a laugh among undergraduates in the Cambridge history faculty (where the book originated), but they seem unworthy of a serious historian. Butterfield gives no documentation,

1

so it is hard to know whom exactly he had in mind, but when we consider that the authors who could have been aimed at include Water Pagel, John Read, E. J. Holmyard, Joseph Needham, and J. R. Partington, it reveals a sadly closed mind. Few people write the history of science like that anymore.

Butterfield is a rather late representative of the first phase of discussion. The second phase had been inaugurated by Lynn Thorndike, in his *History of Magic and Experimental Science*, the genesis of which dates back to his 1902–3 Columbia master's thesis on magic in medieval universities.[2] Published in eight volumes between 1923 and 1958, Thorndike's work, which is more in the nature of a detailed chronological survey than a critical history, did more than any other book to establish the occult as a serious subject of study. In the course of its publication a certain change of emphasis can be seen: an increasingly favorable attitude to the occult, and a corresponding lack of sympathy with its critics (Thorndike's summaries of Renaissance critiques of the occult are not always fair). Also visible, it must be said, is a certain lack of familiarity with the development of nonoccult science, and I note this not in reproach – who could ever hold the whole of such a vast field in his head? – but rather to account for the sense of imbalance that one feels when his history reaches the period of Galileo and Kepler.[3] By faithfully immersing himself in the vast occult tradition, it seems to me, Thorndike lost contact with the many changes that were taking place in the sciences and came to judge them from the point of view of the occult.

If Thorndike inaugurated this second phase by making a more positive claim for the occult, the claim was made still more strongly by scholars who had worked not in the field of experimental science but in more general history of philosophy and history of art. By "experimental science" I mean the traditions – slow to evolve, no doubt, but none the less real – of experiment, empirical observation, quantification, mathematical analysis, as seen in such sciences as physics, statics, dynamics, mechanics, astronomy, mathematics, optics, whose history can be traced back, with interruptions, to the Greeks. Obviously a "pure" science cannot be neatly separated out from a general philosophical context until the late Renaissance, and I do not mean to claim that this was the only, or the most important, part of science or that the history of science in the Renaissance can be studied solely in these terms, for this would be to take us back to the bad old days of phase one. I simply point to the historical fact that when the second phase of discussion reached its peak in the 1950s and 1960s, in which the occult sciences were assigned a greater status, indeed at times were said to have made "the scientific revolution" possible, these claims were not made by those scholars who had worked in the history of the mathematical-experimental sciences, such as Alexandre Koyré, Otto Neugebauer, E. J. Dijksterhuis, Anneliese Maier, Marshall Clagett, Edward Rosen, I. B. Cohen, Owen Gingerich, Edward Grant, and others. Rather, they were made by – or based on the work of – scholars who had studied Renaissance philosophy or the history of art, such as P. O. Kristeller, Eugenio Garin, Paolo Rossi, Frances Yates, Cesare Vasoli, Paola Zambelli, and others. Common to this group is an extensive knowledge of Renaissance philosophy, especially Florentine Neoplatonism, as represented by Ficino and Pico della Mirandola, and the more eclectic traditions represented by Giordano Bruno or Cornelius Agrippa.

In Sonnet 111 Shakespeare writes of the influence of an environment which can be so strong that
> almost thence my nature is subdu'd
> To what it works in, like the dyer's hand.

Similarly, I suggest, those scholars who worked in Renaissance Neoplatonism, a school of thought that had always been welcoming to occult, especially magical, ideas, tended to see the whole of Renaissance science from that viewpoint. The term "hermeticism" came to be used as a holdall for the occult sciences, a way of thinking of which Hermes Trismegistus became a convenient symbol, albeit as anachronistically now as in the Renaissance. For just as Isaac Casaubon (and others before him) exploded the myth that Hermes was as ancient as Moses, so, in 1949, A. J. Festugière showed that the varied collection of texts ascribed to Hermes, and dating from the second to third centuries A.D., represents a fusion of astrology, alchemy, numerology, and magic that is entirely derivative, part of the common mental stock of Hellenistic thought, and has no special distinguishing features.[4] Further, the influence of the hermetic texts was small in comparison with that of the main occult sciences, and their presence in Renaissance philosophy makes for just one more syncretic ingredient in an already syncretic mixture. As Charles Schmitt has said, in a valuable survey of this issue,[5] too many users of the term fail to realize that

> it was Hermeticism which became assimilated into Neoplatonism and seldom, if ever, was Hermeticism itself thought of, even by its Renaissance proponents, as an independent system of ideas. It was Neoplatonism which served as a strong trunk onto which ideas derived from Hermetic, Orphic, Zoroastrian, Neopythagorean, Cabalistic and other sources could be grafted during the Renaissance, continuing a tendency already begun in antiquity. (p. 206)

Although Neoplatonism was the receptive body, it was also the organic whole that sustained these accretions. It was "the Neoplatonic system of metaphysics and epistemology which provided a life-giving sap to hold it all together'' (p. 206). The claim for hermeticism's importance in Renaissance science has been critically reviewed by Robert Westman and J. E. McGuire, who have effected substantial reductions and delimitations of its status.[6] As Schmitt concludes, ''Hermeticism never becomes a real driving force of any significant cultural movement during the Renaissance,'' and where hermetic texts are cited, as they are even by Aristotelians, it is often as a ''rhetorical embellishment to help substantiate arguments put forward for other reasons'' (p. 207). The name of Hermes, one could say, has a pleasantly exotic or glamorous ring about it, like that of Orpheus or Pythagoras; any of them can be used to give an air of ancient wisdom and authority.

The kinds of arguments made on behalf of the occult in this second stage of discussion can be conveniently sampled from an essay published by the late Frances Yates in 1967, called ''The Hermetic Tradition in Renaissance Science.''[7] This begins by affirming that the ''core'' of Neoplatonism was ''Hermetic, involving a view of the cosmos as a network of magical forces with which man can operate'' (p. 255). ''The Renaissance magus . . . exemplifies that changed attitude of man to the cosmos which was the necessary preliminary to the rise of science'' (p. 255). The texts of Hermes Trismegistus, as interpreted by Ficino and Pico, create ''man as magus . . . with powers of operating on the cosmos through magia and through the numerical conjurations of cabala'' (p. 257). More than this, because Miss Yates believed that ''the Renaissance magus was the immediate ancestor of the seventeenth-century scientist,'' she concluded that the Neoplatonism of Ficino and Pico ''prepared the way for the emergence of science'' (p. 258). To these pioneers Miss Yates added other syncretist occultists, such as Cornelius Agrippa, whose advocacy of Pythagorean numerology is said to constitute ''an operative use of number'' (p. 259); Tommaso Campanella, who apparently classified mechanics as '' 'real artificial magic' '' (p. 259); Fabio Paolini, in whose mind she diagnoses ''a basic confusion . . . between mechanics as magic and magic as mechanics'' (he thought the *anima mundi* inspired the movement of clocks: p. 260); and John Dee, whose mixture of traditional mathematics and mysticism in his preface to Billingsley's translation of Euclid (1570) is said to make that work ''greatly superior . . . as a manifesto for the advancement of science'' to Bacon's *Advancement of Learning* (p. 262). Dee is said to have become ''imbued with the importance of mathematics'' precisely because ''he was an astrologer and a conjuror, attempting to put into practice the full Renaissance tradition of Magia and Cabala'' (p. 262). It is this ''Hermetic attitude toward the cosmos,'' Miss Yates wrote, which ''was, I believe, the chief stimulus of that new turning toward the world and operating on the world which, appearing first as Renaissance magic, was to turn into seventeenth-century science" (p. 272).

Even on that brief summary it is evident that Miss Yates blurred fundamental distinctions, as between mathematics and numerology (both share an "interest in number," yet in wholly opposed ways;[8] whether you think the numbers used by Agrippa and Dee were "operative" depends on whether or not you believe they were able to conjure angels or control material forces by the use of arithmology). As for the presentation of the case, it is significant to what extent Miss Yates relied on affirmations of personal belief, assertion not argument, generalization without instance, lack of reference to any counterthesis, and such rhetorical tricks as repetition (the word "new," for instance), denial of other opinions ("impossible to deny"), and a whole series of rhetorical questions that insinuate ideas without ever adequately exploring them ("Is it possible that Bacon avoided heliocentricity because he associated it with the fantasies of an extreme Hermetic magus, like Bruno? And is it further possible that William Gilbert's . . . magnetic philosophy of nature . . . also seemed to Bacon to emanate from the animistic philosophy of a magus, of the type which he deplored?": p. 268). One is reminded of the opening of Bacon's own essay, "Of Truth": "'What is truth?' said jesting Pilate, and would not stay for an answer." We could wish Miss Yates had stayed a little longer, sometimes, for the answers that have been given to the questions she raised have not always been those that she implied.

The reaction to Yates's thesis (restated in her later work, but with little additional evidence) inaugurated what I shall term the third phase of our dispute. On the one hand, it found supporters, such as P. M. Rattansi, A. G. Debus, P. J. French, and others, who agreed that the occult had a formative influence on the new science.[9] On the other, many of her assumptions and arguments were challenged, by M. B. Hesse, Edward Rosen, Paolo Rossi, Charles Trinkaus, and others.[10] The debate has been vigorous, yet it has neither led to any detailed discussion of the main issues nor provoked a thorough reexamination of the texts. Charles Schmitt has commented on the tendency of the proponents to emphasize their case "without bringing to light any compelling new information" (p. 205), being apparently "unwilling or unable to go back to a fresh reading of original sources" (p. 207). Some of Miss Yates's opponents have relied on mere rebuttal, not subjecting the argument to sharp scrutiny, or have linked it with unrelated issues, such as the internalist versus externalist approach or whether the history of science should be written backward or forward. (It has always seemed to me that any form of history must be simultaneously diachronic and synchronic and that the internalist or externalist approaches can be complementary, but need not be so; that is, an internal analysis of a scientific work is just as legitimate an occupation as a study of its sources or of its influence or of its author's social situation, and none of these is at fault unless it claims to present a "total history.")

It would be disingenuous of me not to make clear that I find the Yates thesis almost wholly unfounded. From what we knew of Renaissance Neoplatonism before her book, *Giordano Bruno and the Hermetic Tradition*, it was clear that that philosophy was the last stage in a continuously syncretic tradition and that the only new thing about it was the extent of its synthesis. The cabala could be absorbed by Pico because it was itself an eclectic occult science with features common to Neoplatonism, such as the use of numerology and hierarchical categories. There was nothing new about the use of talismans or magic or correlative thinking: All these can be traced back to Greek sources and earlier. As for the "turning towards the world," since many Neoplatonist texts, and indeed some of those in the hermetic corpus, describe matter as the source of evil, then for many disciples it was a question of turning away from the world rather than toward it. At all events the Neoplatonists, like all occultists, were never interested in matter for its own sake or in general terms. Nature had value to them either as a symbol system, as in hierarchies of descent from the godhead or in degrees of purity, or else as an adjunct to human health or longevity. We do not find the Neoplatonists studying the behavior of falling bodies, taxonomizing plants, or dissecting the human body simply to find out why these things are as they are. Their lack of interest in the physical world goes along with a positive distaste for quantification: Symbolic arithmetic, attributing moral values to numbers, was acceptable, but anything to do with measurement or computation was rejected as mundane or ephemeral. Agrippa, Bruno, and Fludd show this rejection of mathematics and quantification particularly clearly.

Yet, although the issue was wrongly formulated by Frances Yates and her followers, it remains an important and challenging topic. The occult had a long and widely diffused influence, in parallel – as I see it – with the nonoccult sciences, and it seems essential to anyone wanting to understand the Renaissance to try to evaluate what debts, if any, the two traditions owe each other. The title of this book, in the word "mentalities," places the emphasis where I believe it should be put: on two traditions each having its own thought processes, its own mental categories, which determine its whole approach to life, mind, physical reality.

The studies collected here belong, in part, to this third phase of discussion. Some of them are still concerned with putting right issues that were misrepresented by the Yates thesis. For instance, in Chapter 1 Nicholas Clulee takes up the case of John Dee, who has been claimed (by Boas, Yates, and French) to have transformed magical and her-

metical ideas into an "experimental science." Clulee examines Dee's
central science of "Archemastrie" and reveals the several long traditions behind it. Dee's concept of "experimental science" goes back
to Cusanus and Roger Bacon; his knowledge of magic and talismans
goes back to Arabic sources, especially Avicenna; his *Science Alnirangiat* turns out to derive from a most arcane source, the *Ars sintrillia*
of Artephius. In other words, this side of Dee's work belongs to the
occult sciences, which relied on a cumulative philological tradition, a
series of texts whose doctrine was handed down substantially unchanged for hundreds or thousands of years. Clulee's closer study of
these sources reveals that they are occult magical texts, involving divination by the manipulation of reflecting surfaces, catoptromancy,
which is related to what Dee called his "optical science." Dee was a
distinguished mathematician, as is generally recognized,[11] but in this
area he linked himself to various occult traditions whose persistence
unchanged in the Renaissance disproves Frances Yates's claim that
the "magic" that supposedly helped found a new science in the Renaissance was a new form of magic. Dee's magic is here shown to be
"in no way novel in substance"; it "was not any uniquely Renaissance
or 'hermetic' variety but medieval and Arabic." That in turn can be
shown to derive from Hellenistic sources, and since, as Festugière has
shown, there was nothing new in Hermes's magic, we are confronted
again with one of the most interesting, and least studied, phenomena
of the occult, namely, its resistance to change. If Dee's magic was not
helpful to the growth of science, neither was his presentation of it,
since he deliberately cultivated obscurity, another long-standing feature of the occult which also serves to distinguish it from the nonoccult
sciences. Dee, it might be said, attempted to fuse two incompatible
attitudes to reality: Interpreters of his work do not have to follow him.

In Clulee's chapter we find a careful reconstruction of a historical
context, with a concern for the original texts that extends even to
examining Dee's own copies of his books and his marginalia. Mordechai Feingold's work on Renaissance universities is marked by a
similar awareness of the importance of primary texts; indeed, no one
has ever searched the archives of Oxford and Cambridge more thoroughly. His contribution (Chapter 2) sets out to disprove one of the
supporting arguments in Frances Yates's thesis about the "dangerous"
nature of the occult and its liability to persecution. (In her 1967 essay
she had defined what she called the "Rosicrucian type" of scientist,
who "tends to have persecution mania," and this need to discover
persecution and oppression seems to have been one of the organizing
schemes in her very dynamic form of historiography.[12]) In the present
case her argument, based largely on Giordano Bruno's treatment at
Oxford in 1583, was that Oxford and Cambridge were not only sticks-

in-the-mud about the new sciences but were actively hostile to occult sciences, to the extent that they were not taught and that to be found studying them was shameful. As for the universities' attitude to the nonoccult sciences, the old idea that Tudor mathematics derived its strength from the practical arts and not from the universities was disproved years ago by W. P. D. Wightman,[13] and Feingold shows that mathematics did not disappear from the curriculum in the late sixteenth century; rather, the contrary. Bruno's unpopularity may have been due to his contentious manner, and certainly his cavalier handling of Copernican texts – texts well known to his audience – cannot have endeared him to Oxford scholars.[14] Oxford was not hostile to Copernicanism or to Platonism or to the occult; indeed, Feingold's survey of the period 1558–1619 reveals a remarkable open-mindedness and tolerance on this issue. The occult sciences were tolerated for private study, provided students did not cast the monarch's nativity, debase coins, or practice witchcraft. The universities enjoyed the same intellectual freedom as did the country at large, and the occult sciences figured in university disputations, were the subject of informal teaching, with much exchange of books, and even an accepted center at Gloucester Hall. But, as in the wider realm, we find some exceptions: Respondents were expected to argue against the occult sciences, and some students got into trouble for practicing them. We find the same disillusionments, too, as are so poignantly recorded elsewhere. Henry Briggs, a foremost mathematician, as a young man "thought it . . . a fine thing to be of Gods Counsell, to foreknow secrets," and thus prepared himself for "the search of Judiciall Astrology: But there he found his expectation frustrate, there was no certainty in the rules of it." An astrologer whom he consulted admitted that "the Rules of that Art were uncertaine indeed," so Briggs returned to mathematics, eventually becoming Savilian professor of geometry. The fact that the occult could be discussed so openly proves Feingold's contention that the early modern universities did not persecute it, and in view of the absence of evidence that has characterized so many accounts of the occult many will agree with his conclusion that only a prosopographical approach will take us further.

We have some admirable examples of such an approach in recent studies of the Royal Society, expecially those by Michael Hunter and Theodore Hoppen, the latter having shown for the first time the degree to which the leaders of the new science in England could still use concepts and categories deriving from the occult.[15] The problem for modern historians is to understand how such men were able to operate simultaneously within two traditions that have become generally recognized as incompatible since, say, the first generation after Newton. But the existence of two traditions, and their mutual incompatibility, had been an accepted historical fact for some critical minds since the 1580s. The development of a critical rationalistic mentality in the sixteenth century has been studied in great detail by Keith Thomas, in a book yet to be taken notice of by historians of the occult.[16] In my own contribution to this symposium (Chapter 3) I have studied one aspect of the divergence between the two traditions: their attitude toward metaphor and symbol. In the scientific tradition metaphor and analogy are given a subordinate role, whether heuristic or explanatory, in a discourse that is primarily nonmetaphoric and that draws a clear distinction between the literal and the metaphorical. In the occult tradition this distinction breaks down: Metaphors (such as the microcosm and macrocosm) are taken as realities, words are equated with things, abstract ideas are given concrete attributes. This tendency is particularly marked in Neoplatonism, and I have drawn on the admirable studies by P. O. Kristeller, D. P. Walker, and E. H. Gombrich to define the occult's tendency to reify images and to use this as a way of distinguishing it from the nonoccult sciences.

The occult discourse is essentially symbolic:[17] In whatever discipline – astrology, alchemy, numerology, or magic – nature is significant not in itself but as a system of signs pointing to another system of mental categories. Objects, plants, stones, planets are given various attributes (good/evil, pure/impure, male/female) and fitted into a system of operations that, far from being addressed to a disinterested study of nature, returns again and again to a self-centered concern with the individual's welfare. The typical questions the practitioner of the occult asks include: Will I be happy? Will I be rich? How can I avoid bad luck, or ill health? How can I live long? In this desire to anticipate the future, or "be of Gods Counsell," as Henry Briggs put it, the occult symbol system carries a great interpretative responsibility, since the manipulation of the symbols is supposed to correspond with events in reality. When the occult came under sustained critical inquiry in the late sixteenth century the element that received sharpest attention was precisely its reliance on symbols taken as equivalent to realities. Many of these attacks circle around the figure of Paracelsus, who was seen to be extreme in his fusion of literal and metaphorical, he and his followers being frequently attacked for treating analogy as if it were identity. These terms sound modern and anachronistic, but in fact they are the precise formulations of Erastus, Libavius, Francis Bacon, Sennert, and Van Helmont, all of whom attacked the Parcelsians on this score. Such distinctions were perfectly clear to Renaissance thinkers, who had a far more intense education in logic and rhetoric than any modern historian.

The controversies between the occult and the experimental sciences (for each side was aware of the threat to its existence posed by the other) undoubtedly had the effect of making the new science more conscious of itself.[18] They had an effect in the early seventeenth century analogous to that of the manifestoes and research programs produced later, which were still concerned to outflank and disarm the occult.[19] By studying these polemics we can define at first hand the ways in which the Renaissance scientists understood their own project. In Chapter 4 William Hine analyzes the views on magic of Marin Mersenne, who has been recognized since Robert Lenoble's study as a test case for the opposition between science and the occult.[20] In his polemic against Vanini and Francesco Georgio, Mersenne rejects many of the assumptions of the occult sciences, partly in terms of religious orthodoxy, but also from a critical rationalist point of view. He rejects Neoplatonist magic, invokes Casaubon's clinching disproof of the antiquity of Hermes Trismegistus, and attacks the occult's reliance on association or correlation. Ficino had claimed that the effects music has on individuals derive from association with planetary constellations: Mersenne replies that music affects us all in the same way, whatever the constellation. (Mersenne was, of course, an important pioneer in the scientific analysis of music.[21]) The occult had traditionally believed in the association of certain metals and stones with certain planets, but Mersenne replies that neither reason nor experience justifies such correlations. One way in which the occult broke down distinctions between the conceptual and the physical was to claim that written characters added to an object would give it special powers: Mersenne rejects this, too, together with numerology, the cabala, and the idea that the Hebrew alphabet had special powers. For him, manipulating letters and anagrams was a mere permutation, with no symbolic dimension and no operative effect on the world.

Mersenne rejects much of the conceptual structure of occult science, the whole analogical-correlative method, its symbolism, its confusion of mental and physical worlds. These issues figure in many of the polemics against the occult, such as Libavius against Croll, various writers against Paracelsus, and – most famous of all such confrontations – Kepler against Fludd. In Chapter 5 Robert Westman takes up this opposition, well known since the Jungian interpretation by the Swiss physicist, Wolfgang Pauli,[22] and throws new light on it. Fludd's refutation of Kepler – a sentence-by-sentence refutation, all too typical of Renaissance controversy, where the opponent's text had to be chewed up and spat out, a fragment at a time – reveals Fludd's visual, pictorial epistemology, depending in part, as Westman shows, on a consciously formulated esthetic, including hitherto unnoticed debts to Albrecht Dürer. One might add that this tendency to think in images, with a corresponding inability or reluctance to use abstractions, marks other occult scientists, notably Paracelsus,[23] and may be typical of that tradition as a whole. Kepler, by contrast, believed that the principles defining the structure of reality are picturable only in a certain sense. What is entirely lacking from the Fludd mentality is any interest in measurement or in testing an analogy against data derived from experience, and in this respect Kepler's assumptions and methods are wholly different. The crucial issue is the relationship between pictures, words, and things. Fludd starts with ideas and pictures, finds words to describe them, and then links this composite to reality. Kepler, who deals with reality in terms of geometry, rejects Fludd's analogies as visual or rhetorical, never capable of demonstration and often arbitrary. A modern analysis confirms that Kepler has given an accurate description of Fludd's mentality and that his setting of Fludd's tradition as apart from, and antithetical to, his own is justified.

The further attraction of Westman's chapter is that he has reconstructed the background to the modern historian of science who first discussed Fludd, discovering that Pauli's dreams had been analyzed by Jung himself. I write these words in the building where both worked, and where this symposium was held, and I congratulate Mr. Westman on not only discharging the rhetorical topos of "allusion to the place" but of uncovering a fact that sheds light – albeit in a rather special case – on the mentality of those who write the history of science. The further significance of Jung's concept of the mandala or circle containing the quaternities of the human psyche is that it is itself an example of the occult methodology of correlating preexisting categories or superimposing matrices on each other, a practice that can also be seen in Pauli's dreams. Whether this proves the existence of archetypes, or merely shows that Jung and Pauli had steeped themselves in occult literature, it is now clear for the first time how well qualified Pauli was to mediate the dispute between Kepler and Fludd.

Less well known, but also important for our understanding of Renaissance science, was the polemic of Julius Caesar Scaliger against Girolamo Cardano. Cardano, lists of whose accomplishments[24] tend to make him sound like Dryden's Zimri ("A man so various, that he seem'd to be / Not one, but all Mankind's Epitome," who, "in the course of one revolving Moon, / Was Chymist, Fidler, States-Man, and Buffoon"), published *De subtilitate* in 1551, which Scaliger made the subject of mockery in his *Exotericae exercitationes de subtilitate* of 1557. Ian Maclean's study of this polemic (Chapter 6) makes a number of significant points for the historiography of science: first, that modern writers tend to focus on the openly experimental literature and neglect the vast amount of humanist science, as general philosophy, Aristotelian physics and Galenic medicine, which not only formed the minds of nearly all the experimental scientists who had a university education, but still represents – for those with the knowledge and philological expertise needed to make use of it – a virtually untapped source of information about Renaissance intellectual attitudes. In the case of Scaliger's book the neglect is all the more surprising because it was widely used in European universities: Galileo knew it, and Kepler recorded in his *Mysterium cosmographicum* (1621) that he had once been "fascinated . . . by the teachings of J. C. Scaliger on the motory intelligences."[25] Maclean's analysis of the publishing details of the two books also illuminates the sociological implications of scientific publication: Cardano's book appeared in a popular context, printing being shared by half a dozen Paris booksellers, while Scaliger's appeared under the aegis of a reputable publisher of university text books. (More studies are needed on the type of science published by Oxford and Cambridge university presses in this period.)

As for the mental world view, the two books cannot be neatly separated because the writers have more in common than Kepler and Fludd. This is partly a question of temperament (since Cardano is never as wholly mystical as Fludd) and partly the effect of their university educations, which inculcated some standardized habits of thinking and writing. Their central difference concerns the concept of subtlety, which Cardano locates in substances, accidents, representations, as something sensible and intelligible, yet comprehended only with difficulty. To Scaliger, however, subtlety is sited not in nature but in the human mind, a distinction between nature and man's perception of nature that looks forward to the division of primary from secondary qualities so crucial to the new sciences, as developed by Galileo, Descartes, and Locke. Cardano, though seemingly in the vanguard of sixteenth-century science by his rejection of Aristotelianism, is in fact more old-fashioned or less than scientific. He claims that experience is the only trustworthy authority, yet under experience he includes hearsay – he is nearer to Bacon and Henry More in this than to Galileo. In invoking five principles, three elements, and two qualities he seems to be taking an anti-Aristotelian, antioccult line, but in fact he is only making an idiosyncratic selection from the existing categories. Idiosyncratic, too, is his attitude to the correlations so fundamental to occult thought. Where Mersenne rejects them altogether, Cardano rejects some but invents others, in a wholly arbitrary way. He correlates metals, colors, tastes, and planets because, in his system, they happen to have the same number – that is, he defines four or seven items in each set and then interlinks or, rather, interequates them. Much of occult science, if I may sum up the conclusions of my own researches, is built out of purely mental operations, the arrangement of items into hierarchies, the construction of categories that become matrices for the production of further categories. Far from being a science of nature, or even of man, it comes to seem more and more like a classification system, self-contained and self-referring.

From these studies of the disputes between the two traditions, high-level intellectual pugilism, it can be seen that the distinction between occult and nonoccult was widely understood and employed in the sixteenth and seventeenth centuries. Yet, as everyone knows, precisely those scientists who delivered sharp and acute attacks on the occult mentality – Bacon, Kepler, Mersenne, Van Helmont, to name those we have so far encountered – themselves retain many instances of occult beliefs and thought habits. How are we to handle this contradiction? Since this may come to be seen as the major problem in the historiography of Renaissance science, it is not surprising that no one has yet answered it with any certainty. I would suggest three simple responses, all of which are displayed by the contributors to this volume: first, not to ignore or deny its presence (there is no longer any sense in denying that Newton did alchemy, and it is mere self-deception to call what he did chemistry); second, to base our analysis on a firsthand knowledge of the primary texts and to pay attention to unpublished manuscript material; third, to refine and deepen the intellectual models we use in attempting to understand this phenomenon. We are no longer prone to describe Hildegard of Bingen's picture of the universe as the outcome of migraine, as Charles Singer did, and to think, with Arthur Koestler, that Kepler's attempt to correlate the five regular solids with the distance between the planets was a form of paranoia.[26] The fact is that Renaissance scientists were able to operate for a while, at least, in two finally incompatible traditions.

For some historians, however, even to suggest that there were two distinct traditions would be to indict sixteenth- and seventeenth-century scientists as having suffered from schizophrenia. In order to complete the rehabilitation of the occult, its proponents claimed that there was one central, unified tradition, the division of which into two is the product of purely modern attitudes. Walter Pagel has said of William Harvey that while it is agreed that Harvey "laid the foundations of modern physiology and made possible the development of medicine as an applied science, through his discovery of the circulation of the blood," to acknowledge this is to see Harvey from the standpoint of modern science, which implies "a selection of what is relevant today or in the light of the development after Harvey's death."[27] Yet Harvey's discovery of the true nature of the circulatory system was surely the main significance of his work for his contemporaries, or for people in 1700 or 1850, just as much as today. Clearly, what Pagel is really attacking is the reluctance to deal with Harvey's "non-modern aspects":

At best a kind of split mind is admitted: Harvey is presented
as a dweller in two worlds, that of Aristotle whom himself
acknowledged as his master and that of modern science to
which he contributed one of its revolutions.

And yet it would appear that there was a time when what
sounds contradictory today was no contradiction. Unification of what is today sound and relevant with its apparent
opposite must have been possible in the same mind which
yet somehow retained its integrity and power. (Pagel, p. 1)

This is, I think, an entirely accurate account of what seems to us the
peculiarity of some minds in the seventeenth century, their ability to
live in what Sir Thomas Browne called "divided and distinguished
worlds."[28] The very hesitation with which Harvey recorded his hunch
that the blood returns to the heart unconsumed and goes out again all
the time ("I began to think whether there might not be a motion, as it
were, in a circle. Now this I afterwards found to be true")[29] shows
that emancipation from accepted views did not necessarily come easily.
The coexistence of a new system with parts of the old one that it is
designed to replace is surely a normal stage in the process of education
or intellectual change. People do not throw out their ideas or concepts
or categories overnight, as they might clear out a cupboard. It seems
to me unrealistic to expect a black-white separation; indeed, Pagel
himself subsequently refers to "the suture lines which join and unite
the naturalist and the philosophical aspects in Harvey's work."[30]

In his important studies of Paracelsus and Van Helmont, Pagel has
effectively recognized the existence of two traditions, writing that in
many areas "Paracelsus presents a tangle of observations and speculations – partly contradictory and fantastic – from which some sound,
progressive and even modern ideas emerge."[31] He notes in Paracelsus
"the strange and intimate blending of sound scientific principles with
a system of magical and fantastic analogies," a mixture of "sound and
judicious principles" of medicine with astrology, a union of "unrealistic theories" with others having "a sound observational component,"
which makes it "difficult to separate the empirical and these 'cosmological' components of this theory." Yet he feels sure that the "progressive aspects of his work . . . emerge from a mantic and cosmological system which is removed from scientific medicine."[32] This final
metaphor of Paracelsus's progressive ideas "emerging from" a mantic
system – one organized around signs and prophecies – obviously begs
the question of influence and tradition. But this question has been
begged so often, and so often by metaphors, that we need to remember
that it has yet to be demonstrated. Wolfgang Pauli wrote, in terms very
typical of an earlier German tradition in which intellectual history could
be written in terms of signposts, "from . . . to . . ." (as in Nestle's *Vom*

Mythos zum Logos),[33] of a "truly scientific way of thinking" emerging only in the seventeenth century, one which "grew out of the nourishing soil of a magical-animistic conception of nature."[34] Such a simple linear model would be very convenient, if only it were true.

The problem remains, however, of doing justice to the coexistence of "what sounds contradictory today." As Pagel notes, we find it hard to conceive that a seventeenth-century thinker could entertain such opposed views. Harvey, the paragon of scientific medicine, persisted in the Paracelsian-Helmontian practice of applying to a tumor "the hand of a man dead of a lingring disease," which experiment, Robert Boyle records, "the doctor was not long since pleased to tell me, he had sometimes tried fruitlessly, but often with good success."[35] What is at issue is the concept of "unification," as invoked by Pagel. We might prefer the less question-begging term "coexistence," since "unification" implies that these different thought worlds were unified to the degree that neither was aware of the other, which is clearly not the case (at times, one feels, Renaissance critics of the occult could detect its presence in others' minds but not in their own). We need, rather, level-headed discussions of this issue, not in the hectoring vein of P. M. Rattansi, who protests that "to call Newton an alchemist is to split him into 'rational scientific' and 'irrational Hermetic' selves which have nothing in common."[36] And since Newton insisted on "banishing causal hypotheses from 'experimental philosophy' and [conceived] its task as that of formulating quantitative laws by the rigorous analysis of phenomena," then, Rattansi suggests, in addition to this "'hard-headed' and phenomenologically-inclined Newton [we] would have to invent another Newton, a mystical and crankily fundamentalist one . . . in order to explain the theological and alchemical studies which absorbed so much" of his life.[37] Since Newton spent so much of his life studying alchemy, in all the traditional ways, it seems to me entirely proper to call him an alchemist, without getting involved in a value judgment of whether the activity was "rational" or "irrational" – terms that have in any case outlived their usefulness for this discussion. And there is no need to "invent another Newton," whether or not we call him a crank, for the Newton who made over a million words of notes on both alchemy and religious studies certainly existed. That is one fact; another fact is that he chose neither to disclose these occupations nor to publish his findings. These activities are all performed by "the same Newton," but the question is whether the same parts of his mind are engaged in each activity. Professor Rattansi seems to be denying that he could perform any of them without this influencing or determining the others. We must at least entertain the hypothesis that Newton, like other human beings, could devote himself to different activities, each for its own sake, without them all having to be seen as forms of the same activity, or each indelibly affecting the rest.

As against such reductivism I would cite the more measured approach by R. S. Westfall to what he describes as the "interaction of the two traditions," mechanical and hermetic, in Newton's thought, where he diagnosed a constant "degree of tension between the two": "The animism and the active principles of the Hermetic tradition might be disguised; they could not be wholly assimilated into a mechanical system as the 17th century understood it."[38] Whatever our agreement with Westfall's specific argument, the terms within which it is presented seem more reasonable than Rattansi's. Several scholars have noted instances of Renaissance scientists simultaneously inhabiting two distinct traditions. R. P. Multhauf showed that Paracelsus attempted to combine a Neoplatonist cosmology with chemical attitudes, but after Van Helmont "this enterprise seems to have ended with the chemical and mystical aspects of Paracelsian thought taking different directions."[39] Van Helmont was able to combine experimental biology, chemistry, and medicine with Neoplatonism and mysticism. As Lester S. King puts it: "From our viewpoint he combined incompatibles. We today feel an incongruity between chemistry and neoplatonism . . . between mysticism and experimentation. But Van Helmont was not aware of any inconsistency, and indeed from his viewpoint, none existed."[40] True though this may be for Van Helmont's own work, he was abundantly aware of the inconsistencies in Paracelsus's thinking, uttering a running commentary of the most scathing kind against his predecessor's use of analogy and imaginary classificatory schemes in place of empirical observation.[41] The paradox of Van Helmont is that he existed simultaneously in two separate traditions, which at times he played off against each other, invoking mystic experience to denounce the limitations of reason as used in the university logic schools, while invoking clinical experience to destroy occult astrological-botanical medicine. His amphibious nature was evident to his contemporaries, for when Hermann Boerhaave prefixed a "History of Chemistry" to his *Elementa chemiae* of 1732, he divided "the major early chemists into four distinct classes or schools: (1) those he calls the 'systematical Writers,'" who "reduced the operations of chemistry to the form of systems," especially for the preparation of chemical remedies – his list "shows a progressive emancipation from Paracelsus"; (2) "the metallurgical chemists"; (3) "alchemical writers, among them Paracelsus and Sendivogius"; "and (4) the 'chemical improvers of natural philosophy' a class into which he put Robert Boyle." Faced with "the difficulty of classifying Van Helmont, with his amalgam of experimentalism and mysticism, Boerhaave lists him both with the alchemical writers and the 'improvers of natural philosophy.'"[42] Such a pragmatic acceptance of the coexistence of the two traditions seems to me a model modern historians might consider more seriously.

The simultaneous presence of occult and nonoccult – or even antioccult – tendencies in the Renaissance scientist can be seen here in the studies devoted to Kepler, Bacon, and Newton. Kepler's involvement with astrology has attracted very diverse comment, often wholly undiscriminating.[43] One of the virtues of Chapter 7 by Edward Rosen is that it sets out very clearly Kepler's commitment to the practice of astrology, in particular the compilation of horoscopes, with his simultaneous drastic delimitation of the scope of the art. He attacked astrologers for attempting to cast the nativity of a whole year and denied that a person's future could be predicted from his horoscope, since many other factors needed to be taken into account. A similar rationalist attitude governs his denial that the superiority of the wine, say, in one country can be put down to astral influence: The relevant factors here include sun and geographical position. If Kepler sounds like Mersenne in his rejection of occult explanations for the effect of music, he shares with Mersenne the desire to reject arbitrary divisions, such as that of the heavens into twelve houses, or arbitrary correlations, such as those between the zodiacal signs and the human limbs, or between Saturn and the moon as creating cheats. As he wrote to Harriot in 1606, "Ten years ago (1596) I rejected the division [of the heavens] into twelve equal parts, the houses, the dominations . . . keeping only the aspects and transferring astrology to harmonics." As well as rejecting much of the system, Kepler drew attention to the social, psychological, and political facets of astrology, its role as the solver of problems in everyday life: "For since [people] ask many questions, the astrologer thinks of a way to give many answers" – that is, astrology has a way of covering all eventualities. Yet it is not an independent scientific system: In politics, or at court, it can be "induced to say what pleases both sides" and can exploit people's gullibility. Professor Rosen does not attempt to disguise the fact that Kepler retained belief in some parts of astrology and attempted to integrate them into a mathematical cosmology and astronomy, but he brings out the complementary antioccult attitudes, an epistemological and conceptual critique that makes Kepler something quite other than an orthodox Renaissance Neoplatonist.

Even sharper evidence of Kepler's deliberate and conscious distancing of himself from the occult tradition is provided by his response to numerology. In the mystical arithmology derived from Pythagorean and pseudo-Pythagorean sources, some whole numbers (initially 1 to 10) were granted symbolic attributes by a series of manipulations and internal distinctions. If even numbers are symbolized as female, odd numbers as male (the visual thinking behind such symbolism will become clear if one represents 4 by four dots in the form of a square, then makes 5 by an additional dot placed in the center of the square), then the sum, or better still, the product of an odd and even number could symbolize the union of man and woman. So 5, or 6, can be a "marriage number." The number 7, since it cannot be so combined, can then represent chastity and be equated with a virgin goddess, such as Artemis. This whole system, in which numbers function in a self-contained symbolic grid, had a long life in occult and eclectic speculative philosophical traditions, but was never taken up by mathematics proper, and in the revival of mathematics in the Renaissance was the subject of concerted attacks by Italian mathematicians from Tartaglia to Galileo.[44]

Judith Field's study of Kepler's attitude to numerology (Chapter 8) starts from his rejection of Rheticus's justification for there being in Copernican astronomy six, not seven, planets (the moon was now considered a satellite), on the grounds that six is a perfect number according to the Pythagoreans. Kepler's reply is that numerology, being the work of man, is later in Creation than the universe and cannot be used to explain the work of God. Further, he distinguishes pure (that is, abstract or undimensioned) numbers, the symbolic integers of the numerologists, from numbers that derive from measurement, the tools of astronomers and physicists. The first type he called *numeri numerantes* ("counting numbers") as opposed to *numeri numerati* ("counted numbers"), and his attitude to the first type, throughout his life, was total rejection. Only the second type could be used in science because they referred to empirical reality, not to a human-produced symbolism. Kepler's astrology and music theory depend on *numeri numerati* in the form of musical ratios among the arcs into which the circle of the zodiac is divided by bodies that are at aspect to one another, ratios that are to be expressed in geometrical, not arithmetical, form. Kepler wanted to prove God a Platonic geometer, not a Pythagorean numerologist, so the occult science is rejected from the outset. The real gap between Kepler and Fludd is brought out by Dr. Field's analysis of their attitude toward number and harmony. Kepler points out that Fludd's harmonies ignore actual units and use abstract symbolic numerical relationships, whereas he finds musical ratios among quantities measured in the same units, such as the extreme angular speeds of planets as seen from the sun. This is a classic demonstration of the incommensurability of the two traditions, as can be seen from Fludd's reassertion of the occult view, during which it becomes clear to us that Fludd's "geometry" is in fact purely symbolic pattern making, opposing symbols of light and darkness. Fludd also still followed the cosmology of Sacrobosco, in which the spheres of the planets were given an equal thickness, another a priori pattern imposed on physical reality. "Kepler complained that

Fludd was concerned only with his own concept of the world." There is no way in which these attitudes could be reconciled, nor can they be seen as belonging to the same tradition. Kepler consciously allied himself to a scientific tradition deriving from Ptolemy; Fludd asserted his allegiance to Hermes Trismegistus.

Kepler was perfectly aware that Fludd represented the extreme pole of the occult, a system resistant to any of the new ideas arising from sixteenth-century science. Francis Bacon was equally conscious of that extreme, and on several occasions delivered swingeing attacks on the alchemists, magicians, Paracelsians, and others. He also developed a vast program for the new science, a mixture of perceptive criticism of stagnant intellectual traditions and a call for reform involving observation, experiment, cooperation, and the establishment of a scientific method. While historians of science are no longer prone to hail induction as a great tool for scientific research, Bacon's achievement, and above all his timing, as a proponent of scientific reform, was considerable.[45] Yet, as has been evident since 1953 and an essay by Lynn Thorndike,[46] Bacon had much more in common with the occult tradition than we might expect, given the terms in which he attacks it. Graham Rees, who has done more than anyone to clarify this area of Bacon's thought, presents in Chapter 9 the latest of a number of important recent manuscript discoveries that illuminate further this vitalist-animist world view. Because modern students of Bacon have never paid much attention to the chronology of his scientific work, and have concentrated on the more carefully finished parts of his system, such as the *De augmentis scientiarum* (1623) and the *Novum organum* (1620), it used to be thought that the presence of animist ideas in a work like the *Sylva sylvarum* (1626), whose thousand paragraphs (arranged in ten "centuries") include a remarkable mishmash of observation, experiment, and uncritical or hearsay legend and marvel, could be ascribed to the haste with which Bacon had put this work together in the few years remaining after his public disgrace. (Disconcerting evidence of Bacon's prestige as a scientist, or of the early seventeenth century's hesitation over what constituted true experimental method, is that when John Wilkins referred to the sections of the *Sylva* he called them "Experiment 731" and so on.)[47]

Now Dr. Rees has shown that Bacon's speculative, biological-qualitative philosophy of nature goes back much earlier than used to be thought, and that he had worked out a philosophy of terrestrial change as early as 1611–12. The working out, though, was more a question of synthesis than of original experiment or discovery, and the underlying concepts bear obvious similarities to Galenic notions of spirit and Neoplatonist concepts of the astral body. Dr. Rees shows that Bacon's matter theory conceived of two families of quaternions, qualitatively related substances arranged on a scale of opposites, with intermediates being both animate and inanimate. This is obviously a classification scheme imposed a priori, not developed by observation and experiment, and has much in common with the Greek binary systems fundamental to Aristotelian and Galenic biology, as in the four elements/four humors theory.[48] It belongs, equally, to the qualitative methods that continued in the life sciences long after the new physics had successfully replaced qualities with quantities.[49] Bacon conceives of the process of aging as a battle between the vital spirits and the inanimate spirits, the latter being the main agents of change in the terrestrial realm, but in this manuscript only he suggests that vital spirit is elaborated from the inanimate. Such a breakdown of fundamental distinctions between animate and inanimate is typical of the occult tradition, of course, and totally at variance with the new sciences, which insisted on separation and clear boundaries. Dr. Rees suggests that Bacon did not regard his inductive-axiomatic method as the exclusive, omnicompetent tool that it has become for some historians, and I would agree that from the fragmentary nature of much of Bacon's work,[50] its sense of grandiose designs never being fulfilled in real terms, then the *Novum organum*, far from being the crown of his oeuvre, was perhaps only an intermediate methodological excursus. Yet Bacon's failure to use or develop his speculative biological ideas later could also suggest a lack of faith in them, and his leaving them in manuscript may have been a decisive self-criticism. The more we know about this "speculative philosophy," the more accurate our picture of Bacon will be, even though its coexistence with his rigorous logical methods based on observation and experiment suggests not so much a unified whole as a radical incoherence – or the simultaneous acceptance of incompatibles.

With Newton the presence of such diverse strands may present an eternally insoluble problem. Indeed, recent attempts to show that the alchemical ideas can or must be integrated with the physics and optics – as if their coexistence in Newton's mind would otherwise be a threat to our sanity, if not to his – may be fundamentally misguided. Why should Newton be incapable of researching into biblical chronology, composing alchemical treatises, and pursuing the mathematicization of physics, all in the same year or month? This may offend our concept of rationality, but it evidently did not bother him – at any rate, not as pursuits; their publication or publicizing was another matter. The zeal to discover a single organizing key to Newton's activities is, actually, anachronistic, unhistorical, a product of late-twentieth-century belief in a "unified" scientific mentality. We have one – he must have had one.This search for a key to the whole may be just a phase in Newton studies, as is suggested by the development of J. E. McGuire, who in

1966 joined with P. M. Rattansi in arguing for a substantial element of hermetic thought in Newton's scientific achievement.[51] In 1974, however, in a lecture surveying what had become of the hermeticist claim in the interim, McGuire concluded that "the traditions of magic and alchemy did not play a significant role in shaping Newton's conception of nature" and that even when Newton was invoking Christian hermeticism as a possible legitimizing factor in the 1690s, this was only for a short time and was an activity in any case much removed from "the magical world-picture of the Hermetic writings."[52]

A similar cooling off can be seen in the work of R. S. Westfall, one of the leading modern authorities on Newton.[53] In 1972 he could write that "the Hermetic elements in Newton's thought were not in the end antithetical to the scientific enterprise," but were "wedded together." In 1975 he could even question whether modern historians have not "mistaken the thrust of Newton's career," since he devoted over thirty years to his alchemical studies: "To us, the *Principia* inevitably appears as its climax. In Newton's perspective, it may have seemed more like an interruption of his primary labor."[54] This suggested turning of the history of science upside down did not take place, however. Longer acquaintance with Newton's alchemical work, and the discipline of preparing a magisterial 400,000-word biography, led Westfall to a recognition of the frustrations Newton experienced in his alchemical studies, leading to a decisive disillusionment in 1693, near the time of Newton's breakdown.[55] Anyone who has studied Newton's alchemy will record that much of it is entirely traditional, philological, based on extensive knowledge of a wide range of sources that Newton tried to integrate – in vain, one must report, given the ever-increasing size and ever-less-definitive makeup of his "Index chemicus" – into a unified system, and whose teachings he followed out in experiment. Many of the characteristic methods of the occult sciences can be found: the reification of symbols, words turned into things and allegories, numerological classification, the correlation of preexisting categories, the desire for secrecy, and the development of cryptographic systems. Whereas all these are traditional features, Newton showed a more original scientific attitude (as Kenelm Digby had done a few decades earlier)[56] by applying quantitative techniques, yielding far greater accuracy in measurement – an incongruous mixture of occult and experimental traditions, one may feel, analogous to Kepler's rigorously geometrical astrology. One of the main issues still at stake, as Professor Westfall shows with exemplary lucidity in Chapter 10, is the source of Newton's concept of attraction. Westfall rehearses the undeniable evidence that Newton studied alchemy, collected books and manuscripts, performed experiments, applied to this heterogeneous material the systematizing attitude of the new science and its concern for quantitative accuracy. The resulting argument is that in such passages as his description of a "secret principle of un-sociability" in nature, to explain why liquids and solids do not mix, or in his development of a "more concrete notion of force" to explain attraction, Newton was drawing on alchemical ideas or, rather, fusing them with elements from the mechanical philosophy.

Clearly this is a complex issue, and we need more analyses of the concept of force in both traditions. As the argument stands at present I have yet to be convinced. Is it the case, for instance, that Newton's concept of a life force animating matter is necessarily alchemical? It is vitalist, certainly, and belongs to a biological concept of matter that goes back to the Greeks and that we associate more with the occult "panpsychic" tradition than with exclusively alchemical sources. True, Newton rejected the mechanical philosophy's passive concept of matter and aligned himself with an alternative tradition, but does that mean that his specific debt is to alchemy? As for the verbs in which Newton expressed his "perceptions of spontaneous activity" in chemical reactions, rather than proving the specific influence of "the alchemical concept of active agents," it seems to me that this is more the consequence of the anthropomorphism endemic to the whole occult tradition: Substances "lay hold" on each other, "carry up" another, indeed marry, copulate, give birth, die, are resurrected. I am not sure that this can be linked with anything significant in Newton's dynamics.

Above all, as I have already suggested, I wonder whether the question has been properly posed. Professor Westfall in effect asks "whether Newton's alchemy was an activity isolated from the rest of his natural philosophy or whether it exerted an influence on his work in physics." It seems to me that another position is possible; namely, that it was perhaps not "isolated" (note the question begged here) from the rest of his work, but that it did not necessarily exert an influence on that work. The undeniable fact that Newton wrote half a million words on alchemy in the seven or eight years following his *Principia* still does not prove that he "regarded his alchemical endeavours as an harmonious part of his total philosophical programme"; indeed, the very concept of a "total philosophical programme" may be anachronistic. As Professor Westfall has shown, Newton surrounded his alchemical activities with obscurity and exacted secrecy from his collaborators. He was content to leave all his work in manuscript, and his editors and commentators connived at the concealment. Perhaps he was holding back on announcing his alchemical work until he had achieved a success with it that would satisfy his own standards of scientific accuracy, cogency, the mutual cohesion of theory and experiment – a breakthrough that never came. Perhaps disappointment was the cause of his abandonment of alchemy in the 1690s. His silence about his alchemical studies leaves a vacuum from which we cannot extract much certainty other than that of a disavowal. Perhaps alchemy was more of a mystical experience for him, a process of illumination rather than an experimental goal. The tone in which Newton records his achievements resembles the visionary more than the scientist: "May 10 1681 I understood that the morning star is Venus and that she is the daughter of Saturn and one of the doves. May 14 I understood ⟵ [the trident?] . . . May 18 I perfected the ideal solution. That is, two equal salts carry up Saturn. Then he carries up the stone and joined with malleable Jove." On July 10, 1681: "I saw sophic sal ammoniac." "Friday May 23 [1684] I made Jupiter fly on his eagle."[57] These were discoveries, or experiences, that even his closest associate, Humphrey Repton, did not know about. The mystic experiences of the alchemist were meant to be available to the adept alone, not to the vulgar world.

In other words, Newton himself differentiated his alchemy from his mathematics, physics, and optics. The one activity was individual, private, resulting, perhaps, in purification and illumination; the other was social, public, based on demonstrable propositions and mathematical argument, designed to be published for the good of mankind, not the benefit of the adept, and to serve some objective concept of truth. Everyone knows the scorn with which Newton attacked the negative side of hypothesis as a merely personal fantasy; equally biting were his attacks on the principle of infinite regress created by resorting to "occult Qualities" as an explanation for phenomena in nature: "To tell us that every Species of Things is endow'd with an occult specific Quality by which it acts and produces manifest Effects, is to tell us nothing."[58] Perhaps Newton's final silence over alchemy is not unrelated to his failure to find the one fundamental process that could create the metamorphosis of matter. Alchemy and biblical criticism may have been occupations for the diversion of an unresting intellect, but not subjects in which he professed competence and expected to make a living or career. His silence is eloquent, at least, as to what the occult sciences were not, for him.

Yet, although Professor Westfall has not wholly convinced me, his arguments, and the widely culled evidence, are presented so clearly and honestly, without either prestidigitation or bullying, that those who would either develop or challenge the argument know exactly where they must begin.

Of all the aspects of the occult, that connected with demonology and witchcraft seems the most difficult to come to terms with. While alchemy, astrology, and numerology were all self-contained and mutually reinforcing systems, with some social consequences, none of them impinged on life and death in the way witchcraft did. Astrologers or alchemists ran the occasional danger of having their windows smashed or their books burned, but they never had to endure systematic persecution stretching over twenty years and more. Studying the other occult sciences is difficult enough, given the proliferation of that literature and its frequent repetitiveness, but nothing is quite so disheartening, I find, as studying the literature of demonology. The records of the trials are unspeakably depressing, as much for the delusions of the victims as for the prejudices of the prosecutors, a collusion that may have given the witch a sense of importance for a while but that ended horribly. As an intellectual and social phenomenon, too, witchcraft is harder to understand than any of the other occult sciences and has provoked an extensive controversy.[59] Many more factors seem to be involved than with alchemy or astrology – social, legal, religious, psychological – and many of them are rather hard to pin down. In inviting three contributions on witchcraft to this symposium I have tried to balance differing but complementary approaches.

In Chapter 11 Robin Briggs draws on extended first hand knowledge of a remarkable archive of witch trials in Nancy between 1580 and 1630, amounting to over two hundred dossiers. Their particular significance is that they preserve the earlier stages of the trials, the accusations, shedding much light on popular, as opposed to learned, attitudes. The majority of the accused were poor women, most of them over forty, who had had a long local reputation of *maléfice*, actual harm to neighbors and animals. While they had been tolerated for many years, a sudden dispute, followed by a misfortune, could precipitate a prosecution. Frequently the accused began their defense with a pathetic confession of how they had been tempted by the devil, diabolic pacts being a key feature of popular beliefs. Dr. Briggs challenges the distinctions sometimes made between Catholic and Protestant attitudes, finding little difference in terms of ideas of personal responsibility. His analysis supports the model worked out by Alan Macfarlane and Keith Thomas on English material, by which the persecuted person is one to whom charity has previously been refused. Yet, while we can see, in Durkheimian terms, that witchcraft accusations might be a way of bringing errant individuals to heel, social pressure could result in more pressure, as the subject retaliated with malice or violence against his or her oppressors. This vendetta situation, familiar from family conflict in small communities, took on terrible implications when it was moved to a court of law in which the prosecutor had almost unlimited power. One of the saddest points made by Dr. Briggs is that, luckily enough, *curés* did not take a very active part in instigating persecution of witches: Had they done so there would have been many more trials. At least the *curé* could not reveal the secrets of the confessional.

If we take this body of trials from Nancy as a representative sample, it is striking how often the same patterns recur. It is as if the formal possibilities of the witch trial are limited: However diverse the individuals, they all fit into a restricted set of roles. Analysis of trial proceedings in linguistic terms would seem appropriate; indeed Richard Gordon attempted such an analysis, using roughly contemporary English material. (As Robin Briggs says, it is easier "to understand witchcraft beliefs and persecution synchronically than it is diachronically": After some unsatisfactory sweeping surveys, covering several centuries and countries, perhaps what is needed are careful studies of more limited material.) In his paper (not, unfortunately, available for publication) Dr. Gordon discussed the linguistic structure of witch accusations, using English material from the late sixteenth century. The basic narrative contains a certain kind of speech act and records an event of suffering. Drawing on Jeanne Favret Saada, *Les Mots, la mort, les sorts: la sorcellerie dans le Bocage* (Paris, 1977; English trans. *Deadly Words* [Cambridge, 1981]), Gordon stressed that the crucial act in witchcraft is located in the word and that the spoken word constitutes power. The witch accusations are direct recollections of observed utterances that parallel on the verbal plane the sense of abnormality, or breaking of expectations, found in the events themselves. Witchcraft accusations include illicit linguistic utterances, such as asserting the future as a fact, and a suppression of connectives (sentences being joined merely by the copula "and"), so creating a refusal to explain the event. In "paraded limit narratives" (the term is from Pierre Bourdieu, *Outline of a Theory of Practice* [Cambridge, 1977] and refers to the discrepancy between statements and practice), the name of the witch or magic is sometimes not mentioned, setting up an area of indeterminacy. This can be broken by the witch's self-declaration, for what she wants out of others is the ascription of power, and by this ascription she can get social recognition. In a curious act of collusion with her persecutors, the witch, through complicity, gains publicity. In declaring herself the witch enters into a contract with society: "I claim power – you give me recognition – perhaps you kill me."

As for the suffering event narrated, it involves some deviation from the patterns of the natural world, a structural inversion of normal health or prosperity for which no immediately visible reason exists. (Here witchcraft resembles the wider function of the occult sciences, which offer an alternative system of understanding the world. Everyday disasters can be explained, made sense of, as the work of human instruments or a diabolic agency. The pattern is causal, but needs to find a human embodiment of the invisible and intangible.) Witches are seen to exist in a reversed world or to represent the invasion of the wild into the domestic space, as in their "familiars," such as cats, who are both domestic and wild. In socioreligious terms the witch sets up a pollution drama which only the authorities can de-pollute. (This sense of evil as a concrete substance needing purification again links witch-craft to the wider occult system, in which words can be reified into substances, or spirit coagulate as matter, a breakdown of categories that other systems keep clearly distinct.) Paradoxically, the witch gains social credit by an act in which she destroys herself. One might add a further reflection on the curious mutual dependency of witch and ac-cusers on a system of ideas: She needs their suspicion to become the center of attention; they need her collusion to sustain their roles; both parties need the system of ideas to justify the whole activity.

The linguistic structures of witch trials can be related to their larger narrative structures. Using the techniques of Propp or Souriau,[60] one could say that the basic roles are limited: witch, with or without helper(s); victim; victim's dependants or property. Further, many of the trials record a simple sequence of action and reaction: Witch re-quests alms, or services; victim refuses request; witch avenges refusal. A typical instance would be the following, from the trial at Saint Osyth in 1582:

> The said Joan saith, that in summer last, Mother Mansfield
> came unto her house and requested her to give her curds.
> She saith that answer was made that there was none, and so
> she departed. And within a while after some of her cattle
> were taken lame and could not travel to gather their meat.[61]

The denial can be not only of a gift but of a loan: One family refuses to lend Joan Robinson a hayer, since they need to use it themselves, "and presently after there arose a great wind which was like to have blown down their house" (p. 154). The denial can even be of a wish to buy land or animals (pp. 154–5); whatever the favor denied, it ret-rospectively becomes the cause of the disaster. The causation is ex-pressed in a narrative reduced to its absolute minimum, similar to that described by the narrator in a novel by Russell Hoban: "A story is what remains when you leave out most of the action; a story is a co-herent sequence of picture cards. *One*: Samson in the vine yards of Timnal; *Two*: the lion comes roaring at Samson; *Three*: Samson tears the lion apart."[62]

Yet, however minimalist, the narrative structure highlights a human relationship, here that between suppliant and donor. The witch is the suppliant, a normal role in narrative as in life, but the victim is one who has rejected the role of donor. In denying alms, or help, or even refusing to sell a commodity at a market price, he or she violates a principle of social order and is apparently punished by the witch, with the aid of natural and supernatural powers. Violations of any system of reciprocity are almost always problematic. Marshall Sahlins has distinguished three types of reciprocity: "generalized," when the donor gives without expecting a return; "balanced," when the recipient returns the gift; and "negative," when the recipient does not repay the gift or takes instead of waiting to receive.[63] Considering the witchcraft evidence, we might add a further type of negative reciprocity: the denial of the gift by those in a position to help the poor or needy. The roles of witch (frustrated recipient) and victim (the nondonor) are reciprocal, but in a purely destructive sense. The ascription of malice to the witch may be a transference of the unwilling donor's sense of guilt; yet the prosecution of the witch, while condemning witchcraft, does not condemn the denial of the gift. The social order and the moral order are differently interpreted by witch and victim. There is evidently nothing legally wrong in turning away an old woman begging, but it can have evil consequences. To Blake's exhortation, "Then cherish pity, lest you drive an angel from your door,"[64] we can add the rider, "or a witch."

The last of the three papers on witchcraft, Stuart Clark's wide-ranging inquiry into the scientific status of demonology (Chapter 12) links up with Chapters 4 and 6 in reconstructing Renaissance attempts to make distinctions within the occult and scientific traditions. All three chapters show that these distinctions are considerably more complex than might have been expected and that within the areas of natural and demonic magic no simple categories apply. Orthodox demonology could embrace natural scientific explanations of occult phenomena without thereby doubting the existence of witches and demons. The central characteristic of demonic phenomena was that they were extraordinary, often prodigious, puzzling events conceived of as having "no certain cause in nature." Investigation of such events – in which the devil and his agents were believed to have the power of simulating changes in nature – raised a whole series of issues concerning the natural, the marvelous, the difference between illusion and reality, issues that overlapped with scientific concerns, as we see from the continuing interest in witchcraft shown by seventeenth-century scientists, from Bacon to Glanvill and More. Dr. Clark clears up some modern misconceptions about demonology. First, that the original texts concentrated exclusively on the sensational aspects of witchcraft belief, such as the demonic pact and sabbat: In fact the writers examined any dubious phenomenon that might have been demonically caused. The resulting spread of interest, from natural magic to alchemy, astrology, mechanical marvels, and many other occult phenomena, means that we should integrate demonology into its whole cultural context, as an attempt to define the borderline between the natural and the demonic. Second, where some modern historians have divided attitudes to witchcraft into either belief or scepticism, the fact is that most Renaissance writers were able to operate with two models that seem to us incompatible.

From the great variety of texts cited by Dr. Clark it is evident that a wider and more complex range of explanations was open to Renaissance demonologists than has sometimes been thought. Yet their common goal, to expose the limitations and deceptions of the devil, resulted in a shared language and a shared methodology. Despite their widely differing backgrounds and nationalities, these writers shared a common strategy of exposing the almost unlimited powers of the devil to deceive and delude. While the devil was denied the power to create fresh forms or change the essential character of existing forms, he was granted the ability to simulate such changes, and discussions of lycanthropy, for instance, were devoted to pinpointing how such simulations were brought about. The maintaining of this strict division within the devil's powers meant that his acts were limited to the natural realm, could not overrule the powers of nature, and were therefore denied the status of the supernatural or miraculous. (Compare Mersenne's very similar argument.) Thinking in terms of modern concepts of the supernatural can only confuse the issue, which Renaissance writers defined as "quasi-natural" or "preternatural," drawing on a much wider concept of nature than our own. Dr. Clark's comment on this fundamental difference supports my analysis of the problem of "coherence" or "unification" that twentieth-century historians have read back into the minds of seventeenth-century scientists: "The question we have to ask, therefore, is not the one prompted by rationalism – why were intelligent men able to accept so much that was supernatural? – but simply the one prompted by the history of science – what concept of nature did they share?" The concluding section of this chapter shows conclusively that in their concept of nature a concern in demonology was entirely concomitant with an interest in science.

As Dr. Clark shows so well, the intellectual history of the Renaissance cannot be written in wholly modern terms. We have to make a continual effort of historical reconstruction, an imaginative displacement out of our concerns, categories, concepts, even vocabulary, into theirs. So much is evident to anyone who has ever studied the history of the English language, where the form of words persists but their meanings have been transformed beyond naive recognition. The traps involved in assuming that what they meant by "nature" or "science" or "experiment" or "enthusiasm" or "virtue" or "pleasure" is what we mean by these terms (or what some of us mean) ought by now to be universally apparent. As Lotte Mulligan's contribution (Chapter 13) shows, "reason" is another of these protean words. *Recta ratio* is, of course, a Stoic idea, and the English seventeenth century has to be seen in the context of a continuous debate over such concepts that had been carried on since classical antiquity,[65] a debate given additional contemporary significance by the divisions between religious sects and political groups. Like Ian Maclean and Stuart Clark, she finds that the opposed camps defined by modern historians had, in fact, much in common. Many different writers agreed that reason was a God-given faculty of the mind through which man could come to know both the creation and the Creator, and to which faith or revelation were complementary. Attacks on "reason" are often attacks on scholastic syllogistic reasoning or the rigid logic of the university curriculum.

Given these fundamental points Professor Mulligan shows that those modern historians (P. M. Rattansi, Charles Webster, Christopher Hill) who have posited a radical discontinuity in attitudes toward reason between the mid and the late seventeenth century, have misrepresented the issue. Rather than a shift, as Rattansi has it, from an "illuminist, fideistic, hermetic strain" of the 1640s and 1650s to "the empirical, rational, mechanical philosophy" of the Royal Society, she is able to show that both strands persist throughout the period – indeed coexist in the same writers. Walter Charleton, one of Rattansi's test cases, never rejected *right* reason (close analysis of the context shows that Rattansi interprets his quotation from Charleton to mean the opposite of what it actually says), and while Charleton denied that natural ratiocination can provide knowledge of God, he did not reject reason as the proper means to study men and nature. Again, where Rattansi alleges that Charleton shifted from a hermetic to a scientific world view because of his awareness of the social danger of occultism (which would, in any case, put the crudest self-seeking or paranoid motives on his change), the fact is that while he certainly embraced with enthusiasm the principles and discoveries of the new sciences he continued to use the occult concepts of macrocosm and microcosm, signatures, and the alphabet of nature. It is evident that the history of seventeenth-century thought has been overdramatized, turned into a series of momentous changes, such as the scientific revolution, with a whole panoply of apostates and renegades, persecutors and witch hunts. One can only agree with Professor Mulligan's reminder that we should "give due weight to evidence of continuity," for this will help us to understand "how it was possible for seventeenth-century writers to hold at the same time two or more – to us incompatible – models."

The phenomenon of coexistence of incompatibles, the frequently hybrid nature of much seventeenth-century science, can be glimpsed again in the work of John Webster, Baconian and anti-Aristotelian while simultaneously Fluddean and Boehmian. His *Metallographa* [*sic*] of 1671 is an orthodox history of metals, partly observational and partly philological, whose title page also promises to divulge "Mystical Chymistry, as of the Philosophers Gold." Samuel Gott's *Nova Solyma*

(1648) contains several "languages" which might seem incompatible to us, Baconian, hermetic, scholastic, Platonic, Christian. A similar eclecticism runs through the work of other seventeenth-century writers, notably Thomas Vaughan and Henry More, who have been too sharply polarized by modern historians. Vaughan was a professed eclectic, believing that natural philosophy and theology, medicine and alchemy were all inseparable. For him reason needed supplementing by illumination, but knowledge was to be gained by hard work, not by mystic insight. More, whose presentation of himself as the proponent of reason has misled unwary historians into thinking him a "rational" mind in modern terms, while he attacks Vaughan for using metaphor (a common abusive trick in the controversies of this period),[66] himself uses parables and occult analogies (the macrocosm and microcosm), believes in the doctrine of signatures, and espouses Platonic mysticism, visionary enthusiasm, and the Christian cabala.

It is by now clear that despite many real differences, seventeenth-century writers, whatever their political or religious allegiances, spoke much the same language, shared many concepts and categories. Professor Mulligan defines a spectrum of the uses of "right reason," from radical sectarians at one end to orthodox Anglican casuists at the other, and several contributors to this volume have used the metaphor of a spectrum or continuum to describe the spread of attitudes they have found. I believe this to be a more accurate conception than those that put seventeenth-century thinkers into wholly separate groups, or have them experiencing drastic and total conversions, or divide the period into "radical discontinuities." I would go on to make a further point: While it is essential for us to understand the issues at stake, and in the terms in which they were presented and understood, it is important not to accept those terms uncritically. Not everyone accused of lacking reason in this period lacked it; nor did those making the accusation automatically possess it. Reason was not the exclusive property of one group, any more than wisdom or virtue. The late Gregory Bateson suggested that human categories could be divided, on the basis of the difference between analog and digital computers, into two types: those of a yes/no, and those of a more/less, nature.[67] Some of the confusion in modern historiography of the seventeenth century is due to our having taken at face value the pronouncements of controversialists, and understood as yes/no questions some that were in fact more/less ones. For several of the writers studied by Professor Mulligan, reason and illumination, reason and revelation, were not mutually exclusive opposites, but were rather complementary. The polarization into neatly defined opposing groups cannot be sustained by a thorough examination of the historical context, which stresses, rather, continuity and simultaneity.

Reading the chapters of this volume has been a salutary experience for me, and I hope it will be for others. No party line is espoused or was required. Contributors were chosen because from what I knew of their work, published and unpublished, they seemed likely to make a positive contribution to this debate. I think they have, but it is not of a single nature. Two main directions are visible. One group insists on some fundamental differences between occult and – still for want of a better word – scientific (observational-experimental-mathematical) attitudes. They either make distinctions within the work of Renaissance writers or follow out controversies that derived from clearly definable oppositions (as in Chapters 1, 3, 4, 5, 7 and 8). The other group argues that some polarities have been falsely defined, that both attitudes persist simultaneously in the same institutions, throughout the work of one scientist, or within the work of several apparently different writers (Chapters 2, 6, 10, and 13). The two chapters on witchcraft complement each other, but Chapter 12 links up with the second group, and both share the quality which, it seems to me, all these contributions have in common – the quality of going back to the original texts in a critical spirit, ready to challenge received opinion, if necessary.

With history, as with all disciplines, it is essential to combine involvement with the subject with a certain detachment and an awareness of the categories and concepts within which we ourselves think. From the revaluation made here it is evident that the claims for the similarities between the occult and the experimental sciences in the Renaissance have been based on a rather limited range of texts, interpreted in a forceful but one-sided way. Superficial similarities were snatched at; fundamental differences ignored. The state of mind in which future historians need to approach this issue seems to me to be exemplified by Max Weber in 1908, reconstructing "Agrarverhältnisse im Altertum" and criticizing those contemporary historians who ignored the differences between classical and medieval conditions. Weber rejected their reliance on analogies and similarities to produce a spurious causal interpretation of history. The truly critical historian, he urged,

> will put the stress on the *changes* that emerge in spite of all parallels, and will use the similarities only to establish the *distinctiveness* vis-à-vis each other of the two orbits [i.e., the ancient and the medieval] . . . A genuinely critical *comparison* of the developmental stages of the ancient polis and the medieval city . . . would be rewarding and fruitful – but only if such a comparison does *not* chase after "analogies" and "parallels" in the manner of the presently fashionable schemes of development; in other words, it should be concerned with the *distinctiveness* of each of the two developments that were finally so different, and the purpose of the

comparison must be the causal *explanation* of the differences. It remains true, of course, that this causal explanation requires as an indispensable preparation the isolation
(that means, abstraction) of the individual components of the
course of events, and for each component the orientation towards rules of experience and the formulation of *clear concepts* without which causal attribution is nowhere possible.[68]

In the admirable caveat Weber puts his finger precisely on what I see
as our goal as historians of the Renaissance: to isolate, identify the
individual components of the two systems, the occult and the experimental, which persisted side by side for nearly two thousand years.
Above all we must formulate the "clear concepts" needed to define
each system and the relations between them.

It has long been recognized that the history of science aims to recover
and reformulate the concepts and categories used in the past. As one
recent statement has it: "The first task before the historian of science
is to reconstruct the actual thought process of early scientific thinkers
– their goals, their methods, the criteria which they used to judge their
own achievements."[69] While Professor Westman legitimately invoked
the aid of philosophy of science to make the historian aware of "the
conceptual matrix and implicit presuppositions which attend a certain
scientific issue and of which both he and the early scientist may have
been unaware" (ibid.), it seems to me that for the relation between
science and the occult two other disciplines need to be drawn on. One
is the history of thought in general. Too many accounts of "Renaissance hermeticism" have ignored the continuities from medieval
sources, and indeed from earlier sources in late classical and Hellenistic
schools. One cannot study Renaissance Neoplatonism without an
awareness of its synthetic remolding of several traditions – or at least,
if one does ignore its past, one cannot make statements about what is
"new" in it.

The second discipline the historian of science might draw on is social
anthropology. Many of the occult sciences had a magical component.
Their goals were as much religious as worldly, as in alchemy with its
techniques of self-purification and salvation; and their processes resemble rituals as much as they do laboratory experiments. (Indeed, in
the occult tradition "experiment" often meant "experience" of a religious or mystical kind.) Their world view is based on such fundamentally religious concepts as the pure and the impure, and the opposed
states of pollution and purification. These categories are of fundamental
importance in the religion of many societies, and they entered the occult sciences in the West from ancient Greece. From the Greeks, too,
the occult took those anthropomorphic categories (found in many societies, primitive and advanced) based on a supposed qualitative differ-

ence between male and female, right and left. The use of hierarchical classification schemes and the technique of correlating or interequating such grids also came from Greek sources, and Babylonian ones before them, and can be found in Chinese, African, and other cultures.

The organizing structures, mental categories, and thought patterns of the occult sciences are common to many societies, ancient and modern. The historian of the occult can learn a great deal from the work on magic and ritual of such anthropologists as Marcel Mauss, E. E. Evans-Pritchard, and S. J. Tambiah;[70] on the concepts of purity and impurity, from Emile Durkheim, Louis Moulinier, and E. R. Dodds;[71] on the religious dimension of alchemy, from Mircea Eliade and C. G. Jung;[72] on the symbolism of right and left, from Robert Hertz, Rodney Needham, and others;[73] on hierarchical classification in other societies, from Marcel Granet, Derk Bodde, Joseph Needham, and Germaine Dieterlen, to name but a few.[74]

The problems of understanding the occult are many. Equally difficult, it seems to me on the evidence presented by this symposium, is understanding the functioning of the minds of seventeenth-century scientists, who were able to live with mutually incompatible mental categories. The historian of science can have his appreciation of this issue sharpened by reading Robin Horton's analysis of the different ways of thinking in traditional thought and in modern science,[75] together with Ernest Gellner's critique and elaboration of it,[76] or J. D. Y. Peel's essay, "Understanding Alien Belief-Systems."[77] The whole debate by anthropologists over the nature of existence of the "primitive mentality" has great relevance to our topic.[78]

One of the issues anthropologists face every day is the need to interpret cultures, languages, symbol systems that are largely or wholly alien to Western thought. In their continuing discussion of the problems involved in this process, two opposed but complementary positions can be picked out. One, which I choose J. D. Y. Peel to represent, argues that we must understand a magical technique, say, in the same way that its practitioners do: If they believe it to be "instrumental," achieving some clearly defined practical result, we falsify the issue by calling the technique "expressive," that is, uttering some personal emotion or drawing on a symbol system. This "refusal to use the actor's own categories" derives from the observer's ethnocentricity, and the remedy for this grave fault is for social anthropology to "set itself apart from its own social setting – our scientific culture which has given rise to it." Peel concludes that "in the study of alien belief-systems we must aim at a more difficult goal, a temporary suspension of the cognitive assumptions of our own society."[79] I have a great deal of sympathy for this position, but it is countered by another, equally important one, represented by Alastair MacIntyre, for instance, when he argues that "beliefs and concepts are not merely to be evaluated by the criteria implicit in the practice of those who hold and use them," since such criteria "are not necessarily coherent; their application to problems set within that social mode does not always yield *one* clear and unambiguous answer." We should realize that "sometimes to understand a concept involves not sharing it."[80] As he put it elsewhere, "the understanding of a people in terms of their own concepts and beliefs does in fact tend to preclude understanding them in any other terms."[81] The historian, like the anthropologist, has two complementary tasks: to analyze concepts and beliefs within the appropriate social and historical contexts, yet to bring other analytical categories to bear on them.[82] It is not enough to take them at face value.

The existence of these two opposed but complementary demands sharpens the anthropologist's awareness of his own modes of thinking and of the difference between traditional and modern cognitive activities. One of the most suggestive treatments of this issue has been by Robin Horton, and I would like to propose that the distinction he draws between primitive and Western modes of thought is similar in many ways to that between the occult and the scientific traditions.

Horton begins by sketching some of the presuppositions of a modern scientific outlook.[83] The search for explanatory theory in science since Galileo, Kepler, and Newton "is basically the quest for unity underlying apparent diversity," for simplicity, order, and regularity (p. 132), yielding a theoretical scheme that "breaks up the unitary objects of common sense into aspects, then places the resulting elements in a wider causal context" (p. 144). The functioning of this analytical model depends on the scientist's ability to abstract and to integrate at a higher level of abstraction, a process that encourages an awareness of the theorizing activity itself. The "key difference" Horton sees between African thought and Western science is that

> in traditional cultures there is no developed awareness of alternatives to the established body of theoretical tenets; whereas in scientifically oriented cultures, such an awareness is highly developed. It is this difference we refer to when we say that traditional cultures are 'closed' and scientifically oriented cultures 'open.' (p. 153)

In the same way, I would argue, the occult is a closed system and has many of the attributes of traditional thought. It is self-contained, a homogeneity that has synthesized its various elements into a mutually supporting relationship from which no part can be removed. Frances Yates described the "Renaissance Hermetic" view of the cosmos as "a network of magical forces with which man can operate." In a very similar metaphor (evidently "there's magic in the web") Horton quotes Evans-Pritchard's account of Azande witchcraft, in which "all their beliefs hang together," forming a "web of belief" in which "every strand depends upon every other strand," so that "were a Zande to give up faith in witch-doctorhood, he would have to surrender equally his faith in witchcraft and oracles." In the occult tradition, likewise, if a belief in numerology were abandoned, it would destroy the basis for alchemy and astrology; if a belief in astrology were abandoned, it would destroy alchemy, botanical medicine, and much else. The situation of the occult scientist is very similar to that of the Zande. The "web of belief" has an esthetic unity of its own, but one which then conditions the whole of their thinking: "A Zande cannot get out of its meshes because it is the only world he knows. The web is not an external structure in which he is enclosed. It is the texture of his thought and he cannot think that his thought is wrong."[84] In other words, as Horton puts it: "Absence of any awareness of alternatives makes for an absolute acceptance of the established theoretical tenets, and removes any possibility of questioning them." When established tenets are challenged, this is seen as "a threat of chaos," evoking intense anxiety (p. 154).

I do not suggest that Dee or Fludd or Athanasius Kircher were on the same cultural level as a Zande witch doctor in the 1930s. They were erudite men, with a highly sophisticated attitude to philosophical traditions, who were obviously aware of alternatives to the occult sciences. But, equally obviously, they deliberately rejected the alternatives, in the shape of Copernicus or Galileo and the physical-mathematical tradition they represented, and in the vehemence of Fludd's response to Kepler we see the same intense anxiety shown by those members of a traditional society who genuinely lack an awareness of alternatives. This absence of reference to alternative theories, according to Horton, accounts for several related differences between African thought and Western science. The first is that between a magical and a nonmagical attitude to words, a distinction that, as I argue in Chapter 3, separates the occult and the scientific traditions. The "traditional thinker" sees "a unique and intimate link between words and things," such that words are absolutely bound to reality, and to control or manipulate words is to have the same power over the things they stand for (p. 156). Modern science has dismissed such ideas because they would imply that reality did not exist independently of language and that human whim could control the world (p. 157). This is precisely the force of Kepler's objections to numerology and to much of astrology, as projections of human categories onto the physical world. He, like so many scientists since the seventeenth century, believed that reality could be described (at least partially) by language, but not controlled by it. And whereas the Western scientific tradition has long distinguished matter from spirit, in "traditional African cosmologies . . . everything in the universe is underpinned by spiritual forces," and "what moderns would call 'mental activities' and 'material things' are both part of a single reality, neither material or immaterial" (p. 157). This is obviously true of the occult tradition, which consistently fused such categories as animate and inanimate, spiritual and material into a single set.

The awareness in the scientific tradition that ideas and reality exist on different levels leads to a vision of alternatives that, "by giving the thinker an opportunity to 'get outside' his own system, offers him a possibility of his coming to see it *as a system*" (p. 159). It is no accident that the period in which attacks on occult science first become coherently directed against its methodology and cognitive processes is the period, from Ramus to Descartes, of a new consciousness about methods and systems, a debate that inevitably opened up an awareness of alternatives.[85] The absence of self-criticism or an open-ended spirit of inquiry in the occult tradition is paralleled, again, by the tendency of traditional thought "to get on with the work of explanation, without pausing for reflection upon the nature or rules of this work," ignoring such "second-order intellectual activities" (p. 160). Neither there nor in the occult do we find any concern with logic or with such issues as discovering "the general rules by which we can distinguish good arguments from bad ones" or asking "on what grounds can we ever claim to know anything about the world?" (p. 160). Many Renaissance occultists had university educations and knew of the philosophical discussions of these questions going back to Aristotle and the pre-Socratics; but they ignored them. The characteristic linguistic form of the occult tradition (see H. Cornelius Agrippa's *De occulta philosophia* for an unusually clear instance) is a present-tense statement using the verb "to be": This thing is like this, these things are connected with those. The development of scientific thought – already clearly present in Galileo and Kepler – in which "one theory is judged better than another with explicit reference to its efficacy in explanation and prediction" (p. 161) never took place in the occult tradition because it, like African traditional thought, never formulated "generalized norms of reasoning and knowing" (p. 160) and, I would add, never addressed itself to the physical world with nonanthropomorphic, nonsymbolic categories.

Prediction was not the monopoly of the scientific tradition, of course: It is vital to African magic as it is to the occult. But there are marked differences between the scientific tradition and the other two in their reaction to predictive failure.

> In the theoretical thought of the traditional cultures, there is a notable reluctance to register repeated failures of prediction and to act by attacking the beliefs involved. Instead,

other current beliefs are utilized in such a way as to "ex-
cuse" each failure as it occurs, and hence to protect the
major theoretical assumptions on which prediction is based.
This use of *ad hoc* excuses is a phenomenon which social
anthropologists have christened "secondary elaboration."
(p. 162)

A sick man goes to a diviner and is told that he can appease the anger
of the spiritual agency that is worrying him by performing some re-
medial actions. If he does these, but does not get better, he is likely
to go to another diviner and then another. The client "never takes his
repeated failures as evidence against the existence of the various spiritual beings" supposedly responsible for his illness or as "evidence
against the possibility of making contact with such beings as diviners
claim to do." Neither he nor other members of his community "ever
try to keep track of the proportion of successes to failures in the re-
medial actions based on their beliefs," in order to question them (p.
163).

The phenomenon of "secondary elaboration" is familiar in the occult
sciences, as in astrology, where predictive failure can be put down to
inaccurate information about the exact time of birth; or in alchemy,
where failure to perform transmutation can be blamed on the composition of the metals or the temperature of the furnace. A still wider
escape clause is available in both sciences in the form of the explanation
that the adept or his client or both lacked religious purity. (According
to some authorities alchemists were supposed to fast or abstain from
sexual contact before beginning the great work.) Keeping track of successes and failures is a mark of the anti-occult movement, from Pico's
dispute with astrology to such a tract as William Perkins's *Foure Great
Lyers* (1585), which puts side by side four astrological almanacs with
their predictions of the daily weather, showing amazing divergences
from each other.[86]

Where traditional thought, like the occult, has a protective attitude
toward established theory, the scientific tradition is ready to modify a
theory or scrap it altogether because it knows that "the theory is not
something timeless and absolute" (p. 163). As Horton says: "The collective memory of the European scientific community is littered with
the wreckage of the various unsatisfactory theories discarded over the
last 500 years" – the geocentric universe, the circular motion of the
planets, phlogiston, and many more. "This underlying readiness to
scrap or demote established theories on the ground of poor predictive
performance is perhaps the most important single feature of the scientific attitude" (p. 164). The contrast with the occult sciences could
hardly be sharper, since they never threw away anything, and much
of the system elaborated in the Hellenistic period survives intact today.

Modern astrology has absorbed some later planetary discoveries, and there are some sporadic instances of the application of quantitative techniques to mystical goals (as in Leonhard Thurneisser's use of quantitative analysis of urine to identify the three Paracelsian principles, mixing chemical with analogical and metaphorical procedures),[87] but by and large the occult sciences have gone on unchanged.

The contrast between the static nature of the occult and the progressive nature of science is no accident, since it expresses a fundamentally different attitude to time. In African traditional thought the past is "usually valued positively, sometimes neutrally, and never negatively. Whatever the particular scale involved . . . the passage of time is seen as something deleterious or at best neutral." Things were better "in the golden age of the founding heroes," and various activities are evolved by traditional societies "designed to negate" the passage of time "by a 'return to the beginning.'" In just the same way the occult tradition cherishes Orpheus, Hermes, or Pythagoras as its founding father, insists that its knowledge goes back to Moses or the Egyptians, and draws on the concept of *prisca theologia* to legitimize its pursuits.[88] Indeed, since the past is more holy than the present, by reviving the past it believed that it could revive holiness. Hence the desire to rediscover the language of Adam to overcome the consequences of the Fall or the tower of Babel.[89] In the late sixteenth and seventeenth centuries we find a different phenomenon in the occult; namely, an espousal of millenarian beliefs. These are future-oriented, of course, but not to a future achieved by the work of human hands and brains; rather, to one created by some transcendental religious or mystical experience. Otherwise, it seems true that the occult was past-oriented, with a conception of a golden age, that version of primitivism which Boas and Lovejoy labeled "soft," imagining a state of perfection from which mankind has been steadily declining.[90] The scientific tradition, in sharp contrast, sees the first age as "hard," a state of deprivation out of which we have painfully emerged, thanks to inventors, technologists, scientists. As Horton puts it: "Where the traditional thinker is busily trying to annul the passage of time, the scientist" is "trying frantically to hurry time up. For in his impassioned pursuit of the experimental method, he is striving after the creation of new situations which nature, if left to herself, would bring about slowly if ever at all" (p. 169). As Francis Bacon said, "Nature exists in three states": free; "forced out of her proper state" by natural causes; or "constrained and moulded by art and human ministry," binding nature to new production: "without man, such things would never have been made."[91] Further, given the scientist's "open" attitude to theories, in which a currently held idea is only one possibility among many, and given his experience of the way in which overthrown theories "are replaced by ideas of ever greater predictive and explanatory power," it is inevitable that the scientist should have "a very positive evaluation of time" (ibid.). Once the "idea of Progress" is formed, it "becomes in itself one of the most powerful supporters of the scientific attitude generally" (p. 170). That idea was first formulated in a coherent way by the propagandists for science in the Renaissance, and in the work of Francis Bacon we have one of the earliest recognitions of the increasing fruitfulness of scientific discovery – in a sentence from the Vulgate that he formed into a motto for the new science: *Multi pertransibunt et augebitur scientia.*[92]

The concept of science as being allied to progress, so crassly trumpeted and aligned with materialism as it has been for the last two centuries, has become an embarrassment to modern historians of science. They will admit to it reluctantly, since the tradition in which only positive achievements were deemed worthy of study has been so comprehensively discredited. Yet the fact remains that the occult tradition did not constitute what Charles Schmitt has described as "a genuine science," one "which is progressive, productive, and in some way susceptible to empirical verification or corroboration." Making the appropriate qualifications, he goes on:

> I would be among the last to deny that history of science must include bad and superseded science as well as good and successful science, but we must also realize that there comes a point at which science – and I take this to be one of the characteristic ways in which it differs from art, literature, political thought, or philosophy – must be progressive.[93]

This seems to me perfectly true and to offer a valid mode of discriminating science from the closed system of the occult. One reason for the continuously evolving nature of modern science has been given by Horton, namely, its readiness to consider alternatives and to revise theories and models. Another reason, as I see it, has been its willingness to admit the limits of its knowledge, to state clearly what it does not know. The occult sciences, by contrast, claimed to be omniscient, able to account for all phenomena, and were, as a result, strictly irrefutable. Their system was sufficiently flexible, using secondary elaboration when necessary, to ignore criticism. The process by which the claim of comprehensiveness produces the claim of irrefutability has been underlined by two modern philosophers, I. C. Jarvie and Joseph Agassi, who use it to distinguish magic from science. Their remarks also apply to the totality of the occult: "The strength of the magical world-view is that it is a complete world-view, one that explains anything and everything in terms of magic." Modern scientific thought, by contrast, finds it hard to accept a world view that accounts for

everything, since we have abandoned that goal as impossible or un-
desirable.

> We allow large roles of coincidence, accident, luck and fate.
> All of these categories are vague and introduced *ad hoc*; our
> world-view does not try to explain everything, if it did it
> would be irrefutable and we have ceased to regard irrefuta-
> bility as a desirable quality. The unique and disconcerting
> thing about the western scientific world-view is that it is
> progressive: it is more interested in the question than the an-
> swer; it puts a premium on overthrowing and improving pre-
> vious answers by means of severe criticisms. Among these
> severe criticisms is that of irrefutability: immunity to all pos-
> sible experience.[94]

In this sense the occultists of the seventeenth century were immune
to the experience of the new work that had been done in mathematics,
physics, mechanics, and optics. Their system accounted for everything
already, did not need rethinking, and was in any case directed to other
goals.

Robin Horton's distinction between African traditional thought and
modern science has been developed, in a characteristically incisive
paper, by Ernest Gellner.[95] Gellner accepts Horton's criterion of
"open" and "closed" systems and agrees about "the existence and
observability of an external reality other than the social perceptions
of it, such that styles of thought can be classified in terms of their
stance vis-à-vis that external reality," a reality, further, which is "such
as to render the 'open' outlook sounder, or at least cognitively more
effective, than the closed visions" (p. 166). He disagrees about the lack
of alternatives, since he finds a degree of pluralism in traditional
thought: Members of primitive societies "do transcend their condition
not by reaching out to science, but simply through syncretism" or
"doctrinal pluralism" (p. 166). Certainly the occult sciences in the
Renaissance were nothing if not syncretist; yet the various doctrines
they drew on were fitted into a totality that achieved an epistemological
and methodological unity. They also did not "reach out to science,"
and their awareness of alternatives was limited to those that did not
challenge their fundamental beliefs and methods. For them, as for traditional societies, Gellner's first qualification applies: "that Horton's
crucial differentia be credited not to individuals, nor even groups, but
to systems of thought" (p. 168), or, we might say, mentalities.

Taking the occult sciences as a system of thought, we can apply to
them several of Gellner's own differentia, beginning with what he calls
"the use of idiosyncratic norms": "A traditional belief-system contains
at least one general vision of 'what is normal.' The normal differs from
the abnormal in that it either requires no explanation," or, "if explained

at all, is explained wholesale, by the general myth . . . This normality is both cognitive (in the sense of having these implications for explanatory strategy), and moral . . . Normality is very specific and concrete" (p. 170). "By contrast, the crucial feature of scientific thought-systems is that the notion of normality is not conspicuously present in them." When they distinguish between what does and what does not require explanation, "the base-line for explanation . . . is relative, temporary and problem-bound rather than socially entrenched." What we might see as the most important difference, bearing in mind that paucity of abstract thinking and of second-order theoretical activity in the occult sciences, is that the explanatory baseline of scientific thought systems

> can generally be specified only in terms of the formal properties of explanation, rather than in terms of concrete properties of the thing explained. The most widely favoured baseline of this kind is what is popularly conceived as mechanism or materialism: the existence of a structure, built of publicly available materials with no unsymmetrical, locally idiosyncratically defined properties, and repeatable in accordance with a publicly stateable and socially neutral recipe or formula, such that the behaviour to be explained follows from the properties of that structure. (p. 170)

And Gellner adds in parentheses that the materialism is actually irrelevant: "As long as the criterion of publicity and repeatability is satisfied, it matters little whether the structure invoked is built of tangible materials, or remains abstract" (pp. 170–1).

Many elements of the occult sciences are touched on by that characterization. First, their lack of abstraction, that is, their reliance on the "concrete properties of the thing explained," which is seen as a unity rather than a system of relations. Second, their "occult" or hidden nature. Where the scientific tradition is "built of publicly available materials," the occult has always been secretive, restricting knowledge to adepts or initiates, communicating only in hermetic forms or in messages designed to sabotage themselves (such as alchemical recipes in cipher or exotic foreign languages – Ethiopian, say – or with the names of crucial substances or quantities omitted).[96] Where scientific experiments are repeatable and public, occult experiments, or experiences, are personal and notoriously not repeatable (above all not in alchemy, where the absence of any established criteria for determining the purity or concentration of substances, solid or liquid, or of standardizing temperature, made for insuperable difficulties in emulation). Where the occult sciences continued to use anthropomorphic, socioreligious, or ethical categories (male/female, right/left, pure/impure, with all the allied concepts of "spirit," "matter," and "base residues"), scientific thought systems are "socially neutral" and are thus "ill suited for the underpinning of moral expectations, of a status- and value-system" (p. 171). Where alchemy continues, for much of the seventeenth century, to be for many adepts a religious pursuit, science in that period separated itself from the status of a rival magic: It did not challenge the tenets of religious belief, but it was not in itself a religious activity.[97]

Throughout his essay Gellner makes the very necessary break with that tradition which chose to define and delimit science "not in terms of the type of *explanation* it tolerates, but in terms of its sources of *information*" (p. 171). This fits my own conviction that, if we wish to understand them, we should lay the stress not so much on the content of the occult sciences as on their thought processes and categories, the ways in which they arrive at explanations. The occult sciences represent a long-established tradition of trying to make sense of the world we live in in homocentric, symbolic, and religious terms, and attention to that tradition's epistemological and cognitive processes will show more clearly the respects in which it differs from the experimental, mensurating, quantifying, scientific tradition. One important distinction between the two is their attitude to what we might call the whole and its parts. While traditional thought systems, like the occult, form totalities in which everything mutually coheres, yet where differing criteria of evaluation apply to differing classes of objects – concrete reality being the determining factor, not any system of relationships that can be handled at a higher abstract level – the modern scientific tradition depends on a classification of knowledge and language into various types. This process entails using "criteria of validity," such as classifying propositions into "those which stand or fall in virtue of factual checking" or "of formal calculation" or "of consonance with the speaker's feelings" (all of which procedures can be found in Galileo's *Dialogue Concerning the Two Chief World Systems*). As Gellner points out, these theories make their greatest impact not through their specific detail but through their shared approach, "the assumption of specificity of function. By habituating people to the idea that there is a single, simple criterion and function, governing the evaluation of any one given cognitive or verbal act, they profoundly modify their outlook" (p. 173). That is, whereas in life and language as they actually exist, "various purposes or functions are conflated and confused," modern philosophy and science teach us to see these various functions as "'really' distinct" (pp. 173–4). Gellner's account of the effect of this invocation of "functional specificity" clarifies the distinction I have myself tried to make between the occult, which seems to conflate and fuse parts into the whole, and the scientific attitude, which seems to distinguish and separate them. The criterion of functional specificity

> in fact favours the mechanistic, disenchanted vision of the
> world as against magical enchantment. The enchanted vision

> works through the systematic conflation of descriptive, eval-
> uative, identificatory, status-conferring etc. roles of lan-
> guage. A sense of the separability and fundamental distinct-
> ness of the various functions is the surest way to the
> disenchantment of the world. (p. 174)

It is by disentangling the threads of the web that one becomes aware
of them. As Agassi and Jarvie put it, magic is "a substitute for systematic and analytic thinking," and as several anthropologists have
shown, primitive thought systems are able to tolerate logical contradictions that would be unthinkable to a modern European.[98]

Gellner draws two other important distinctions from this point. One
concerns the way in which traditional societies make no clear-cut distinction between concepts "which have an empirically operational role,
and those whose reference is transcendent." They work with "concepts that are, so to speak, semioperational, which have both empirical
and transcendent reference, invoked according to a locally recognized
sliding scale" (p. 176). While it would be helpful to see the magic of
Pico or Ficino in those terms, Gellner's analysis here seems to me
more relevant to the scientific tradition than to the occult:

> The really important job done by three centuries or so of
> empiricist propaganda has not been the proscribing or the
> discouragement of the transcendent: it has been the system-
> atic inculcation of a sensitivity to the existence of the
> boundary between that which is testable and that which is
> not, and above all the consequent inhibition of such bound-
> ary-hopping. (p. 176)

That is clearly true of science since Galileo, Mersenne, Descartes, and
Locke. Perhaps Newton's withholding of his researches into alchemy
and biblical history and chronology shows a tacit awareness that the
existence of "such a boundary discourages systematic conceptual
boundary-hopping" (p. 176). One could add that much earlier science,
and philosophy, imposed or recognized analogous boundaries or distinctions and that in much of Aristotle's work, say, "orderly and regular
conduct is exacted from concepts"; but while extending the historical
scale back in time, one would still grant the validity of the distinction.

A second distinction is related to it; namely, that traditional thought
systems depend a great deal on "entrenched constitutional clauses,"
convictions that, if destroyed, will bring down the whole of a system.
They are "cross-tied by so many firm links to all other institutions that
they cannot be shaken without everything being shaken" (pp. 178).
The concept of "the sacred or the crucial," say, in a traditional thought
system, "is more extensive, more untidily dispersed, and much more
pervasive" than in a modern thought system, where it is "tidier, narrower, as it were economical, based on some intelligible principle,"

tends to be not "diffused among the detailed aspects of life," and therefore is much less helpful in reinforcing "the fabric of life and society" (p. 178). In modern science and philosophy "the entrenched clauses have been reduced to a kind of formal minimum," as in empiricist theory, which "describes our view of the world as a kind of mosaic, in which all individual pieces are independent of each other and can be replaced without disturbing any of the rest" (p. 179). While this is a feature of modern philosophy – as Gellner shows, not the only one, or particularly accurate, despite its insistent self-propaganda – it can be taken as a fair description of the scientific tradition in the Renaissance, or indeed in ancient Greece, where suppositions could be criticized without toppling the whole system and where developments in one scientific area could be fruitfully applied in another. Far from permitting this degree of individual autonomy and interplay, the occult sciences sustained and defended their ideas of the macrocosm and the microcosm, sympathy, correlation of categories, and a numerological concept of harmony because their whole system depended on them. Their protective attitude to their world view was rather like Shakespeare's Ulysses, arch-politician, ready to invoke "entrenched clauses" for the purposes of political manipulation, appealing to the concept of social hierarchy as the bond that holds the universe together:

> Take but degree away, untune that string,
> And hark what discord follows. Each thing meets
> In mere oppugnancy,

a clash of parts which will end in appetite, "an universal wolf," eating up itself.[99] Yet when the microcosm and macrocosm and the other components of the occult system ceased to be widely accepted, no such dramatic consequences ensued. Science went one way, and the occult, discredited as a serious or valid intellectual activity, went another – as, I believe, they always had done. For me the occult is worthy of historical study, and in the same way in which an anthropologist like Evans-Pritchard started from the assumption that "people of alien cultures think neither more nor less intelligently and efficiently than ourselves, but merely live out their lives in the light of different initial premises."[100] The error, as I see it, lies in arguing that the occult sciences in the Renaissance were productive of ideas, theories, and techniques in the new sciences.

Not all the contributors to this symposium share my interpretation of the relationship between the two traditions; nor do they necessarily share my placing of the occult in a wider anthropological context. We do share the conviction that the issue has been inadequately discussed so far, that the questions have been wrongly defined or supported with flimsy historical evidence. Criticism ought to be constructive: We hope our contribution to this debate will push it on to a further stage of discussion, where some lasting solutions may be found.

Notes

1 Herbert Butterfield, *The Origins of Modern Science: 1300–1800*, rev. ed. (London, 1957; New York, 1965), p. 141. For other dismissive comments on occult science, see pp. 65, 75, 84f., 132. I need hardly add that as a general history of selected topics in orthodox history of science the book continues to be helpful.

2 Lynn Thorndike, *A History of Magic and Experimental Science*, vol. I. (New York, 1923), p. ix.

3 For examples of Thorndike's inaccuracies on Renaissance science, see Edward Rosen, "In Defence of Kepler," in *Aspects of the Renaissance*, ed. Archibald R. Lewis (Austin and London, 1967), pp. 141–58, and Alexandre Koyré, *The Astronomical Revolution*, trans. R. E. W. Maddison (London, 1973), p. 449 n. 125.

4 A. J. Festugière, *La Révélation d'Hermès Trismégiste*, 4 vols. (Paris, 1944–54), esp. vol. I, *L'Astrologie et les sciences occultes*, e.g., the conclusion, borne out by the whole of the preceding analysis: "En ce qui concerne l'hermétisme populaire, c'est-à-dire les écrits de sciences occultes qui ont fait l'objet du présent travail, il apparaît tout d'abord que cette littérature hermétique n'eut rien de propre et d'original. Le nom d'Hermès a couvert un mouvement qui se recontre, tout identique, sous le patronage d'autres prophètes. L'alchimie d'Hermès . . . ne diffère pas de l'alchimie d'Ostanès. La botanique astrologique d'Hermès ne se distingue en rien de celle qui est attribuée à d'autres maîtres (Salomon, Alexandre, Ptolémée, etc.) . . . Les recettes magiques d'Hermès sont en tout semblables à celles, par exemple, d'Apollonius. Bref, dans ce domaine, Hermès paraît n'avoir été qu'un des prête-noms dont on se sert à l'époque hellénistique pour contenter le besoin de révélation qui travaillait alors un si grand nombre d'esprits." Festugière states that he would have arrived at exactly the same result if he had chosen one of the "mages hellénisés," such as Zoroaster, or even Apollonius of Tyana. "D'un mot, il n'y a pas d'occultisme proprement hermétique, en ce sens que les écrits du Trismégiste sur ces matières n'apportent rien de neuf" (pp. 355–6). I find it strange that the findings of this fundamental study should have been ignored by proponents of the hermeticist thesis.

5 Charles B. Schmitt, "Reappraisals in Renaissance Science," *History of Science*, 16 (1978), pp. 200–14. In addition to Schmitt's caveat against the vague use of the term "hermeticism," see Eugenio Garin, "Divagazioni ermetiche," *Rivista critica di storia della filosofia* 31 (1976), pp. 462–6, against "troppo facili sintesi," which finish "per non significare più nulla." Garin ends by stating that if the history of science ever becomes a scholarly discipline this is only possible when it "osserva le regole della ricerca storica" (p. 466).

6 R. S. Westman and J. E. McGuire, *Hermeticism and the Scientific Revolution* (Los Angeles, 1977), consisting of Westman, "Magical Reform and Astronomical Reform: The Yates Thesis Reconsidered" (pp. 1–91) and McGuire, "Neoplatonism and Active Principles: Newton and the *Corpus Hermeticum*" (pp. 93–142). Charles Schmitt's article, "Reappraisals," is an essay review of this volume. For a different, pro-Yatesian review, see Brian Copenhaver in *Annals of Science*, 35 (1978), pp. 527–31, who believes that "the Renaissance revival of the Hermetic tradition contributed to the development of new ideas about will, operation, power and human capacity which culminate in the work of Francis Bacon" (p. 530). It would be

interesting to see that claim argued out with full acknowledgment of the counterevidence.

7 Frances Yates, "The Hermetic Tradition in Renaissance Science," in *Art, Science, and History in the Renaissance*, ed. Charles S. Singleton (Baltimore, 1967), pp. 255–74. See also her *Giordano Bruno and the Hermetic Tradition* (London, 1964).

8 On the radical differences between mathematics and numerology, and on the Renaissance mathematicians' rejection of the latter, see Edward W. Strong, *Procedures and Metaphysics: A Study in the Philosophy of Mathematical-Physical Science in the Sixteenth and Seventeenth Centuries* (Berkeley, 1936; Hildesheim, 1966).

9 See, for example, P. M. Rattansi, "The Intellectual Origins of the Royal Society," *Notes and Records of the Royal Society*, 23 (1968), pp. 129–43, and "Some Evaluations of Reason in Sixteenth- and Seventeenth-Century Natural Philosophy," in *Changing Perspectives in the History of Science*, ed. M. Teich and R. Young (London, 1973), pp. 148–66. For some of A. G. Debus's more occult-sympathetic work, see his review of Yates's *Giordano Bruno* in *Isis*, 55 (1964), pp. 389–91; "Renaissance Chemistry and the Work of Robert Fludd," *Ambix*, 14 (1967), pp. 42–59; *The English Paracelsians* (London, 1965); "Mathematics and Nature in the Chemical Texts of the Renaissance," *Ambix*, 15 (1968), pp. 1–28. In the biographical dictionary edited by Debus, *World's Who's Who in Science* (Chicago, 1968), "Robert Fludd is given 82 lines, more than Lavoisier (35) and Dalton (46) added together," as W. A. Smeaton pointed out in his review in *Ambix*, 17 (1970), p. 133. Debus's introductory textbook, *Man and Nature in the Renaissance* (Cambridge, 1978), is slanted wholly toward hermetic and occult science and gives a very one-sided view that is likely to mislead the kind of undergraduate or general reader at whom it is directed. Elsewhere, of course, Debus has made valuable contributions to the history of chemistry and has indeed criticized the Yatesian approach: A certain duality of outlook is evident. A less scholarly and indeed largely derivative book is P. J. French, *John Dee: The World of an Elizabethan Magus* (London, 1972).

10 For critiques of the Yates claims, see Mary B. Hesse, "Hermeticism and Historiography: An Apology for the Internal History of Science," in *Historical and Philosophical Perspectives of Science*, ed. R. H. Stuewer (Minneapolis, 1970), pp. 134–60, and "Reasons and Evaluation in the History of Science," in Teich and Young (eds.), *Changing Perspectives in the History of Science*, pp. 127–47, although her counterarguments are not wholly satisfactory: see, for critical comment, Arnold Thackray in the Minnesota volume, pp. 160–2; Rattansi, "Some Evaluations of Reason"; Westman, "Magical Reform"; and Rossi, cited below. Professor Hesse seems to want to stifle study of the occult sciences altogether; I would rather try and understand them. For more reliable appraisals, see Edward Rosen, "Was Copernicus a Hermeticist"? in Stuewer (ed.), *Historical and Philosophical Perspectives*, pp. 163–71; Paolo Rossi, "Hermeticism, Rationality and the Scientific Revolution," in *Reason, Experiment, and Mysticism in the Scientific Revolution*, ed. M. L. Righini Bonelli and W. R. Shea (New York, 1975), pp. 247–73, an outstanding essay that deserves to be required reading; and Charles Trinkaus, *In Our Image and Likeness*, 2 vols. (London, 1970), for a more balanced account of Renaissance Neoplatonism that implicitly calls in question much of Yates's interpretation and often explicitly rejects it: e.g., pp. 396 n. 21, 487–525, 800 n. 41, and the index.

11 See E. G. R. Taylor, *The Mathematical Practitioners of Tudor and Stuart England* (Cambridge, 1954); N. R. Clulee, "John Dee's Mathematics and the Grading of Compound Qualities," *Ambix*, 18 (1971), pp. 178–211; and Clulee, "Astrology, Magic, and Optics: Facets of John Dee's Early Natural Philosophy," *Renaissance Quarterly*, 30 (1977), pp. 632–80, with further bibliography of works on Dee as a mathematician at p. 676 n. 140.

12 Yates, "Hermetic Tradition," p. 263. See also Brian Vickers, "Frances Yates and the Writing of History," *Journal of Modern History*, 51 (1979), pp. 287–316, at 289–301.

13 W. P. D. Wightman, *Science and the Renaissance: An Introduction to the Study of the Emergence of the Sciences in the Sixteenth Century*, 2 vols. (Edinburgh, 1962), I, 145f. For a well-informed study of the whole issue, see Charles B. Schmitt, "Philosophy and Science in Sixteenth-Century Universities: Some Preliminary Comments," in *The Cultural Context of Medieval Learning*, ed. J. E. Murdoch and E. D. Sylla (Dordrecht, 1975), pp. 485–537.

14 But however controversial Bruno's interpretation of Copernicus was on this occasion, it seems unlikely that to him "the Copernican diagram" was no more than "a hieroglyph, a Hermetic seal hiding potent divine mysteries" (Yates, *Giordano Bruno*, p. 241). Miss Yates has misrepresented both Oxford and Bruno: see Westman, "Magical Reform," pp. 12–34; J. K. McConica, "Humanism and Aristotle in Tudor Oxford," *English Historical Review*, 94 (1979), pp. 291–317; and A. D. Weiner, "Expelling the Beast: Bruno's Adventures in England," *Modern Philology*, 78 (1980), pp. 1–13, who shows that Mornay was not a hermeticist but a staunch Protestant, like Sir Philip Sidney, whom Bruno (anti-Protestant, pro-French) addressed in the hope of converting him. (Miss Yates made Sidney out to be a Brunonian.) For more balanced accounts of Bruno, see Paul-Henry Michel, *La Cosmologie de Giordano Bruno* (Paris, 1962), English trans. R. E. W. Maddison, *The Cosmology of Giordano Bruno* (London, 1973); and Hélène Védrine, *La Conception de la nature chez Giordano Bruno* (Paris, 1967).

15 See Michael Hunter, "The Social Basis and Changing Fortunes of an Early Scientific Institution: An Analysis of the Membership of the Royal Society, 1660–85," *Notes and Records of the Royal Society*, 31 (1976), pp. 9–114; K. Theodore Hoppen, "The Nature of the Early Royal Society," *British Journal for the History of Science*, 9 (1976), pp. 1–24, 243–73.

16 Keith Thomas, *Religion and the Decline of Magic: Studies in Popular Beliefs in Sixteenth-Century England* (London, 1971).

17 The role of symbol systems in astrology has been well discussed by Gérard Simon, *Kepler astronome astrologue* (Paris, 1979). Charles Schmitt, referring to the Yatesian-Warburgian importance attached to symbolic images, has commented on "the radically different methodologies of art historians and historians of science. While symbols may well play a role in scientific discovery from time to time, they have little to do with the finished formulations of science, whether 'science' be taken as based on an Aristotelian, a medieval (*i.e.*, rational theology), or a modern model." A symbol may "serve a useful analogical function" in science, and the "abstractions used in scientific discourse" can be treated as symbols, but the "peculiarly scientific aspects" of the history of science "do not play host to symbols." Symbols are "ambiguous and often obscure," and their use, therefore, "goes specifically against the ideal of precision which has always been one of the chief criteria of any valid science," where "every

effort is made to obtain precise and unequivocal linguistic formulation"
("Reappraisals," p. 203). For some other studies pointing up the constraints
placed upon analogy in the nonoccult scientific tradition, see Agnes Arber,
"Analogy in the History of Science," in *Studies and Essays in the History
of Science and Learning Offered in Homage to George Sarton*, ed. M. F.
Ashley Montagu (New York, 1946), pp. 221–33, repr. in slightly abridged
form in Arber, *The Mind and the Eye* (Cambridge, 1964), pp. 32–44; Owsei
Temkin, "Metaphors of Human Biology," in *Science and Civilization*, ed.
R. C. Stauffer (Madison, Wis., 1949), pp. 167–94; E. Gilson, "Le
Raisonnement par analogie chez T. Campanella," in Gilson, *Etudes de
philosophie médiévale* (Strasbourg, 1921), pp. 125–45; Johanna Bleker,
"Chemiatrische Vorstellungen und Analogiedenken in der Harndiagnostik
Leonhart Thurneissers (1571 und 1576)," *Sudhoffs Archiv*, 60 (1976), pp.
66–75. The reified use of analogies and symbols by the occult seems to me
to have more in common with "mythical thought" as defined by Ernst
Cassirer in *The Philosophy of Symbolic Forms*, vol. 2 (New Haven, 1955) –
without accepting Cassirer's evolutionary scheme from "mythical" to
"logical" thought – and with "pre-operative thinking" as defined by C. R.
Hallpike, *The Foundations of Primitive Thought* (Oxford, 1979).
18 Mary Hesse diagnosed the "conscious self-definition of the new science in
the course of vigorous repudiation of the hermetics and all their works"
("Hermeticism and Historiography," p. 153).
19 See P. B. Wood, "Methodology and Apologetics: Thomas Sprat's *History
of the Royal Society*," *British Journal for the History of Science*, 13 (1980),
pp. 1–26, and Brian Vickers, "The Royal Society and the Reform of
English Prose Style: A Reassessment," forthcoming in Brian Vickers,
Defence of Rhetoric (Oxford).
20 See Robert Lenoble, *Mersenne ou la naissance du mécanisme* (Paris, 1943,
1971).
21 See D. P. Walker, *Studies in Musical Science in the Late Renaissance*
(London, 1978), and Penelope M. Gouk, "Music in the Natural Philosophy
of the Early Royal Society," Ph.D. dissertation, Warburg Institute, London
University, 1982.
22 Wolfgang Pauli, "The Influence of Archetypal Ideas on the Scientific
Theories of Kepler," in *The Interpretation of Nature and the Psyche*, ed.
Carl Jung and Wolfgang Pauli; English trans. Priscilla Silz (New York, 1955;
first pub. Zurich, 1952), pp. 149–240.
23 See, for example, H. Kayser, "Das Formdenken des Paracelsus," *Nova
acta paracelsica*, 1 (1944), pp. 103–8; O. Temkin, "The Elusiveness of
Paracelsus," *Bulletin of the History of Medicine*, 26 (1952), pp. 201–17.
24 The subtitle of Markus Fierz's recent *Girolamo Cardano (1501–1576)* (Basel
and Stuttgart, 1977) is "Arzt, Naturphilosoph, Mathematiker, Astronom und
Traumdeuter." In *Doctor Cardano, Physician Extraordinary* (London,
1969), Alan Wykes describes him as a mathematician, "also an inventor, an
astrologer, an astronomer, a philosopher, and a doctor. On the side, as it
were, he was skilled in divination by palmistry and geomancy" (p. xi).
Robert Lenoble's account, *Mersenne*, pp. 122–8, 135, stresses the fantastic
nature of Cardano's occult ideas and practices; R. P. Multhauf, by contrast,
in *The Origins of Chemistry* (London, 1966), pp. 239–41, 284, 286, 316–19,
gives his theories of matter serious, indeed respectful, attention. It is worth
noting, in support of Dr. Maclean's outline of the similarities between
Cardano and Scaliger, that John Wilkins, referring to them both in his

Mathematicall Magick (1648), often does so side by side, in illustration of the same point: see, for example, *The Mathematical and Philosophical Works of the Right Rev. John Wilkins*, 2 vols. (London, 1802), II, 120, 195.

25 Max Jammer, *Concepts of Force* (Cambridge, Mass., 1957), p. 90, quoting from Johannes Kepler, *Gesammelte Werke*, ed. Walther von Dyck, Max Caspar, Franz Hammer et. al. (Munich, 1937–), VII, 319. See also Max Caspar, *Kepler*, trans. C. D. Hellman (London and New York, 1959), p. 45: Scaliger's book "at that time was passed from hand to hand among the students and was eagerly read. This work, so Kepler informs us, awakened in him all possible thoughts about all possible questions, about heaven, souls, spirits, the elements, the nature of fire, the origin of springs, the tides, the shape of the earth and surrounding seas, and so forth."

26 Charles Singer, "The Scientific Views and Visions of Saint Hildegard (1098–1180)," in *Studies in the History and Method of Science*, ed. Charles Singer (Oxford, 1917; London, 1955), pp. 1–55, e.g., p. 53: "We give examples from the more typical of these visions, in which the medical reader or the sufferer from migraine will, we think, easily recognize the symptoms of scintillating scotoma"; Arthur Koestler, *The Sleepwalkers* (London, 1959).

27 Walter Pagel, "William Harvey Revisited: I," *History of Science*, 8 (1969), pp. 1–31, p. 1.

28 Thomas Browne, *Religio Medici* (1643), bk. I, sec. 34.

29 Translated by Pagel, p. 4, correcting the translation of Robert Willis (1847), which has: "When I surveyed my mass of evidence."

30 Walter Pagel, "William Harvey Revisited: II," *History of Science*, 9 (1970), pp. 1–41, p. 38.

31 Walter Pagel, *Paracelsus: An Introduction to Philosophical Medicine in the Era of the Renaissance* (Basel, 1958), p. 152.

32 Ibid., pp. 189, 200, 201f.

33 For some comments on this way of periodizing history, see Brian Vickers, *Towards Greek Tragedy* (London, 1973), pp. 167–9, and S. J. Tambiah, "The Magical Power of Words," *Man*, n.s. 3 (1968), pp. 175–208, at pp. 187–8.

34 Pauli, p. 154.

35 Pagel, "Harvey Revisited: II," p. 24.

36 Rattansi, "Some Evaluations of Reason," p. 155. Cf. Also Rattansi, "Newton's Alchemical Studies," in *Science, Medicine and Society in the Renaissance*, ed. A. G. Debus, 2 vols. (London, 1972), II, 167–82: e.g., p. 168 on "the problem of reconciling his supposedly alchemical commitments with his published views on celestial topics . . . it seems inconceivable that there should be no connection between" Newton's alchemy and chemistry: a candid *petitio principii*, at least. He concludes, again, by stressing the "danger" of "splitting Newton into irreconcilable 'scientific' and 'mystical' selves" (p. 179). The question has been loaded from the outset by the metaphors of "splitting" and "irreconcilable": What is assumed by them is a total unity – or perhaps better – an essential monism in Newton's thought.

37 Rattansi, "Some Evaluations of Reason," p. 165.

38 R. S. Westfall, "Newton and the Hermetic Tradition," in Debus (ed.), *Science, Medicine and Society in the Renaissance*, I, 183–198, at 186, 190. Westfall also notes (p. 185) the consistent opposition of scientists from Galileo to Newton against the explanatory concept of "occult qualities."

This division was evidently an operative category in the scientific Renaissance. A reviewer of Westfall's essay (Kathleen Ahonen, in *Ambix*, 21 [1974], p. 247) expressed the hope that he would "reconsider his overzealous conclusion that in Newton's science the Hermetic and the Mechanical were wedded together."

39 Multhauf, p. 241.

40 L. S. King, *The Road to Medical Enlightenment, 1650–1695* (London, 1970), pp. 60f.

41 See Chapter 3 of this volume; see also Brian Vickers, "On the Functions of Analogy in the Occult," *Renaissance Tradition,* ed. Allen G. Debus and Ingrid Merkel (Associated University Presses, forthcoming).

42 Henry Guerlac, "Guy de la Brosse and the French Paracelsians," in Debus (ed.), *Science, Medicine and Society in the Renaissance*, I, 177–99, at 177–8, 192 n. 3.

43 P. M. Rattansi, for instance, aligned Kepler with Paracelsus because he "shared with him not only the basic conception of the study of nature as tracing correspondencies, but also the Neoplatonic view in which they were rooted. Kepler cannot be turned into a 'modern' by citing his strong criticism of Robert Fludd; he was steeped in the Renaissance Neo-Platonism of which Hermeticism was an extension, and the Kepler–Fludd debate is better understood as a debate between various lines of development within sixteenth-century Neo-Platonism than as one between a 'rational scientist' and an irrational Hermetic magician" ("Some Evaluations of Reason," p. 153). To empty out the scientific, mathematical, computational, observational, and theoretical aspects of Kepler's work in this way seems somewhat rash. For more reliable estimates, see Gérard Simon, *Kepler*, and his "Kepler's Astrology: The Direction of a Reform," trans. J. V. Field, *Vistas in Astronomy*, 18 (1975), pp. 439–48. Franz Hammer, "Die Astrologie des Johannes Kepler," *Sudhoffs Archiv*, 55 (1971), pp. 113–35 stresses Kepler's use of "scientific" techniques in astrology, especially geometry.

44 The best short introduction to classical arithmology is by F. E. Robbins, essays prefixed to M. L. D'Ooge, trans. of Nichomachus of Gerasa, *Introduction to Arithmetic* (New York, 1926), esp. pp. 88–110, 121–3. On the Pythagorean tradition, see the admirable study by Walter Burkert, *Lore and Science in Ancient Pythagoreanism*, trans. E. L. Minar, Jr. (Cambridge, Mass., 1972); for the medieval and Renaissance continuity of the occult tradition, see Vincent Hopper, *Medieval Number Symbolism* (New York, 1938), and Christopher Butler, *Number Symbolism* (London, 1970); for the attack by the mathematical tradition, see Strong, *Procedures and Metaphysics*. Strong's account was challenged by Cassirer in an article published in German in 1940, and translated in *Galileo Man of Science*, ed. E. McMullin (New York, 1967), pp. 338–51, with a reply by Strong, pp. 352–64. The dispute is mostly about whether Galileo was a "Platonist." In any case Cassirer concedes that number mysticism did not play a role in the mathematics of the new natural science: pp. 342f., with Strong's comments pp. 353–8.

45 For more positive evaluations of induction as used by Bacon, see Mary Hesse, "Francis Bacon's Philosophy of Science," repr. in *Essential Articles for the Study of Francis Bacon*, ed. Brian Vickers (Hamden, Conn., 1968; London, 1972), pp. 114–139, and Mary Horton, "In Defence of Francis Bacon: A Criticism of the Critics of the Inductive Method," *Studies in the*

History and Philosophy of Science, 4 (1973), pp. 241–78. For a chronology of Bacon (the first, it would seem), see Brian Vickers, *Francis Bacon* (London, 1978), pp. 38–44. This is based mostly on Spedding and will need revision in the light of Dr. Rees's researches.

46 Lynn Thorndike, "Bacon and Descartes on Magic," in *Science, Medicine and History*, ed. E. A. Underwood, 2 vols. (Oxford, 1953), I, 451–4.

47 John Wilkins, *Mathematicall Magick* (1648), in *Works*, II, 109, 160, etc.

48 See, for example, Erich Schoner, *Das Viererschema in der antiken Humoralpathologie* (Wiesbaden, 1964; *Sudhoffs Archiv*, Beiheft 4).

49 See E. A. Burtt, *The Metaphysical Foundations of Modern Physical Science* (London, 1924; rev. ed. 1932).

50 See Brian Vickers, *Francis Bacon and Renaissance Prose* (Cambridge, 1968), pp. 202–4.

51 J. E. McGuire and P. M. Rattansi, "Newton and the 'Pipes of Pan,'" *Notes and Records of the Royal Society of London*, 21 (1966), pp. 108–43, something of an *omnium gatherum* of possible influences from the *prisca theologia* tradition, Cambridge Platonism, and much else, which never quite engages with any demonstrable chain of influence on Newton's scientific development. The claim is made that the "apparent contradiction between . . . a traditional Neo-Platonic philosophy, and the stern inductivism of the *Principia*, dissolves when we examine more closely how Newton modified the 'mechanical' philosophy of nature which was current earlier in the century" (pp. 124f.), but this claim is hardly substantiated in the remaining section of the article (pp. 125–38), which prefers, instead, to survey a great range of intellectual history, with, at times, inconclusive results, such as that "Newton did not borrow a great deal from Cudworth's learned account" (p. 135), that he in fact rejected several arguments of this school (pp. 135f.) and specifically did not adopt "Cudworth's identification of Moschus and Moses" (p. 142 n. 68). Much of the impressive scholarship behind this article failed to focus on the issues it raised.

52 McGuire, "Neoplatonism and Active Principles," pp. 132–3.

53 See, for example, R. S. Westfall, *Force in Newton's Physics: The Science of Dynamics in the Seventeenth Century* (London, 1971), and *The Construction of Modern Science: Mechanisms and Mechanics* (New York, 1971). Both of these study exclusively the nonoccult work, while an early article, "Isaac Newton: Religious Rationalist or Mystic?" *Review of Religion*, 22 (1957–8), pp. 155–70, rejected Newton's "mysticism." The first sign of a shift of interest is perhaps the essay "Newton and the Hermetic Tradition," in Debus (ed.), *Science, Medicine, and Society in the Renaissance*, I, 183–98. The shift is taken further in "The Role of Alchemy in Newton's Career," in Bonelli and Shea (eds.), *Reason, Experiment, and Mysticism in the Scientific Revolution*, pp. 189–232, 305–16, and in "Isaac Newton's Index Chemicus," *Ambix*, 22 (1975), pp. 174–85. In his magnum opus, *Never at Rest: A Biography of Isaac Newton* (Cambridge, 1980), Professor Westfall has integrated the alchemical work into Newton's whole career: see esp. pp. 281–309, 357–77, 407, 443f., 491–3, 510–11, 524–33, 537–9, 645, 793. In these recent studies Westfall has acknowledged a debt to Betty Jo Dobbs, *The Foundations of Newton's Alchemy* (Cambridge, 1975), to which should be referred the important review article by Karin Figala, "Newton as Alchemist," *History of Science*, 15 (1977), pp. 102–37.

54 Westfall, "Newton and the Hermetic Tradition," p. 195; "Role of Alchemy," p. 196. See other essays in the Bonelli-Shea volume critical of

his presentation of alchemy in Newton: by Paolo Casini (pp. 233–8) and by Marie Boas Hall (pp. 239–46).

55 Westfall, *Never at Rest*, pp. 530–1.

56 See B. J. Dobbs, "Studies in the Natural Philosophy of Sir Kenelm Digby," *Ambix*, 18 (1971), pp. 1–25; 20 (1973), pp. 143–63; 21 (1974), pp. 1–28, esp. pt. II, pp. 143f., 162, and pt. III, pp. 3, 24ff.

57 Westfall, *Never at Rest*, pp. 367–9.

58 Ibid., p. 645; cf. pp. 373, 375. See also Keith Hutchison, "What Happened to Occult Qualities in the Scientific Revolution?" *Isis*, 73 (1982), pp. 233–53.

59 For preliminary reading on witchcraft, the books I have found most helpful are Alan Macfarlane, *Witchcraft in Tudor and Stuart England* (London, 1970); Keith Thomas, *Religion and the Decline of Magic* (London, 1971); Sydney Anglo (ed.), *The Damned Art: Essays in the Literature of Witchcraft* (London, 1977); A. C. Kors and E. Peters (eds.), *Witchcraft in Europe, 1100–1700: A Documentary History* (Philadelphia, 1972); R. H. West, *The Invisible World; A Study of Pneumatology in Elizabethan Drama* (New York, 1969; 1st ed., 1939); B. Rosen (ed.), *Witchcraft* (London, 1969). H. R. Trevor-Roper's work, *The European Witch-Craze of the Sixteenth and Seventeenth Centuries* (New York, 1969; a part reprint of *The Crisis of the Seventeenth Century: Religion, the Reformation and Social Change*, 1968), is to be used with caution: see, for example, the important review by Robin Briggs in the *Times Literary Supplement*, 30 October 1970, repr. in *TLS 9* (London, 1971), pp. 121–31.

60 See, for example, Vladimir Propp, *Morphology of the Folktale*, trans. L. Scott, rev. ed. (Austin and London, 1968); E. Souriau, *Les Deux Cent Milles Situations dramatiques* (Paris, 1950).

61 Quoted from *Witchcraft*, ed. Barbara Rosen (London, 1969), p. 142. For other instances of the request denied having disastrous consequences, see this anthology, pp. 96–8, 108–9, 117–19, 130, 132, 139, 142–3, 154–6, 185–6, 188, etc.

62 Russell Hoban, *Pilgermann* (London, 1983), p. 38.

63 Marshall Sahlins, "On the Sociology of Primitive Exchange," in Sahlins, *Stone Age Economics* (London, 1974), pp. 191–6.

64 "Holy Thursday," *Songs of Innocence*; in *The Complete Writings of William Blake*, ed. Geoffrey Keynes (London, 1966), p. 122.

65 See, for example, E. Zeller, *The Stoics, Epicureans, and Sceptics*, trans. O. J. Reichel (London, 1870), pp. 75, 141, 339, 346; and M. C. Horowitz, "The Stoic Synthesis of the Idea of Natural Law in Man: Four Themes," *Journal of the History of Ideas*, 35 (1974), pp. 3–16.

66 See Vickers, "Royal Society."

67 See Gregory Bateson, *Towards an Ecology of Mind* (London, 1973), and Anthony Wilden, *System and Structure* (London, 1972).

68 Max Weber, "Agrarverhältnisse im Altertum," *Gesammelte Aufsätze zur Sozial- und Wirtschaftsgeschichte* (Tübingen, 1924), pp. 257, 288; trans. Guenther Roth, in Weber, *Economy and Society,* ed. G. Roth and C. Wittich (Berkeley and Los Angeles, 1978), p. xxxvii.

69 Robert S. Westman, "Kepler's Theory of Hypothesis and the 'Realist Dilemma,'" *Arbor Scientiarum*, Reihe A, Band I, *Internationales Kepler-Symposium, Weil der Stadt 1971* (Hildesheim, 1973), pp. 29–54, at p. 29.

70 M. Mauss, *A General Theory of Magic*, trans. R. Brain (London, 1950); E. E. Evans-Pritchard, *Witchcraft, Oracles and Magic Among the Azande*

(Oxford, 1937; abridged ed. E. Gillies, Oxford, 1976). Also S. J. Tambiah, "The Magical Power of Words," *Man*, n.s. 3 (1968), pp. 175–208; "Form and Meaning of Magical Acts: A Point of View," in *Modes of Thought: Essays on Thinking in Western and Non-Western Societies*, ed. Robin Horton and Ruth Finnegan (London, 1973), pp. 199–220; and "A Performative Approach to Ritual," Radcliffe-Brown Lecture of the British Academy, *Proceedings of the British Academy*, 65 (1979), pp. 113–69.

71 Emile Durkheim, *The Elementary Forms of the Religious Life* (London, 1915), trans. J. W. Swain from *Les Formes élémentaires de la vie religieuse* (Paris, 1912); Louis Moulinier, *Le Pur et l'impur dans la pensée des Grecs d'Homère à Aristote* (Paris, 1952); E. R. Dodds, *The Greeks and the Irrational* (Cambridge, 1951).

72 Mircea Eliade, *The Forge and the Crucible* (London, 1962), trans. S. Corbin from *Forgerons et alchimistes* (Paris, 1956); Eliade, "The Forge and the Crucible: A Postscript," *History of Religion*, 8 (1968), pp. 74–88; C. G. Jung, *Psychology and Alchemy* (London, 1953), trans. R. F. C. Hull from *Psychologie und Alchemie* (Zurich, 1944; rev. ed. 1952); Jung, *Alchemical Studies*, trans. R. F. C. Hull (London, 1968); and Joseph Needham, *Science and Civilisation in China*, vol. 5, pt. 2 (Cambridge, 1974).

73 Robert Hertz, *Death and the Right Hand* (London, 1960), trans. R. and C. Needham from "La Prééminence de la main droite: étude sur la polarité religieuse," *Revue philosophique*, 68 (1909), pp. 553–80; Rodney Needham (ed.), *Right and Left: Essays on Dual Symbolic Classification* (Chicago, 1973); and V. Fritsch, *Left and Right in Science and Life* (London, 1968), trans. from *Links und Rechts in Wissenschaft und Leben* (Stuttgart, 1964).

74 Marcel Granet, *La Pensée chinoise* (Paris, 1934); Derk Bodde, "Types of Chinese Categorical Thinking," *Journal of the American Oriental Society*, 59 (1939), pp. 200–19; Joseph Needham, *Science and Civilisation in China*, vol. 2, *History of Scientific Thought* (Cambridge, 1956), pp. 232–303; Germaine Dieterlen, "Les Correspondances cosmo-biologiques chez les Soudains," *Journal de psychologie normale et pathologique*, 43 (1950), pp. 350–66, and "Classification des végétaux chez les Dogon," *Journal de la Société des Africanistes*, 22 (1952), pp. 115–58; Rodney Needham, *Reconnaissances* (Toronto, 1980), chap. 2: "Analogical Classification," pp. 41–62.

75 Robin Horton, "African Traditional Thought and Modern Science," *Africa*, 37 (1967), pp. 50–71, 155–87; repr. in slightly abridged form in *Rationality*, ed. B. R. Wilson (Oxford, 1970), pp. 131–71.

76 Ernest Gellner, "The Savage and the Modern Mind," in Horton and Finnegan (eds.), *Modes of Thought*, pp. 162–81; see also Gellner, "Concepts and Society," in Wilson (ed.), *Rationality*, pp. 18–49.

77 J. D. Y. Peel, "Understanding Alien Belief-Systems," *British Journal of Sociology*, 20 (1969), pp. 69–84.

78 See, for example, Lucien Lévy-Bruhl, *How Natives Think* (London, 1926), trans. L. A. Clare from *Les Fonctions mentales dans les sociétés inférieures* (Paris, 1912), an interpretation hotly disputed by recent anthropologists and one that Lévy-Bruhl himself came to disown: see Lévy-Bruhl, *The Notebooks on Primitive Mentality* (Oxford, 1975), esp. pp. 47–9: "Final Renunciation of the Prelogical Character" of the primitive mentality, dated 27 June 1938. A penetrating evaluation of Lévy-Bruhl was made by E. E. Evans-Pritchard in various essays and books: see esp. his *Theories of Primitive Religion* (Oxford, 1965). See also C. Lévi-Strauss, *La Pensée*

sauvage (Paris, 1962), English trans. *The Savage Mind* (London, 1964). The stimulating recent study by C. R. Hallpike, *The Foundations of Primitive Thought* (Oxford, 1979), based on the theories of Jean Piaget, seems to me to offer a very promising model for understanding the occult.

79 Peel, pp. 75, 76, 79, 82.

80 Alastair MacIntyre, "Is Understanding Religion Compatible with Believing?" in Wilson (ed.), *Rationality*, pp. 62–77, at 67, 69.

81 Alastair MacIntyre, "The Idea of a Social Science," in Wilson (ed.), *Rationality*, pp. 112–30, at p. 130.

82 See also Steven Lukes, "Some Problems about Rationality," in Wilson (ed.), *Rationality*, pp. 194–213, for the argument that "beliefs are not only to be evaluated by the criteria that are to be discovered in the context in which they are held; they must also be evaluated by criteria of rationality that simply *are* criteria of rationality" (p. 208).

83 Quotations are from the abridged version in *Rationality*.

84 Evans-Pritchard, *Witchcraft*, p. 194.

85 See N. W. Gilbert, *Renaissance Concepts of Method* (New York, 1960).

86 Pico della Mirandola, *Disputationes adversus astrologiam divinatricem* (1496), ed. E. Garin, 2 vols. (Florence, 1946–52). See also Paolo Rossi, "Considerazioni sul declino dell'astrologia agli inizi dell' eta moderna," in *L'opera e il pensiero di Giovanni Pico della Mirandola nella storia dell' Umanesimo*, 2 vols. (Florence, 1965), II, 315–31; D. J. Fitzgerald, "Some Notes on Pico's Dispute with Astrology," *Arts libéraux et philosophie au Moyen Age* (Montreal and Paris, 1969), pp. 1049–55; J. D. North, "Astrology and the Fortunes of Churches," *Centaurus*, 24 (1980), pp. 181–211 – an important study of conjunctions and prognostication. On William Perkins's authorship of *Foure Great Lyers*, see the note by H. G. Dick in *Library*, 4th ser., 19 (1939), pp. 311–14.

87 See Johanna Bleker, "Die Harndiagnostik des Leonhard Thurneysser zum Thurn," *Deutsches Aerzteblatt*, 41 (1970), pp. 3202–09, and the article cited in note 17 above.

88 See D. P. Walker, *The Ancient Theology: Studies in Christian Platonism from the Fifteenth to the Eighteenth Century* (London, 1972).

89 See Paul Cornelius, *Languages in Seventeenth- and Early Eighteenth-Century Imaginary Voyages* (Geneva, 1965), who brings out well the motive of a return to a prelapsarian language; James Knowlson, *Universal Language Schemes in England and France, 1600–1800* (Toronto, 1975); and Vivian Salmon, essays on universal language schemes collected in *The Study of Language in 17th-Century England* (Amsterdam, 1979).

90 See George Boas and A. O. Lovejoy, *Primitivism and Related Ideas in Antiquity* (Baltimore, 1935).

91 *The Works of Francis Bacon*, ed. J. Spedding, R. L. Ellis, and D. D. Heath, 14 vols. (London, 1857–74), IV, 253.

92 This phrase (from Daniel 12:4) appears on the engraved title page of the *Novum organum*, reproduced in Brian Vickers, *Francis Bacon and Renaissance Prose* (Cambridge, 1968), frontispiece; see chap. 6, "Philosophy and Image-Patterns," pp. 174–201, for the metaphors of fruitfulness promised by the new science, esp. pp. 194ff.

93 Schmitt, "Reappraisals,", p. 209.

94 I. C. Jarvie and J. Agassi, "The Problem of the Rationality of Magic," *British Journal of Sociology*, 18 (1967), pp. 55–74, at pp. 70, 74 n. 33; repr. in Wilson (ed.), *Rationality*, pp. 172–93, at p. 192.

95 Gellner, "Savage and the Modern Mind," pp. 162–81.
96 The best guide to this aspect of alchemy is Maurice Crosland, *Historical Studies in the Language of Chemistry* (London, 1962).
97 See R. S. Westfall, *Science and Religion in Seventeenth-Century England* (New Haven, 1958).
98 Jarvie and Agassi, "Problem of the Rationality of Magic," p. 192. On the ability of traditional cultures to tolerate contradiction, see Wilson (ed.), *Rationality*, pp. 25, 143, 201–4 (Evans-Pritchard on the Azande, who do not perceive the contradiction in their system "as we perceive it, because they have no theoretical interest in the subject"); Horton and Finnegan (eds.), *Modes of Thought*, pp. 42 (Lévy-Bruhl's concept of "participation" in primitive thought, which fuses objects and attributes that we would separate) and 252–4 (on "indifference to logical contradictions" as one of Lévy-Bruhl's criteria for defining traditional thought).
99 William Shakespeare, *Troilus and Cressida*, 1.3.109ff.
100 Horton and Finnegan (eds.), *Modes of Thought*, p. 37.

1

At the crossroads of magic and science: John Dee's Archemastrie

NICHOLAS H. CLULEE

John Dee has often figured significantly in discussions of the interconnections of occultism and science in the Renaissance. While his interest in the occult, ranging from astrology and alchemy to ceremonial magic, remained strong, his abilities and interests in mathematics, navigation, and computational astronomy are also undeniable. Yet disagreement prevails on the exact interrelationship of the occult to the scientific aspects of Dee's efforts, as it does on the nature and interrelationship of occultism and science generally in the Renaissance. A central text in these discussions has been Dee's "Mathematicall Praeface" to the 1570 English translation of Euclid's *Elements of Geometry*. Early discussions by Johnson, Taylor, and Calder focus on the "Praeface" as a manifesto of modern science by emphasizing Dee's understanding of experimental method combined with quantitative and mathematical theory.[1] Recent scholars have been more cautious regarding Dee's use of the term "scientia experimentalis" in the section of the "Praeface" on "Archemastrie," pointing out that it often meant no more than experience, not the controlled testing of hypotheses as in its modern connotation, and could easily be applied to occult experiences.[2] Nonetheless, Marie Boas thinks that Dee's "Archemastrie" meant "genuine observation of nature" to the extent that "magic was near to becoming experimental science."[3]

On the other hand, Frances A. Yates and Peter French have argued that Dee's movement toward science did not come at the expense of magic and the occult, but was fostered by his adherence to an occult philosophy, based on Renaissance cabala and hermetic sources, which emphasized an operative magic as the key to understanding nature.[4] French accepts without qualification that Dee "proposed a viable theory of experimental science" in his idea of Archemastrie and expressly links this with Dee's interest in magic as showing a "hermetically in-spired desire to control nature."[5]

More specifically than the entire "Mathematicall Praeface," Dee's discussion of Archemastrie has become a key text standing behind these various interpretations. As the last of the numerous mathematical arts described in the "Praeface," Dee considers Archemastrie the sov-ereign science because it builds upon and extends all other arts and sciences.[6] As he defines it, Archemastrie

> teacheth to bryng to actuall experience sensible, all worthy conclusions by all the Artes Mathematicall purposed, & by true Naturall Philosophie concluded: & both addeth to them a farder scope, in the terms of the same Artes, & also by hys proper Method, and in peculier termes, precedeth, with helpe of the foresayd Artes, to the performance of complet Experiences, which of no particular Art, are hable (For-mally) to be challenged.[7]

Archemastrie is both theoretical and practical. It both certifies and makes useful the conclusions of all the mathematical arts and of natural philosophy, and also leads to experiences or accomplishments beyond the scope of other sciences. What has attracted the attention of com-mentators is Dee's emphasis on experience and experiment. In the crucial passage Dee says: "And bycause it procedeth by *Experiences*, and searcheth forth the causes of Conclusions, by *Experiences*: and also putteth the Conclusions them selues, in *Experience*, it is named of some, *Scientia Experimentalis*. The *Experimentall Science*."[8] Dee then refers to Nicolaus Cusanus's *Idiota de staticis experimentis* and to Roger Bacon's works for Clement IV for earlier uses and discussions of the term.[9]

Clearly, Dee's idea of Archemastrie should be invaluable in under-standing Dee's concept of science and the occult and in settling whether he moved toward science because of, or in spite of, his interest in the occult. The surprising thing is how little discussion the actual content of Dee's idea of Archemastrie has received. Like many authors, Dee is more often cited than read. We tend to notice that which is familiar to us and easily accessible to interpretation, such as the term "exper-imental science" and the references to Cusanus and Roger Bacon, while the obscure and unfamiliar slip by us unnoticed because our interpretive nets lack the appropriate catagories. Thus, there has been little discussion of the second half of Dee's passage on Archemastrie, which is written in awkward and turgid prose and abounds in obscure references. I have found that Dee's idea of experimental science is far from novel, that it harks back to Roger Bacon and contains a significant occult and magical dimension little noticed until now.

While Dee's explanation of Archemastrie is not lucid, the references
to Cusanus and Bacon can shed considerable light on its meaning.
Cusanus uses the term "experimentalis scientia" in his *Idiota de staticis experimentis*, in which he argues the importance of the comparison
of empirical quantitative measurements for the investigation of nature.[10] While this is in general conformity with Dee's emphasis on
experience in his idea of Archemastrie and his stress on the usefulness
of mathematics throughout the "Praeface," it is not a close parallel.
Dee's intent becomes more apparent when it is realized that he is paraphrasing not Cusanus but Roger Bacon's discussion in the sixth book
of the *Opus majus*, entitled "De scientia experimentalis."[11] Bacon distinguishes two ways of knowing: one by argument and persuasion; the
other by experience or "scientia experimentalis," which alone removes all doubts. Experimental science has three prerogatives over
all other sciences. First, it investigates by experience (*per experimen-
tiam*) the conclusions that other sciences reach by reasoning. Second,
it is the method for reaching those truths in the other sciences that they
cannot arrive at by their own methods. And third, it has the power
that no other science has to investigate the secrets of nature; namely,
the ability to acquire knowledge of the future, the past, and the present
through wonderful works, by which it forms judgments better than
ordinary judicial astrology.[12] This third aspect of Bacon's experimental
science gives it overtones of the occult, reflecting Bacon's interest in
natural magic as shown in his *De secretis operibus artis et naturae, et
de nullitate magiae*.[13] Dee's explanation of Archemastrie emphasizes
aspects of the first two of Bacon's points, but seems to ignore the third,
magical aspect. For Dee, Archemastrie certifies the conclusions of
other arts completely and fully by sensible experiences, whereas the
arts themselves use only words and arguments that persuade but do
not prove. This "doctrine Experimentall" leads to truths beyond those
of which the other arts are capable.[14] Thus Dee's concept of Archemastrie is not at all novel or original, although it is undoubtedly significant that Dee revived, called attention to, and popularized this idea
in a vernacular work intended for "vnlatined people, and not Vniuersitie Scholars."[15]

Bacon's investigation of the cause of the rainbow, which he gives
as a working illustration of his concept of experimental science, has
led Crombie to argue that Bacon's method involved empirical investigation leading to the formulation of a mathematical model, which is
then verified or falsified through arranged experimental tests.[16] In this
formulation, Bacon's experimental science would be very close to the
idea of experimental-mathematical investigation in "classical 'scientific method,'"[17] making Bacon's idea a genuine contribution to the
creation of modern experimental science when this "ordered and ra-

tional experimentation'' was revived in the sixteenth century by Dee, among others.[18] Dee's adoption of Bacon's concept of experimental science, along with his proposal for empirically testing his mathematical theory of astrology in the *Propaedeumata aphoristica* of 1558, might thus appear to support both the interpretation of Dee's Archemastrie as an important statement of a modern idea of scientific method and Crombie's claim for significant continuity between medieval and early modern science.[19]

A close inspection of the remainder of Dee's text, however, calls in question both this interpretation of Dee's Archemastrie and its positive contribution of medieval to sixteenth-century science. This concluding section is extremely puzzling. Here Dee says that ''to this Science,'' meaning Archemastrie,

> doth the *Science Alnirangiat*, great Seruice. Muse nothyng of this name. I chaunge not the name, so vsed, and in Print published by other: beyng a name, propre to the Science. Vnder this, commeth *Ars Sintrillia,* by *Artephius*, briefly written. But the chief Science, of the Archemaster, (in this world) as yet known, is an other (as it were) OPTICAL Science: wherof, the name shall be told (God willyng) when I shall haue some, (more iust) occasion, therof, to Discourse.[20]

Here the references are so obscure that they have been either overlooked or ignored in most published discussions, and the few who have tried to track them down have not succeeded.[21] I believe it is now possible to identify these references and to establish their meaning in Dee's usage. On this basis Archemastrie takes on a strongly magical dimension, and through these references Dee completes his paraphrase of Bacon by obliquely including the third, magical, prerogative of Bacon's ''scientia experimentalis.''

Although Dee says we should ''muse nothyng of this name,'' ''alnirangiat'' is far from a common term in the Latin West. ''Nirangiyat,'' in various forms (nīrangăt, nāranğăt, nāranğīyăt, nārinğiyyăt, nīranğiyyăi) is a plural form of ''nīrang,'' a word of Persian derivation, and was used by Arabic authors to refer to various kinds of magic involving tricks, talismans, conjuring, and so forth.[22] In the Pseudo-Mağrītī Gaya (the Arabic basis for the *Picatrix*), ''nīranğ'' means a magical charm or spell, involving often complex recipes, which is useful in achieving all the usual aims of magic.[23] Dee claims to have found the word in print, and the only Arabic author whose works were known and translated in the West who uses the term is Ibn Sînâ (Avicenna).[24] In a small work, *On the Division of the Sciences* (*De divisionibus scientiarum*), translated by Andrea Alpago and published in 1546, Avicenna lists a ''scientia *alnirangiat*'' among the subalternate branches of the principal natural sciences.[25] Dee cited this particular edition earlier in the "Prae-face" when he quoted Avicenna's definition of algebra, or algiebar as Avicenna calls it, and Dee's copy of this work has the section on al-nirangiat annotated in his hand.[26] In Alpago's Latin version, "alniran-giat" is defined as the science of the magic art that joins together the virtues of earthly things to produce strange and extraordinary effects.[27] In Dee's copy this passage is annotated "magicae" in the margin, and "alnirangiat" and "artis magicae" are underlined. Thus, alnirangiat is a form of natural magic for the manipulation of the hidden virtues of things in order to perform the wonderful works of nature and art that Bacon included in the third aspect of his "scientia experimentalis."[28] Avicenna's definition undoubtedly also appealed to Dee because it made magic a science derived from the theoretical sciences of nature whose attribute of certain knowledge corresponds to Dee's idea of Archemastrie.[29]

Artephius's *Ars sintrillia* is a more difficult reference to track down with assurance. Artephius (also found as Artefius, Arthephius, Arte-pius, Artesius), whose identity remains obscure and who may never have existed, was occasionally cited by medieval and Renaissance authors, being granted a reputation for deep and extensive knowledge in the occult, particularly alchemy and magic.[30] The earliest mention of him has been found in a twelfth-century manuscript, and two Latin works ascribed to him, a *Clavis sapientiae* or *Clavis maioris sapientiae* and a *Liber secretus*, are found in a number of manuscripts and were printed several times in the seventeenth century, some in vernacular translations.[31] Although he was in some instances confused with Or-pheus or Apollonius of Tyana and in others represented as a student of the latter, the ultimate provenance of his ideas and some of his works was most likely a Muslim author, since an Arabic original of the *Clavis sapientiae* has been identified.[32] Perhaps significantly for this inves-tigation, Roger Bacon mentions Artephius a number of times as a nat-ural philosopher who gained exotic knowledge through travel to the Orient, used methods of concealing philosophical secrets from the mul-titude, and, in conjunction with his treatment of "scientia experimen-talis," acquired through experience such a knowledge of the occult properties of nature that he was able to prolong life.[33]

The most likely source for Dee's reference to Artephius is a man-uscript codex Dee owned in 1556 that contained an *Ars sintrillia* among its constituent works.[34] While the codex has been identified, this por-tion of the manuscript is missing and no other work with the title *Ars sintrillia* has, to my knowledge, been found either in manuscript or in print.[35] The *Clavis sapientiae* and the *Liber secretus* are of potential interest, the *Liber secretus* being an alchemical handbook, and the *Clavis* an alchemically inspired cosmology with strong magical overtones in its prescriptions for drawing celestial forces into the human spirit; but neither contains material that would allow it to be identified with an *Ars sintrillia*.[36] Other titles have been attributed to Artephius in various bibliographies, only one of which has a secure provenance extending back earlier than the sixteenth century.[37] It is possible to reconstruct the contents of this work and to show that this is most likely the same work as Dee's *Ars sintrillia*.

In the *De rerum varietate*, Girolamo Cardano gives a lengthy description, including apparent quotations, from an "Ars magica Artefii et Mehinii" that he found in an old parchment manuscript also containing works by Euclid and Campanus.[38] The eight divisions Cardano describes cover (1) characters of the planets and images; (2) the significance of the motions of birds; (3) the interpretation of the voices of birds and animals; (4) the virtues of herbs; (5) the philosopher's stone; (6) the knowledge of the past, present, and future by three vases; (7) experiments with these vases; and (8) the prolongation of life.[39] Several of these conform in one way or another with teachings attributed to Artephius by earlier authors, including Bacon, William of Auvergne, and Ristoro d'Arezzo.[40] Of particular interest are the sixth and seventh divisions, which describe the use of three vases or vessels of different materials containing different liquids with semiprecious stones at the bottom. These are to be arranged in various ways with candles and, by the reflection of the rays of the sun, moon, and stars from a polished sword into the liquids, attended with the utterance of ceremonial formulae, make possible the various kinds of divination, especially knowledge of the past, present, and future.[41] Gianfrancesco Pico della Mirandola, along with reporting the legendary accounts of Artephius's longevity and ability to understand the language of birds, also mentions an "ars Artephij" involving both knowledge of the past, present, and future and prophecies of hidden things through the gathering of celestial rays in a mirror.[42]

There is evidence that this part of the work Cardano describes circulated as a separate book with no definite title much earlier than the sixteenth century.[43] The earliest reference to Artephius is in a twelfth-century manuscript entitled "Alchamia," which cites an Artesius in connection with divination by the reflection of rays of the sun or moon in liquids or mirrors.[44] Later, William of Auvergne (ca. 1180–1249) mentions, in connection with a discussion of revelations through the inspection of lucid objects, the practice of an Artesius for obtaining visions of all hidden things through the glittering of water placed below a polished sword.[45] Both of these are in keeping with what Cardano describes. The most revealing detail is that William refers to this practice as the "ars triblia vel syntriblia."[46] I have found no precise meaning for these terms, but they are possibly based on the Greek root for three, which would link up with the three vases of Cardano's report and also with Dee's *Ars sintrillia* by means of a trivial clerical error through which "syntriblia" became "sintrillia."[47]

Clearly, Artephius was associated with a magical technique involving divination by means of reflecting surfaces and celestial radiations. Although the manuscript tradition of this work of Artephius's is obscure, there is so much similarity in the various descriptions of this "ars" that more than likely it is the same work that is referred to variously as the "Ars magica Artefii & Mihinii," the "ars triblia vel syntriblia," the "arte suttrillia," or Dee's *Ars sintrillia*. This identification of Dee's mention of *Ars sintrillia* also makes sense in the context of the passage in the "Mathematicall Praeface," for several reasons. First, the magical nature of Artephius's art makes it a particular instance of the general science of magic, or alnirangiat, as Dee describes it.[48] Second, as a procedure for obtaining knowledge of the past, present, and future, it conforms to the other part of Bacon's third prerogative of his "scientia experimentalis."[49] Third, the central mechanism in Artephius's art is optical – the reflection and refraction of rays projected from celestial bodies – which provides a possible connection between this and Dee's final reference to the "chief Science, of the Archemaster, . . . is another (as it were) OPTICAL Science."[50]

My suggestion that Dee's "other Optical Science" is some form of magic related to Artephius's divination is supported by Dee's own conception of optics. Dee, following the medieval tradition in which optics and the science of perspective were identical, defined perspective earlier in the "Praeface" as

> an Art Mathematicall, which demonstrateth the maner, and
> properties, of all Radiations Direct, Broken, and Reflected
> . . . It concerneth all Creatures, all Actions, and passions,
> by Emanation of beames perfourmed. Beames, or naturall
> lines, (here) I meane, not of light onely, or of colour (though
> they, to eye, giue shew, witnes, and profe, whereby to
> ground the Arte vpon) but also of other *Formes*, both *Sub-
> stantiall*, and *Accidentall*, the certaine and determined actiue
> Radiall emanations.[51]

In this conception perspective was not limited to vision, but was a general science of radiated influences and a foundation for natural philosophy and astrology as well as optics and catoptrics.[52] Like Bacon in his investigation of the rainbow, Dee asserts that through optics the true and natural causes of various visual phenomena can be discovered. Even more broadly, he claims that the art of perspective provides the means of fully understanding, verifying, and extending natural philosophy, astronomy, and astrology, which conform to the first two powers of Bacon's *Scientia experimentalis* and Dee's Archemastrie. Astrology was the major area in which Dee elaborated a theory based on the models of perspective, closely following both Roger Bacon's theory of the multiplication of species and a similar theory of al-Kindī.[53] In common with Bacon and al-Kindī, Dee also saw magical implications in this theory of astrology in the possibility of manipulating celestial influences by means of optical devices.[54] Further on, Dee says: "The whole Frame of Gods Creatures, (which is the whole world,) is to vs, a bright glasse: from which, by reflexion, reboundeth to our knowledge and perceiuerance, Beames, and Radiations: representing the Image of his Infinite goodness, Omnipotency, and wisedome."[55] This implies that his broad conception of optics and perspective could provide the key to the ultimate secrets of creation.

The "other" optical science to which Dee refers as the chief science of the Archemaster would thus involve some additional method of investigating radiated influences going beyond the first two prerogatives of Archemastrie. The most likely related science would be catoptromancy, or divination by mirrors or other reflecting surfaces. In this regard it is worth noting that Bacon included among the experiences pertaining to *scientia experimentalis* interior divine illumination, which he considered more certain than external sense experience.[56]

Divination by mirrors, crystals, gems, and other reflective surfaces was frequently mentioned in the occult and magical literature of Renaissance authors.[57] Interest in these techniques was not confined to theory; there are records of such practices among associates of Dee in England in 1567.[58] Whether Dee practiced such things as early as the "Praeface" is not certain, but his skrying activities with Edward Kelly beginning in 1581 are a form of such divination. Thus, Dee's "spiritual exercises" were not an isolated aberration, but were related to his earlier ideas and his concept of Archemastrie, and there is a hint that he was involved in attempts at such divination as early as 1569.[59]

Although this reconstruction of what Dee meant by *Ars sintrillia* and "an other Optical Science" must remain tentative in the absence of more conclusive evidence, the evidence nonetheless is highly suggestive. Archemastrie was to him a master science for investigating nature because it confirmed through experience the conclusions of natural philosophy and other sciences, and alone offered a knowledge of the innermost secrets of creation through its ability to understand and manipulate the virtues and radiated influences that, Dee believed, are the ultimate mechanism of natural causation and medium of divine revelation.

Some of the implications of this discussion bear elaborating. First, Archemastrie cannot be seen as a practice by which, as Boas suggests, "magic was near to becoming experimental science" through the rejection of occult or mystical experience in favor of the "genuine ob-

servation of nature.'' While isolated aspects of Dee's ''doctrine experimentall'' may perhaps represent a fruitful method for the
investigation of nature and look forward to modern experimental methods, these aspects are inextricably tied to the magical aspects of Archemastrie. This magic is not a narrow practical or instrumental natural
magic that rejects occult virtues or the special esoteric and mystical
insight of the sage. Rather, it is a magic related to Dee's occult and
esoteric interests as found in the earlier *Monas hieroglyphica* and in
later spiritual exercises. Thus it points in the direction of a spiritual
knowledge so opposed to natural science as later understood that it is
impossible to cite Dee's concept of Archemastrie as evidence that Renaissance magic and occultism unambiguously contributed to the evolution of a new science. The effort to find a dividing line between magic
and genuine science – a crossroads where magic either transforms itself
into science or is left behind and true science taken up – is, in regard
to Dee, mistaken because it pushes a later conceptual distinction between magic and science, involving a narrowed definition of legitimate
science, back onto Dee, for whom it is inappropriate. I think Dee considered Archemastrie a unique and autonomous activity, not original
to him but unknown and unpracticed at his time. Dee's Archemastrie
combines the practices of the mathematical arts and of magic, and the
goals and theories of natural philosophy, with methods of experimental
verification. The resulting fusion cuts across traditional disciplinary
divisions and, on his terms, consummates these disciplines by verifying
them, grounding them in experience, while extending them toward the
fulfillment of the highest objective of all the sciences: knowledge of
the most fundamental and hidden principles of the Creator's work. It
is not merely a composite of these components, nor can it be reduced
to primarily one or another of them, such as experimental science or
magic.

Second, even if we admit that magic served as the stimulus that
attracted Dee's attention to experimental method and that it was in
this way that magic in the Renaissance contributed to the foundations
of a new science, still Dee's Archemastrie does not support the argument of Yates and others that the magic responsible for this was the
new style of magic that was developed in the Renaissance in conjunction with the hermetic texts, the cabala, and late antique Neoplatonic
writings. Whatever the significance of Dee's ''doctrine experimentall,''
it is in no way novel in substance, part being a paraphrase of Roger
Bacon and the remainder being inspired by him. If Dee was attracted
to the concept of empirically testing theories of nature by an interest
in magic, the magic was not any uniquely Renaissance or ''hermetic''
variety, but medieval and Arabic. For all that Dee has been considered
a representative of the ''Renaissance hermetic tradition,'' the magic

of the Archemastrie passage derives from no Renaissance hermetic sources and is founded upon no uniquely hermetic ideas. I think it is important to realize that occultism in the Renaissance was neither univariant nor coeval with hermeticism, but more various in content and pluralistic in its sources than is often recognized.

Third, the first two observations suggest a possible role for Archemastrie in our attempt to understand the extremely diverse products of Dee's intellectual career. Since the work of I. R. F. Calder in the early 1950s, it has been customary to try to unify the varied and apparently contradictory products of Dee's thought through some single philosophical inspiration that might provide a unifying thread tying together all his activities and works and serving as their interpretive key for the historian. It has also been customary to import this philosophical key fully developed from some tradition external to Dee. For Calder, the source was Renaissance Neoplatonism; for French and Yates in her early writings, it was the attitude of the Renaissance magus of the "hermetic tradition," which subsequently Yates variously modulated into the "hermetic-cabalist tradition," the Rosicrucian phase of Renaissance hermeticism, Christian cabala, and "the occult philosophy."[60] Archemastrie as a programmatic practice for investigating nature may well offer the unifying method at which Dee was aiming in his various activities and writings and which he finally articulated in the "Praeface." This would provide an interpretive device intrinsic to Dee, one which would not erode the vast substantive differences among many of Dee's works.

Finally, in addition to the issue of what influence magic had upon Dee's science, there is another historiographic issue involving the relation of magic, science, and Dee. This is the claim of French and Yates that Renaissance magic offered a stimulus to others – particularly artisans, mathematical practitioners, and mechanicians – in developing a scientific attitude because of Dee's linkage of magic, science, and practical mathematics in the "Praeface," which was influential into the seventeenth century.[61] This suggestion can be evaluated only by studying the readers of Dee, how they responded to the "Mathematicall Praeface," and their subsequent achievements; this I have not done. While it is not unlikely that Dee's "Praeface" may have stimulated others to scientific pursuits, it is more doubtful that Dee's magic in the "Praeface" played any role in this. Could the artisans and practitioners, "being vnlatined people, and not Vniuersitie Scholars," whom Dee claims for his audience, have perceived the magical dimension, which is not only cryptic but, I would argue, deliberately obscure? Artephius's *Ars sintrillia* could not have been well known, and alnirangiat is equally obscure without a reference to Avicenna's text. Since Dee gives an explicit reference to Avicenna earlier in the "Praeface,"

I am inclined to think he deliberately concealed the alnirangiat reference by leaving the term in its Arabic form and giving no source. This
intentional obscurity about the magical aspect of Archemastrie is not
surprising, considering that this section immediately follows a long
apologia in which Dee attempts to lay to rest his reputation, whether
imagined or real, as a conjuror.[62] I think Dee presents two faces to his
public, maintaining a disjunction between the ways he presented different facets of his thought. He was open when dealing with a general
audience about mathematics and his practical pursuits in the mathematical sciences, such as navigation, but was guarded about his interests in natural philosophy and more esoteric subjects. These he intentionally obscured, such as Archemastrie; published only in Latin, and
even then cryptically, such as the *Propaedeumata aphoristica* and the
Monas hieroglyphica; or kept entirely private, such as his spiritual
exercises with Kelly. While Dee perhaps articulated a potentially fruitful concept of method including magic, it may have been his ironic fate
both to have contributed to the progress of science among those who
were ignorant of the magical dimension and to have encouraged a less
modern notion of science among those who ignored everything but the
magical and occult dimension. As students of Dee, we need to be careful lest we assume an unwarranted correspondence between different
aspects of Dee's own work and make unwarranted generalizations
about the influence of his ideas.

Notes

1 Francis R. Johnson, *Astronomical Thought in Renaissance England*
(Baltimore, 1937), pp. 151–2; E. G. R. Taylor, *Tudor Geography, 1485–1583*
(London, 1930), p. 103; I. R. F. Calder, "John Dee Studied as an English
Neoplatonist," 2 vols., Ph.D. thesis, Warburg Institute, London University,
1952, I, 640–5.
2 Allen G. Debus, Introduction to John Dee, *The Mathematicall Praeface to
the Elements of Geometry of Euclid of Megara (1570)* (New York, 1975),
pp. 21–2; Debus, *The Chemical Philosophy: Paracelsian Science and
Medicine in the Sixteenth and Seventeenth Centuries* (New York, 1977), I,
38–43; Marie Boas, *The Scientific Renaissance, 1450–1630* (New York,
1962), pp. 184–5.
3 Boas, p. 185.
4 Frances A. Yates, *Giordano Bruno and the Hermetic Tradition* (Chicago,
1964), pp. 148–50; Yates, "The Hermetic Tradition in Renaissance
Science," in *Art, Science and History in the Renaissance*, ed. Charles S.
Singleton (Baltimore, 1967), pp. 259, 261–2; Peter J. French, *John Dee: The
World of an Elizabethan Magus* (London, 1972), pp. 160–2.
5 French, pp. 162–3.
6 Dee, *Mathematicall Praeface*, sig. Aiij[r].
7 Ibid.
8 Ibid., sig. Aiij[v].
9 Ibid. Dee refers to Cusanus's work as the *Experimentes statikall*, by which

he obviously means the *Idiota de staticis experimentis*. Dee does not mention Roger Bacon by name, but the "RB" in the margin, in conjunction with the description of him as a philosopher native to England who wrote at the request of Clement, leaves no doubt whom he meant. Dee's text gives "Clement the Sixt," but clearly Clement IV (Pope 1265–8) is meant, for whom Bacon wrote the *Opus majus*, the *Opus minus*, and the *Opus tertium*.

10 Nicolaus Cusanus, *Idiota de staticis experimentis*, in *Werke*, ed. Paul Wilpert, 2 vols. (Strasbourg, 1488; repr. Berlin, 1967), I, 284, 277–9.

11 The most thorough discussion of Dee's debt to Roger Bacon for the Archemastrie passage is W. R. Laird, "John Dee on Archemastrie in the *Mathematicall Praeface* (1570)," a paper submitted as part of a course in the Centre for Medieval Studies, University of Toronto, which came to my attention only after I had completed the research for this chapter. Our discussions differ little on the contribution of Bacon to Dee's formulation.

12 Roger Bacon, *The "Opus Majus" of Roger Bacon*, ed. John Henry Bridges, 2 vols. (Oxford, 1897), II, 167–8, 172–3, 202, 215–16.

13 Printed in Roger Bacon, *Opera quaedam hactenus inedita*, ed. J. S. Brewer (London, 1859), pp. 523–51.

14 Dee, *Mathematicall Praeface*, sig. Aiij^v.

15 Ibid.

16 A. C. Crombie, *Robert Grosseteste and the Origins of Experimental Science, 1100–1700* (Oxford, 1953), pp. 139–43, 155–62. For more reserved discussions of the meaning of Bacon's "scientia experimentalis," see Stewart C. Easton, *Roger Bacon and His Search for a Universal Science* (Westport, Conn., 1970), pp. 7, 113, 181; and Lynn Thorndike, *A History of Magic and Experimental Science*, 8 vols. (New York, 1922–1948), II (1923), 658. Crombie only briefly mentions Bacon's third prerogative with its occult overtones and does not discuss Bacon's illustrations of the practical meaning of this prerogative at all, let alone with anywhere near the care he devotes to Bacon's work on the rainbow as an example of Bacon's first prerogative. Likewise, despite his comment that Bacon includes within his idea of experience a range of "interior illuminations" which are of greater certainty than the experience of the exterior senses (Crombie, p. 141 n. 2), he does not explore the implications of this. The significance of these two aspects of Bacon for Dee's Archemastrie will be discussed later in this chapter.

17 Charles B. Schmitt, "Experience and Experiment: A Comparison of Zabarella's View with Galileo's in *De Motu*," *Studies in the Renaissance*, 16 (1969), 134–5, 137; repr. in Schmitt, *Studies in Renaissance Philosophy and Science* (London, 1981).

18 Crombie, pp. 278–9, 290–2, 296.

19 Nicholas H. Clulee, "Astrology, Magic and Optics: Facets of John Dee's Early Natural Philosophy," *Renaissance Quarterly*, 30 (1977), 645, 674–5.

20 Dee, *Mathematicall Praeface*, sig. Aiij^v.

21 Attempts have been made to track down these allusions. Calder (I, 479, 779–81; II, 461) concentrates on Artephius's *Ars sintrillia*, and this only in the context of a discussion of Dee's magic and spiritual exercises, without suggesting any significance in relation to the issue of "experimental science." He mentions Naudé's comment that Artephius's philosophy was based on al-Kindī and suggests that this work is the same as one of Artephius's described by Girolamo Cardano in *De rerum varietate* (lib. XVI, cap. XC), which is similar to a magical art attributed to Artephius in

two French manuscripts described by Emile A. Grillot de Givry in *La Musée des sorciers* (Paris, 1929), p. 307. He also suggests that this is related to the "other Optical Science" and relates both to Dee's skrying activities. Calder has assembled some of the important clues, as we shall see below, but he has no evidence indicating any linkage of these things with a specific *Ars sintrillia*. Laird, "John Dee," discovered the Arabic origin and meaning of the term nīrangīyat, but did not find Dee's source and was largely unsuccessful with the *Ars sintrillia*, finding in addition to the well-known but in this case unhelpful published works of Artephius a reference to a manuscript containing an "Arthephij de Opere Solis" claiming to be an extract from "de Arte Sutrillia." Since he was unable to examine the contents of this, he was not able to determine its significance, although the "Arte Sutrillia" sounds similar to *Ars sintrillia*.

22 Manfred Ullmann, *Die Natur- und Geheimwissenschaften im Islam*, Handbuch der Orientalistik, Ergänzungsband, VI, 2 (Leiden, 1972), pp. 360–3; Toufic Fahd, *La Divination arabe: études réligieuses, sociologiques et folkloriques sur le milieu natif de l'Islam* (Leiden, 1966), p. 40.

23 Ullmann, p. 363; H. Ritter and M. Plessner (trans.), *"Picatrix:" Das Ziel des Weissen von Pseudo-Maǧrītī,* Studies of the Warburg Institute, vol. 27 (London, 1962), pp. 10, 155, 253.

24 Fahd, p. 40; *Dictionary of Scientific Biography*, XV (1978), 497.

25 Avicenna, *De divisionibus scientiarum*, in Avicenna, *Compendium de anima et al.*, trans. and ed. Andrea Alpago (Venice, 1546; repr. Farnborough, 1969), p. 142.

26 Dee, *Mathematicall Praeface*, sig. *ij^v: "And secondly, the very name, is *Algiebar*, and not *Algebra*: as by the Arabian *Auicen*, may be proued: who hath these precise wordes in Latine, by *Andreas Alpagus* (most perfect in the Arabik tung) so translated." Dee's copy is in the Bodleian Library, Oxford, with the pressmark Selden 4° A40 Art. Seld. This, and the information on Dee's annotations, were generously shared with me by Andrew Watson of University College, London.

27 Avicenna, p. 142: "Et ex illa [divisiones, vel partes scientiae naturalis ramificatae, vel subalternatae] est scientia alnirangiat, idest scientia artis magicae. & intentio in ea est permiscere virtutes, quae sunt in substantiam mundi terreni, vt adueniat in eo virtus a quo proueniat actio extranea vel mirabilis."

28 Roger Bacon, *Opus tertium*, in Bacon, *Opera quaedam hactenus inedita*, p. 44; cf. Bacon, *"Opus Majus,"* II, 215–16.

29 Avicenna, pp. 139^v–40; see also A.-M. Goichon, *Lexique de la langue philosophique d'Ibn Sīnā (Avicenna)* (Paris, 1938), pp. 240–7, s.v. Ilm.

30 George Sarton, *Introduction to the History of Science*, 5 vols. (Washington, 1927–31), II, 1, 219, and Thorndike, II, 353–4, give some information, not all of which can be accepted without question. G. Levi Della Viola, "Something More about Artephius and his 'Clavis Sapientiae,'" *Speculum*, 13 (1938), p. 80 and n. 6, has questioned the validity of identifying Artephius with any historical personality. For Artephius's reputation, see Levi Della Viola, pp. 80–1; Herbert Douglas Austin, "Accredited Citations in Ristoro d'Arezzo's *Composizione del Mondo*," *Studi Medievali*, 4 (1913), pp. 368–76; Austin, "Artephius-Orpheus," *Speculum*, 12 (1937), pp. 251–4; Thorndike, I, 774, II, 353–4, 655; and Armand Delatte, *La Catoptromancie greque et ses dérivés*, Bibliothèque de la Faculté de Philosophie et Lettres de l'Université de Liége, fasc. XLVIII (Liége and Paris, 1932), pp. 18–23.

Although not necessarily complete, the following is a list of medieval and
Renaissance authors who mention Artephius: Roger Bacon, William of
Auvergne, Ristoro d'Arezzo, Gianfrancesco Pico della Mirandola, Heinrich
Cornelius Agrippa von Nettesheim, Girolamo Cardano, Francis Bacon, and
Gabriel Naudé.

31 See Thorndike, I, 774, and Delatte, p. 21 n. 1, on the twelfth-century
manuscript in Berlin. On the *Clavis sapientiae* and the *Liber secretus*, see
Sarton, II, 1, 219; J. Ferguson, *Bibliotheca Chemica*, 2 vols. (Glasgow,
1906), I, 51; and Dorothea Waley Singer, *Catalogue of Latin and
Vernacular Alchemical Manuscripts in Great Britain and Ireland*, 3 vols.
(Brussels, 1928), I, 128–30.

32 See Austin, "Artephius-Orpheus," pp. 251–4, and Levi Della Viola, pp.
80–1, on Artephius, Orpheus, and Apollonius; Levi Della Viola, pp. 82–5,
reports the discovery of an Arabic original of the *Clavis sapientiae*.

33 Roger Bacon, *Epistola de secretis operibus artis et naturae*, pp. 540, 541,
545; Bacon, *The "Opus Majus,"* II, 208, 209, 212–13.

34 Dee, "Libri antiqui scripti quos habeo anno 1556," Corpus Christi College,
Oxford, MS. 191, fol. 77ᵛ.

35 This is now Corpus Christi College MS. 233, the contents of which date
from the thirteenth through the fifteenth centuries. Once again, I am
indebted to Andrew Watson for this information. Lynn Thorndike and Pearl
Kibre, *A Catalogue of Incipits of Medieval Scientific Writings in Latin*, rev.
and aug. ed. (London, 1963), cols. 9, 297, 774, gives several citations, none
of which appears to be the work in question.

36 Artephius, *Clavis maioris sapientiae*, in *Bibliotheca chemica curiosa*, ed. J.
J. Manget, 2 vols. (Geneva, 1702), I, 503–9.

37 Austin, "Accredited Citations," pp. 371–2.

38 Girolamo Cardano, *De rerum varietate*, lib. XVI, cap. XC, in *Opera omnia*
(1662; repr. New York, 1967), III, 316. This is discussed in Austin,
"Accredited Citations," p. 372, and Delatte, pp. 19–20. Grillot de Givry, p.
307, describes two French manuscripts with the title "L'Art magique
d'Artephius et de Mihinius" (Bibliothèque de l'Arsenal, nos. 2344 and 3009)
whose content appears to be identical with the work described by Cardano.
I have not seen these, and it is not clear whether these are translations of
Cardano or of his source.

39 Cardano, III, 312.

40 Roger Bacon mentions Artephius in connection with the powers of herbs,
stones, and metals to promote health ("*Opus Majus*," II, 208); the ability to
prolong life (ibid., II, 209, 212–3; *Epistola*, pp. 540–1); and his method of
concealing secrets with letters and characters (*Epistola*, p. 545). William of
Auvergne mentions a work of Artephius "de virtute verborum, &
characterum" in both the *De universo* and the *De legibus* (*Opera omnia*, 2
vols. [Paris, 1674], I, 91, cols. 1D–2A; 1064, col. 2F). Ristoro d'Arezzo
reports Artephius's legendary ability to understand the voices of birds and
animals (Austin, "Artephius-Orpheus," pp. 251, 253).

41 Cardano, III, 314–15. See also Austin, "Accredited Citations," pp. 373–4,
and Delatte, pp. 19–20.

42 Giovanni Francesco Pico della Mirandola, *De rerum praenotione*, lib. IX, in
Opera omnia, 2 vols., Monumenta Politica Philosophica Humanistica
Rariora, series I:14 (Turin, 1972; repr. of Basel, 1573), I, 426–7, 469.

43 Austin, "Accredited Citations," pp. 374–5, argues this. He found that
Ristoro's citation of Artephius is based on the work or part of the work
described by Cardano.

44 Thorndike, I, 774; Delatte, p. 21 n. 1. This is Berlin MS. 956.
45 William of Auvergne, *De universo*, II, 3, p. 20, in *Opera omnia*, I, 1057, cols. 2C–D.
46 Ibid.
47 Although no complete manuscript of this work survives that I know of, there is a manuscript fragment (British Library, MS. Sloane 1118) described in the explicit as "Arthephii de opere solis capitulum extractum de arte suttrillia." See Singer, *Catalogue*, I, 130. This "suttrillia" may be a similar corruption of "syntriblia" or "sintrillia."
48 Dee, *Mathematicall Praeface*, sig. Aiij^v. See the text quoted earlier in this chapter (p. 60).
49 Bacon, "*Opus Majus*," II, 215; *Opus tertium*, in *Opera quaedam hactenus inedita*, p. 44; *Part of the Opus Tertium of Roger Bacon*, ed. A. G. Little, British Society of Franciscan Studies Publications 4 (Aberdeen, 1912), p. 48.
50 Dee, *Mathematicall Praeface*, sig. Aiij^v.
51 Ibid., sig. b.j^r; see also sig. b.j^v where Dee speaks of a demonstration according to the rules of perspective as a "demonstration Opticall."
52 Ibid., sig. b.j^v.
53 Ibid., sig. b.iij$^{r/v}$. See also Clulee, "Astrology," pp. 652–77.
54 Clulee, pp. 672–6.
55 Dee, *Mathematicall Praeface*, sig. b.ij^r.
56 Bacon, "*Opus Majus*," II, 169–72.
57 Delatte, pp. 63, 69, 73, refers to a number of sixteenth-century discussions of catoptromancy.
58 Calder, I, 775, reports a manuscript, British Library Add. MS. 36, 674(4), dated 1567, that contains an account of divinatory visions involving Sir Thomas Smith.
59 Dee's accounts of his skrying in the 1580s has a brief reference to what may have been an episode of the same thing in 1569 (British Library MS. Sloane 3188, fol. 5^r).
60 Calder, I, i, 7–18; Yates, *Giordano Bruno*, p. 148; French, pp. 1–3.
61 French, pp. 171–6, 186, states this most directly, but it is based on suggestions in Frances A. Yates, *Theatre of the World* (London, 1969), pp. 5, 17; and Yates, "Hermetic Tradition," pp. 259, 262.
62 Dee, *Mathematicall Praeface*, sigs. Aj^v–Aiij^r.

2

The occult tradition in the English universities of the Renaissance: a reassessment

MORDECHAI FEINGOLD

Whether for better or for worse, it is no longer possible for historians interested in the "scientific revolution" to regard the movement solely in terms of the victory of true and rational scientific ideas over the scholastic and magical modes of thought circulating in the sixteenth and seventeenth centuries. Not only have the attitudes of various men of science toward scholasticism and Aristotelianism been scrutinized, but the extent to which these men created a solely rational construction of reality has also been questioned.[1] Scholars such as Cassirer, Garin, Kristeller, and Yates have redirected our attention to the importance of the "occult tradition" in generating and disseminating the new scientific modes of thought.[2] Their claim is that Neoplatonism, hermeticism, astrology, alchemy, and the cabala – individually or as a unified ideology – had as great an influence on Kepler, Galileo, or Newton as they did on Ficino, Agrippa, and Bruno.

To be sure, not all historians of science share this perspective. Even those who accept the importance of the occult tradition vary in the degree of their commitment. Paolo Rossi, one of the earliest proponents of the occult tradition, has recently voiced certain reservations:

> What started off as a useful corrective to the conception of
> the history of science as a triumphant progress, is becoming
> a retrospective form of historiography, interested only in the
> elements of continuity [between the hermetic tradition and
> modern science] and the influence of traditional ideas.[3]

My purpose here, however, is not to pass qualitative judgments on the impact of the occult tradition on the genesis of modern science. Instead, I wish to question an assumption held both by those who believe that the occult tradition was seminal to the emergence of the new science and by those who believe that the new science triumphed *despite* the occult tradition. Both groups are united in their criticism

73

of the English universities as bastions of backwardness. Historians of science charge that the universities' lingering commitment to a scholastic and Aristotelian framework resulted in their unwillingness to teach and contribute to contemporary scientific modes of thought. The historians of the occult tradition maintain that this identical commitment to scholasticism and Aristotelianism also led to the universities' hostility to the black arts, including numerology; hence the universities' alleged suppression of occult studies.

The evidence for such assumptions is well known. There is scarcely a study of the development of English science that fails to cite the autobiographical account of John Wallis concerning his student days at Cambridge as proof – and occasionally the sole proof – of the lack of mathematics in the university curriculum before 1640.[4] Historians of the occult tradition have made equally wide use of the account of Giordano Bruno's celebrated visit to Oxford in 1583 to prove Oxford's rejection of both the Copernican and the hermetic world views. According to this account Bruno, the prophet of innovative and true ideas, reasoned in vain with the local pedants, who refused to concede defeat, even after each of their arguments was refuted upon their own scholastic ground, using their own scholastic jargon. In Bruno's own words:

> And if you don't believe it, go to Oxford and have someone tell you what befell the Nolan when he disputed publicly with the doctors of theology in the presence of Prince Albert Laski the Pole and representatives of the English nobility. Have them tell you how learnedly he answered their arguments and how fifteen times, for fifteen syllogisms, the poor doctor, whom they put before the Nolan on this grave occasion as the Coryphaeus of the Academy, felt like a fish out of water. Have them tell you with what uncouthness and discourtesy that pig acted, and about the extraordinary patience and humanity of the Nolan, who showed himself to be a Neopolitan indeed, born and raised under a more benign sky. Have them inform you how they put an end to his public lectures and those *de immortalitate animae* and *de quintuplici sphaera*.[5]

Largely on the basis of this incident Frances Yates claimed the existence of a hostility of the early modern English universities to scientific and occultist studies. According to her, Oxford turned its back on the medieval tradition of Roger Bacon and his contemporaries, an act "which generated and increased Aristotelian rigidity." Henry Savile and Richard Hakluyt were "individual exceptions to the predominantly grammarian and unscientific character of Tudor Oxford . . . [in which] the general tone was set by the contentious 'Aristotelian Party' which despised the mathematical sciences."[6]

Yates's verdict on Oxford and Cambridge still holds today. Allen Debus has elaborated on this theme in his many studies of the English alchemical tradition.[7] Peter French, the author of the most recent biography of John Dee, claims that "Oxford and Cambridge, rejecting their heritage, turned to Ciceronianism, which ultimately degenerated into grammatical pedantry." He adds: "Just as scholastic theologians and humanists would have none of Ficino, Pico and Agrippa, so pedants at the English universities came to disapprove of Dee," the result being that "Dee chose to dissociate himself from the developments taking place at the English universities when he found them inimical to his interests." For French, Dee certainly was not a product of sixteenth-century Cambridge, where the occult philosophy was "largely scorned by the new generation of humanists."[8] Similar claims have been made by Nicholas Clulee: "There is no indication that Dee was introduced in any formal way either to Neoplatonism, Hermeticism and the Cabala or to have been instructed in mathematics in general."[9] Finally, a historian interested in detecting Ficino's influence on Shakespeare has agreed with Yates that the absence of any scholarly translations or commentaries on the Florentine are, at least in part, the result of the "deliberate suppression at this time of Neo-Platonism at Oxford."[10]

It is not within the scope of this chapter to examine the complex relations between the "new humanism" and science in general, and the occult sciences in particular. Far too much work remains to be done on the new ideals of education that emerged in post-Reformation England. However, I would like to add that my own research fails to corroborate the popular claim that the mathematical sciences disappeared from the university curriculum in the latter half of the sixteenth century; if anything, they were stronger by the end of this period.[11] Nor did the heirs to Linacre, Colet, and Grocyn despise the mathematical and occult traditions and discourage students from pursuing them. John Caius, Sir John Cheke, Sir Thomas Smith, and even Roger Ascham were all heirs to this tradition. This impression of the English universities as sterile intellectual climates has arisen because the testimony of a John Wallis or a Giordano Bruno has been stressed out of all proportion to the facts, while other relevant evidence has been almost totally neglected. In the course of this chapter I hope to bring forth evidence to suggest a somewhat different picture of the occult tradition in the English universities. I should like to argue that (1) the opposition to Bruno was not necessarily the result of his advocation of Copernicanism and Platonism; (2) regardless of this "notorious" episode, neither Cambridge nor Oxford had any official vindictive or prosecutive ideology against Platonism and the occult tradition; and (3) numerous university men studied and practiced the various components of the occult tradition. Furthermore, those who left the universities after taking a degree – and even those who voiced doubts about certain aspects of their education – still maintained ties with their former colleagues. Thus there evolved a large, somewhat homogeneous intellectual community composed of the generally obscure – and now forgotten – university men as well as the more celebrated men familiar to historians. For the sake of continuity, I shall confine most of my examples to the quarter of a century before and after Bruno's visit to Oxford, that is, from the time of the retraction of the allegedly progressive Edwardian statutes in 1558 until the foundation of the Savilian professorships in geometry and astronomy at Oxford in 1619.

Both admirers and critics of Giordano Bruno basically agree that he was pompous and arrogant, highly valuing his opinions and showing little patience with anyone who even mildly disagreed with him. And yet no one has ever suggested that it might have been Bruno's manner, his language and his self-assertiveness, rather than his ideas, that so offended the reserved Englishmen. Twenty years after the event, George Abbot, the future archbishop of Canterbury, recalled the disputation that had occurred while he was a young Balliol student:

> When that Italian Didapper . . . had in the traine of *Alasco* the Polish Duke, seene our Vniversity in the year 1583, his hart was on fire, to make himselfe by some worthy exploite, to become famous in that celebrious place. Not long after returning againe, when he had more boldly than wisely, got up into the highest place of our best & most renowned schoole, stripping vp his sleeues likes some Iugler, and telling vs much of *chentrum & chirculus & circumferenchia* (after the pronunciation of his Country language) he vndertooke among very many other matters to set on foote the opinion of Copernicus, that the earth did goe round, and the heavens did stand still; whereas in truth it was his owne head which rather did run round, & his braines did not stand stil.[12]

Abbot's disparaging remarks concentrate on Bruno's pretentiousness and conceit and should not be construed as reflecting Abbot's own ideas about science or the occult. His career provides some evidence for his lifelong interest in both. Abbot was the author of a very popular geographical treatise, *A Briefe Description of the Whole Worlde* (1599), and served as patron to such men of science as Samuel Purchas and John Greaves. His splendid library, which included many astronomical and mathematical books as well as occult tracts, still survives and indicates his wide range of interests.[13]

Abbot was not alone in his irritation with Bruno's pretentiousness. In 1584 one N.W. addressed a letter to Samuel Daniel in which he described Bruno as "that man of infinite titles among other phantastical toyes."[14] Indeed, it appears that Bruno's name became synonymous with contentiousness. Such, at least, was the judgment of Richard Hooker when he described the eccentric and ever-combative Hebrew scholar Hugh Broughton as "an English Jordanus Brunus."[15]

There are other reasons why the Englishmen became frustrated with Bruno. Bruno himself concedes that when he argued on specific points of Copernican theory, his opponents were able to fetch the text and show him that his interpretation did not conform with the text.[16] However, neither Bruno, nor for that matter Yates, is troubled by this incommensurability of discourse: "The truth is," Yates soberly writes, "that for Bruno the Copernican diagram is a hieroglyph, a Hermetic seal hiding potent divine mysteries of which he has penetrated the secret."[17] Hence, Bruno could easily initiate a discussion of the heliocentric theory only to shift the argument and introduce a multitude of factors – some of them extraneous – to the issue at hand. Thus even a friend could write: "I heard from the greatest of men assertions strange, absurd and false, as of a stony heaven, the sun bipedal, that the moon doth contain many cities as well as mountains, that the Earth doth move, the other elements are motionless and a thousand such things."[18] Bruno's opponents, then, certainly misunderstood him, but not always because they were unfamiliar with, or opposed to, the ideas he presented.

Indeed, it should be stressed that the audience addressed by Bruno was not necessarily hostile. At a later date Bruno acknowledged the kindness of Tobie Matthew and Martin Culpeper, the heads of Christ Church and New College, respectively, during this Oxford visit. Culpeper may well have been the person who identified Bruno's reliance on Ficino during the disputation.[19] Certain other friends, such as Gwyne and Gentili, were also present on this occasion. As for the issue of Copernicanism, many in the audience could still remember Henry Savile's Oxford lectures a decade earlier, which included a long and detailed account of the Copernican theory.[20] Savile himself was present on this occasion, as were many of the people we shall have occasion to mention when we discuss the occult tradition at Oxford.

University records suggest that the attitude of the universities to occult pursuits was similar to that of the state; private study was tolerated as long as it did not involve any unlawful casting of the nativities of monarchs or debasing of coins and did not result in any scandalous accusations of cheating or witchcraft. In general, the university and college statutes are extremely reticent about the limits of intellectual inquiry, and the little evidence we have about the official attitude toward occult pursuits comes from the records of the occasional visit-ations.[21] For example, during the 1520 visitation of Oriel College, Ox-ford, Walter May, a fellow, was accused of publicly practicing "judiciaria astronomica."[22] Similarly, during the 1566 visitation of New College, Oxford, Thomas Hopkins, a junior member of the house, admitted to possessing "a book of conjurations" that had been given to him by John Fisher, another New College member. Fisher, in turn, had been given the book by an M.A. of Christ Church. The authorities admonished Hopkins "not to use the art of magic."[23] The various halls, which were subject to less rigorous discipline than the colleges, were particularly concerned about the study of the "unlawful arts" during the visitations. Among the Oxford University archives there is a vol-ume containing a series of articles of visitation of Oxford halls between 1580 and 1649. These articles contain the standard question: "Item, whether there be anie that do studie anie unlawfull studie or science in your house and who they be?"[24] Finally, it is worthy of note that the only explicit distinction between the lawful and the unlawful was made by Sir Henry Savile when he founded his professorship of as-tronomy at Oxford in 1619. The professor, Savile stipulated, "must understand, however, that he is utterly debarred from professing the doctrine of nativities and all judicial astrology without exception."[25]

Additional evidence concerning the attentiveness of the university officials to the interest in the occult sciences is to be found in the large number of questions relating to the occult approved each year by con-vocation for disputation. There exists an uninterrupted succession of questions dealing with astrology, alchemy, and magic. The topics range from general questions about the lawfulness of such studies and whether they are sciences at all, to such narrow topics as the possibility of transmuting base metals into gold and of using spells to cure dis-eases.[26] Frequently the respondents were expected to argue against the occult sciences, thus reflecting a general cautiousness on the part of the university officials and their hesitance to allow a relative laxness to extend into the important and widely attended public exercises in July. But occasionally some freedom for divergence was allowed and the respondents were not categorically ordered to refute the tenets of the occult studies.

An analysis of student notebooks containing mock disputations in preparation for the public disputations substantiates this interest in the occult. A few examples follow. An Oxford student of the late sixteenth century filled an entire notebook with a mixture of theological and occult issues; a contemporary at Christ Church made similar notes.[27] A notebook of 1607 contains notes of a Cambridge student on math-ematical and astronomical issues and indicates an interest in Roger Bacon and John Dee.[28] Two years earlier another Cambridge student had devoted an entire commonplace book to Aristotle, with the exception of the following passage on the Platonic idea that "the soul of the man is the man":

> That men are nothing else than their souls, only Plato among all the philosophers dared assert. This opinion is acceptable to me, not because Platonic, though Plato's authority carries more weight with me than that of any other philosopher, but because his opinion seems to me to approach nearer the truth.[29]

The above evidence suggests that the universities were not as inimical to the study of the occult sciences as is often believed. Despite an official stance against the black arts, only in official circumstances, such as during the visitations, were students of the occult rebuked, and then it appears not very seriously. In this connection the overtures of the universities to the sons of the upper classes should be noted. Elsewhere I have argued that the universities sometimes underwent cosmetic surgery to correct the view they presented to the upper classes. Some of this surgery involved their attitude to the mathematical, and certainly to the occult, sciences. For the upper classes these studies carried dangerous connotations, as is made clear by Francis Osborn when he described the state of education at the turn of the seventeenth century:

> My Memory reacheth the time, when the Generality of People thought her [Mathematics] most useful *Branches, Spels* and her *Professors, Limbs of the Devil*; converting the Honour of *Oxford*, due for her (though at that time slender) Proficiency in *this study*, to her shame: Not a few of our then foolish *Gentry*, refusing to send their Sons thither, lest they should be smutted with the *Black-Art*.[30]

A letter of advice addressed by James, Lord Ogilvy, in 1605 to his grandson confirms this prejudice:

> And seeing, now a days, many young scholars give themselves curiously to understand magick and necromancy, whilk are the greatest sins against God that can be, and has been the destruction of both body and soull of many and their houses, I will beseech you in the name of God never to let that enter your mind.[31]

Sensitive to such criticism and eager to attract the affluent and well-born, the university officials sometimes carried out changes in the official curriculum, although the actual teaching was rarely affected.[32] However, for a better idea of the nature and extent of the study of the occult sciences, it is necessary to determine the identity of those university men interested in the occult.

We might begin our survey with the most celebrated English magicians, John Dee and Robert Fludd. Dee, who matriculated at St. John's College in 1542, almost certainly spent much of his time at Cambridge in the study of alchemy and astrology. In his *Monas hieroglyphica* (1564), for example, Dee mentioned that in Paris two years earlier he had delivered a lecture which had incorporated "whatever twenty years' hard work in the study of alchemy had taught him."[33] If indeed this is true, these studies date back to the year Dee went up to Cambridge. Similarly, according to Dee's own testimony, his *Propaedeumata aphoristica* (1558) also dates back at least a decade.[34] Dee's reputation as a conjurer also originated during his student days at Cambridge. Having been elected the underreader of Greek as well as a fellow of the newly established Trinity College, Dee went on to produce Aristophanes' *Pax*; according to Dee's account, "with the performance of the *Scarabeus* his flying up to Jupiter's pallace, with a man and his basket of victualls on her back: whereat was great wondering, and many vaine reportes spread abroad of the meanes how that was effected."[35]

Fifty years later, in 1592, Robert Fludd matriculated at St. John's College, Oxford. As was the case with Dee, Fludd's published work was the labor of many years. Thus we learn from Fludd's own testimony that while a student of Thomas Allen he had gained sufficient reputation for his astrological skill for his tutor to ask his assistance in discovering a thief who had robbed him. The incident, Fludd relates, occurred "when [he] was so deeply engrossed in [his] treatise on music that [he] had hardly left [his] room for a week." Clearly then, not only were the foundations for Fludd's astrological studies laid at Oxford; his musical theories, which "are the foundation ideas of [his] voluminous works," also date back to this period.[36]

Dee and Fludd were not alone in their pursuit of mathematical and occultist studies at Oxford and Cambridge, and it is unwise to take occasional statements to this effect at their face value. For example, writing to Lord Burghley in 1563, Dee claimed that although the universities had many men in divinity and the learned tongues, yet "Our cuntry hath no man (that I ever yet could here of) hable to set furth his fote, or shew his hand; as in the Science *De Numeris formalibus*, the Science *De Ponderibus mysticis*, and the Science *De Mensuris Divinis*."[37] Dee's self-aggrandizement must be viewed in the context of his attempt to procure patronage and financial security and not as a critique of the universities. And even if very few of these contemporaries could rival either Dee or Fludd in prominence and reputation, many shared the same interests. However, I should like to emphasize that this is not intended to be an exhaustive account; more material awaits historians willing to sift through the masses of manuscripts in British and Continental libraries.

Dee's mentors and patrons from his Cambridge days onward were Sir John Cheke and Sir Thomas Smith, both of whom were interested in astrology and alchemy as well as in mathematics and astronomy.[38] However, Cheke and, especially, Smith served as patrons to other scholars as well. Cheke exerted a strong influence over Roger Ascham's intellectual development and initiated and encouraged the mathematical studies that culminated in Ascham's appointment as university lecturer in mathematics between 1539 and 1541. Cheke and Smith both served as his patrons throughout his career, and on at least one occasion Ascham applied for Smith's advice on astrology.[39] Even more important was Smith's influence on the young Gabriel Harvey. Smith guided Harvey through his studies, entertained him in his house, and in 1570, the year Harvey graduated B.A., obtained for him a fellowship at Pembroke Hall. For the rest of his life Harvey devoted much time to the mathematical as well as the occult sciences. Harvey's relationship with Smith almost certainly contributed to these interests. Direct testimony to this effect is to be found in a manuscript of 1567 entitled "Visions," given to Harvey by Smith shortly afterward. Among Harvey's annotations to the text we read: "Certaine straung visions, or apparitions of memorable note. Anno 1567. Lately imparted unto mee for secrets of mutch importance. A notable journal of an experimental magitian."[40]

The extensive marginalia in many of Harvey's books and manuscripts allow us to follow the course of his studies in mathematics, astronomy, alchemy, and astrology, as well as to identify contemporary students interested in such subjects. John Caius, for example, is known to have employed astrology in his medical studies, but Harvey's notes make clear that Caius's range of interests was wider and included magic. Harvey obtained a manuscript "found amongst the paper bookes, & secret writings of Dr Caius" containing extracts from Agrippa and Petrus de Abano as well as miscellaneous conjurations in Caius's handwriting.[41] Following the death of Caius in 1573, this manuscript passed into the hands of John Fletcher, fellow of Caius College, a talented mathematician, and an astrologer. Although Fletcher graduated B.A. only in 1581, already the previous year Harvey had named him – together with Thomas Blundeville, Thomas Hood, and Christopher Heydon – as one of the most promising mathematicians of the day.[42] Until his death in 1613, Fletcher lived in Cambridge, where he taught and collaborated with such mathematicians as Henry Briggs and Edward Wright. However, Fletcher's reputation suffered as a result of his occult studies. Described as "in arcana naturae penetrare ausus est" in the annals of Caius College, on at least one occasion his astrological practices involved him in a lawsuit. Fletcher also greatly assisted Sir Christopher Heydon in the composition and publication of his *Defence of Judicial Astrology* (1603).[43]

Another of Harvey's notes in the Caius–Fletcher manuscript refers to Dr. William Butler of Clare Hall. Butler was considered one of the best physicians of the day and was consulted by King James I as well as by many of the nobility. But Butler's advice on astrological and alchemical matters was also eagerly sought. Known to have saved the life of Nicholas Ferrar, he probably exerted some influence on Ferrar's astrological studies. He also received applications to impart his knowledge of alchemy, and some of his alchemical and astrological notebooks still survive. Hence the significance of the following comment by Harvey: "The best skill that Mr Butler physician had in nigromancie, with Agrippa's Occulta philosophia, as his coosen Ponder upon his oathe after repeated seriously intimated unto me."[44]

Quite often individuals have provided us with firsthand testimonials about their occult studies while at university in the form of recantations. One such man was Henry Briggs. Briggs matriculated at St. John's College, Cambridge, in 1577, proceeded M.A. in 1585, and went on to become one of England's foremost mathematicians of the first half of the seventeenth century. We know that Briggs studied with John Fletcher, collaborated in astronomical observations with Edward Wright, and delivered highly successful lectures on Ramus's geometry at Cambridge. However, owing to the testimony of the Puritan controversialist John Geree, who had been Briggs's student and friend while the latter served as Savilian professor of geometry at Oxford, we also learn of Briggs's youthful studies in astrology:

> This loving friend of mine, upon a question moved to him by me, touching judiciall Astrology, told me this remarkable story touching himselfe, when he came to *Cambridge*. First, he thought it was a fine thing to be of Gods Counsell, to foreknow secrets, and resolved to have that knowledge what labour soever it cost him: And so early applyed himselfe to the Study of the Mathematicks, beginning with Arithmetick, and so to Geometry and Astronomy, and to lay a good foundation, he left none of these Arts till he had attained exactnesse in them. The foundation thus layed, he then applyed himselfe to his maine scope, the search of Judiciall Astrology: But there he found his expectation frustrate, there was no certainty in the rules of it; when he had tired his body and wits in vaine, he was much dejected with the frustrating of his expectation. At last he repayred to a man in *Cambridge* famous in this Art, and a practitioner in Prognostications by it; to him he made his mone what paines he had taken to be an expert Astrologer, and how the uncertainty of

the Rules in that Art, did now defeat his hopes. The Astrologers reply was, that the Rules of that Art were uncertaine indeed, neither was there any cure for it: whereupon Mr. *Brigs* relinquisht that study.[45]

An exact contemporary of Briggs at Cambridge was the famous Puritan polemicist William Perkins, who matriculated at Christ's College in 1577 and took his M.A. in 1584. During these years Perkins was deeply involved in the mathematical and occult sciences, pursuits that somewhat soiled his character. As Thomas Fuller expressed it when he came to defend Perkins,

> When first a Graduate, he was much addicted to the study of naturall Magicke, digging so deepe, in natures mine, to know the hidden causes and sacred quallities of things, that some conceive that he bordered on Hell it selfe in his curiosity. Beginning to be a practitioner in that *black Art*, the blacknesse did not affright him but name of Art lured him to admit himselfe as student thereof.

Perkins went on to denounce his occult studies, publishing an attack on astrology, *Foure Great Lyers, Striving Who Shall Win the Silver Whetstone*, in 1585, the year after he proceeded M.A. In this tract he admitted:

> I have long studied this Art, and was never quyet, untill I had seene all the secrets of the same: But at ye length, it pleased God to lay before me ye prophanenesse of it, nay, I dare boldly say, Idolatry, although it bee covered with fayre and golden shewes, therefore that which I will speake with griefe, I will desire thee to note with some attention.[46]

Such criticism of the occult does not necessarily mean that the critic abandoned all interest in the subject. A good example is William Fulke. Fulke matriculated at St. John's College, Cambridge, in 1555, graduated B.A. in 1558, and spent the following four or five years at the Inner Temple. In 1560 Fulke published an attack on astrology, *Antiprognosticon*, which reflects his disillusionment with his studies ever since his university days. Yet, as Fulke's recent biographer has noted, after his return to Cambridge around 1563 – where he eventually became master of Pembroke Hall – Fulke "allowed himself to come rather more under the influence of Neoplatonic elements in contemporary scientific thought," though his position shows some inconsistencies. During the 1570s Fulke published an astrological game, and at least one paragraph of an unpublished theological manuscript is "of strongly hermetic character on astrological talismans." Similarly, during a theological disputation in 1581 Fulke probably used Hermes Trismegistus as an authority, a fact that was suppressed from Fulke's published version of the debate but is revealed by the Catholic account.[47]

Of the alchemists who are known to have studied and practiced at Cambridge, mention should be made of Samuel Norton, great-grandson of the celebrated Thomas Norton. Although Samuel never took a degree, he spent some years at St. John's College, where in 1574 he translated George Ripley's "Bosome Book" into English. During this time he was also occupied in the composition of his "Key of Alchemie," dedicated to Queen Elizabeth and bearing the date, 10 July 1577, and the place, St. John's College. Norton probably remained at Cambridge until 1584, at which time his father died and he inherited the family estate. He nevertheless continued to devote much of his time to alchemy, and many of his treatises were published posthumously.[48]

Totally forgotten today is John Tichborne, who matriculated at Clare Hall in 1584 but migrated to Trinity College shortly afterward. Tichborne proceeded M.A. in 1592, was elected a fellow of Trinity, and was created D.D. in 1605. We know of the nature of his studies from a few of his surviving manuscripts. These "contain the complete Latin text and a complete English translation of [Paracelsus's] *De natura rerum* and *De natura hominis* . . . both taken from the Forberger edition of 1573, as well as the English translation of texts contained in Bodenstein's 1572 edition of *Metamorphosis, seu, de natura rerum.*"[49]

Occasionally an inventory offers a glimpse into alchemical interests. Thus John Rodeknight, a fellow of Queens' College who died in 1615, left behind such experimental apparatus as "a glass Limbeck, a still, six long glasses, three stones and one brass mortar, some glasses of distilled waters."[50]

University members were sometimes known to have participated in occult practices. There exists a manuscript recording a spiritual seance that took place at Cambridge in 1557 and was attended by members of the university. This account subsequently passed into the hands of two magicians practicing at Oxford.[51] Some years later two young scholars of King's College, Cambridge, associated with John Heron, who gained notoriety as a conjurer and necromancer. Both were subsequently recorded in the annals of their college as "juniores socii recessere a mathematicis, et ad artes daemonicas se contulerunt."[52]

A similar picture emerges from a study of the occult at Oxford. Perhaps the most eminent Oxford figure from 1570 until his death in 1632 was Thomas Allen of Gloucester Hall. Described as a second Roger Bacon, Allen was also generally regarded as a magician. He was perhaps the most influential teacher of the mathematical sciences of the day and collaborated closely with generations of students and practitioners, including Thomas Harriot, Sir Kenelm Digby, Sir Thomas Aylesbury, and Sir John Davies. Allen was also a close friend and associate of John Dee. The two served as consultants to Robert, Earl of

Leicester, and were frequently spoken of as the earl's conjurers. Dee also gave Allen many of his manuscripts, together with his famous mirror, which had the property of casting and inverting images. Allen, in turn, seems to have made over to Dee the services of the notorious Edward Kelly, who then served as Dee's alchemist. Like Dee, Allen possessed a large and rich library of which only a portion survives. Nonetheless, even this portion suggests the scope of his astrological, alchemical, and magical studies, including as it does many treatises and fragments by Roger Bacon, Raymond Lull, and Hermes Trisme-gistus.[53] Allen's library, like Dee's, was available to anyone who wished to consult it. And many did. Brian Twyne, the Oxford math-ematician and antiquarian, used Allen's extensive collection and was given certain of Allen's manuscripts.[54] Similarly, while a student at Christ Church in 1616–17, Robert Payne – mathematician and future collaborator with Sir Charles Cavendish and Thomas Hobbes – made copious notes from various manuscripts of Roger Bacon scattered among the collections of Allen, Twyne, and John Prideaux, the rev-erend president of Exeter College.[55]

Gloucester Hall appears to have been an important center for occult studies. In his autobiographical account, Thomas Hodgson, who at the age of thirty-five converted to Catholicism, recorded that he had stud-ied at the Hall for seventeen years (1581–98), devoting his time mainly to astronomy and judicial astrology, and later to medicine.[56] Thomas Gent, a member of Gloucester Hall from the 1580s until his death in 1613, was a close friend of Allen. He was also a member of the circle that revolved around William Gilbert and included Dudley Carleton, John Chamberlain, and Mark Ridley. Gent's pursuit of the sciences is also suggested by his donation of four hundred scientific and medical books to the Bodleian Library in 1600. The collection included the works of Pico della Mirandola, Bonatus, Della Porta, Hermes Tris-megistus, Roger Bacon, and Ficino; numerous astrological volumes; and, among the mathematical and astronomical books, the works of Co-pernicus, Clavius, Commandino, Tartaglia, Regiomontanus, and Peur-bach.[57]

John Delaber of Christ Church studied medicine in Basel during the 1510s, before serving as principal of Gloucester Hall from 1581 until his resignation and return to Christ Church in 1593. Delaber appears to have been one of the first to establish a chemical laboratory at Ox-ford. Writing to one of his patients in 1596, Delaber complained of his inability to obtain various chemical remedies in London: "I am forced now to bylde a Laboratorie or Styllhouse of myne owne and am at this present setting upp of my furnasses."[58] Delaber was certainly not alone in his chemical interests. In short succession Oxford enjoyed two Regius professors of divinity who were noted more for these secular studies than for their theological teachings. Edward Cradock, who proceeded M.A. in 1559 from Christ Church, served as Margaret professor from 1565 to 1594. Described by Wood as addicted "much to chymistry," Cradock "spent many years in obtaining the Elixir . . . and was counted one of the number of those whom we now call Rosycrucians." Cradock was also a friend of John Dee, and in his diary the latter recorded a three-day visit to Cradock at Oxford in October 1581. The nature of Cradock's alchemical interests is suggested by some surviving compositions, two of which are dedicated to Queen Elizabeth. To my knowledge, no one has yet studied these manuscripts.[59]

Even less is known about Cradock's successor, John Williams, who was elected a fellow of All Souls College in 1569, Margaret professor in 1594, and Principal of Jesus College in 1602. He also shared the widespread Oxonian interest in Roger Bacon, and in 1590 published an edition of the latter's *De retardandis senectutis accidentibus & sensibus conservandis*.[60]

As mentioned previously, we sometimes hear about a student's astrological or alchemical pursuits only when, and if, he ran into trouble with the authorities. Thus, for example, in 1570 John Bulkeley of New Inn Hall was arrested and accused of assisting one William Bedo in attempting to debase silver. Bulkeley, who is known to have been a keen mathematician and a correspondent of Thomas Harriot, testified that he had read to Bedo out of "a booke made by John Baptista Porta Neappolitanus who wretyth of naturall magyge wherein there were soundry experyments as well of metalles as of other thinges." Following Bulkeley's arrest, all "such bookes as . . . [he had in his chamber] towching the art of estromancy gematry and alcamistrye" were confiscated.[61]

In a similar manner Adam Squier, who served as Master of Balliol College from 1571 until 1580, was almost expelled from his mastership as a result of his having sold familiar demons "to help the purchaser to win at dice."[62] In 1591 even Thomas Allen was charged with having assisted some Catholics with astrological predictions.[63]

Oxford had a number of astrologers as well as critics of astrology who were exceptionally versed in contemporary astronomical and astrological literature. Thomas Heth, a fellow of All Souls College from 1567 until about 1583, was a skillful mathematician and astrologer, highly regarded both by Allen and Dee. In 1583, the year of Bruno's visit, Heth published a small treatise directed against the predictions of Richard Harvey concerning the effects of the conjunction of Jupiter and Saturn to occur later that year. He corrected certain of Harvey's mathematical errors and went on to complain of "simple" astrologers who were ignorant of "Copernicus his hypotheses, Reinholts observations, or Peurbachius."[64] A slightly older contemporary was Richard

Forster, a fellow of All Souls College from 1562 to the late 1570s, who also combined astronomical and astrological studies. Forster's interests are evident from various sources. In 1573 he heavily annotated a copy of Eschenden's *De stellarum conjunctionibus*, and a year later he published his *Ephemerides meteorographicae*, dedicated to the Earl of Leicester. He also wrote an unpublished commentary on Ptolemy, assisted Sir Christopher Heydon in his astrological studies, and corresponded with Clavius and Magini about astronomy and astrology.[65]

Another student who developed an early interest in astrology as well as in astronomy was William Camden, who studied at Oxford from 1566 until 1573. Certain of Camden's astrological notes from this period still survive, as does his copy of Cyprian Leowitz's *De coniunctionibus*, acquired on 18 April 1573. Later in life Camden testified to the diligence with which he observed the new star of 1572, and it is quite possible that he acquired his copy of the 1566 edition of Copernicus's *De revolutionibus* at about this time. Quite possibly Camden's acquaintance with Dee also began during this Oxford stay, for by August 1574 Dee had written Camden a long letter in which he defended, among other things, his *Propaedeumata aphoristica*.[66]

We might conclude our brief survey with the two most famous "Aristotelians" produced by sixteenth-century Oxford: John Case and John Rainolds. In his biography of Case, Charles Schmitt concludes that Case "shows himself heir both to Aristotle and to the Pico–Ficino–della Porta tradition which will culminate in Bacon's thoughts on the same subject a few years later." Indeed, Case's work demonstrates the eclecticism of a man whose "primary allegiance was to Aristotle." A firm believer in astrology and inclined toward alchemy, Case believed in the possibility of the transmutation of metals. He "sang the praises of Roger Bacon" and was willing to accept certain aspects of the corpus of Haly, Bonatti, and Agrippa. Finally, he "adhere[d] to the *prisca* tradition," while at the same time he was able to emphasize "the creative aspect of man's abilities to formulate new knowledge and techniques." Case was one of the more influential and popular teachers at Oxford and was allowed to continue teaching and preparing students for the B.A. despite the fact that he had to relinquish his fellowship at St. John's College at the time of his marriage. Given Case's eclectic beliefs, it is interesting to speculate upon what he might have taught his students.[67]

John Rainolds is another example of a man who tried to resolve the apparent inadequacies of Aristotelianism without falling into the extremism of such innovators as the Paracelsians. In his discussion of Rainolds's 1570 lectures on Aristotle's *Rhetoric*, James McConica notes that for neither Rainolds nor Case was Aristotle an "ossified legacy, but a convenient vehicle, entrenched in the arts curriculum, to mobilize the vast and heterogeneous access of new material that invaded the university in the sixteenth century.'' Rainolds publicly proclaimed the possibility of criticizing Aristotle and commended Ramus's method, but at the same time he sought the via media because of his "love of moderation." Hitherto neglected manuscripts shed even more light on Rainolds's views. Only two years before delivering his lectures on Aristotle, Rainolds delivered an oration, "In Praise of Astronomy," as part of his M.A. exercises. Remarkable in the oration is the extent and variety of sources consulted by Rainolds. Classical in orientation, the text is nevertheless heavily influenced by the Platonic and hermetic traditions; Plato, Ficino, and Hermes Trismegistus are frequently quoted as authorities. Rainolds also dwells on the relevance of astronomy to all aspects of life, from navigation and agriculture to medicine and astrology. Indeed, Rainolds's astrological philosophy is visibly tinged with hermeticism, affirming as it does both the influence of astrology on all aspects of life and its predictive powers. Unlike Case, however, who published philosophical works, Rainolds limited himself to theological publications. Hence we are unable to elaborate upon his position. Our only additional evidence is the unpublished catalogue of his vast library. Although the library consists largely of theological books, it includes a large number of mathematical, astronomical, and medical works. In addition, the important figures of the occult tradition are all heavily represented: Plato, Plotinus, Pyrro, Lucretius, Hermes Trismegistus among the ancients; Nicholas of Cusa, Ficino, Della Porta, Reuchlin, Agrippa, Trithemius, Paracelsus, and Libavius among the moderns. By no means was Rainolds an occultist. But he certainly was well acquainted with the occult literature, and in his early career at least, he was apparently influenced by it.[68]

This long list of examples may have tried the reader's patience. However, without these examples my claim that Dee and Fludd were not singular in their broad knowledge of, or commitment to, the occult sciences would be invalid. If we are to view Dee and Fludd as part of a larger English phenomenon, then only a prosopographical approach will enable us to arrive at meaningful conclusions about the nature and scope of this occult tradition.

Traditionally, historians have been interested in the champions of any given cause, be they such successful scientific innovators as Kepler, Galileo, and Newton, or such colorful radicals as Bruno, Dee, and Fludd. But between these vanguards blasting forward in new, but different, directions lies a vast and virtually unexplored terrain occupied by middle-of-the-road contemporaries. Like the more extreme proponents of the sciences, these men to a large extent shared the knowledge of, and overall commitment to, the "new" as well as the "occult" sciences. What they lacked was not so much originality – for there were contemporaries as ingenious as Dee or Fludd – but some elusive quality such as conviction or courage: whatever it is that makes a person consciously choose to publish his ideas and defend them at all costs. Throughout their entire lives Dee and Fludd strove to legitimize occult studies, make them distinguishable from magic, atheism, and popery, and convince a skeptical audience of their truth and superiority. Patronage remained a necessity in their careers because without it they would have been totally helpless against their critics. Most contemporaries were not so daring. Perhaps they lacked the ego mandatory for such an undertaking or perhaps their natural combativeness was harnessed by religious constraints and public opinion. In one respect the almost compulsive need of Dee and Fludd to publish might have hampered the contemporary study of the occult sciences. Too much public exposure was at direct odds with the traditional secrecy associated with the occult, while polemics about such delicate issues as religion, atheism, and magic provided critics with ammunition and quite often forced the authorities into taking a public stance against the black arts.

If we accept this claim, that despite differences of temperament, Dee and Fludd shared their occult interests with a significant number of contemporaries, then a study of the events taking place within the English universities becomes particularly relevant. At least until the middle of the seventeenth century the occult tradition was essentially an intellectual tradition. Most of its practitioners were university-educated men, first introduced to occult literature at Oxford and Cambridge, where they formed what often evolved into lifelong friendships with other practitioners. In the course of this chapter I have tried to identify some of these university members and trace their occult interests back to their student days. Further research into largely neglected archival material should give us a better basis to decide to what extent the universities played an important role in the occult tradition in England.

Notes

1 See, for example, Charles B. Schmitt, "Philosophy and Science in Sixteenth-Century Universities: Some Preliminary Comments," in *The Cultural Context of Medieval Learning*, ed. J. E. Murdoch and E. D. Sylla (Dordrecht, 1975), pp. 485–530; "Towards a Reassessment of Renaissance Aristotelianism," *History of Science*, 11 (1973), pp. 159–83; James K. McConica, "Humanism and Aristotle in Tudor Oxford," *English Historical Review*, 94 (1979), pp. 291–317.

2 For a useful survey of the literature, see Robert S. Westman, "Magical Reform and Astronomical Reform: The Yates Thesis Reconsidered," in R. S. Westman and J. E. McGuire, *Hermeticism and the Scientific Revolution*,

(Los Angeles, 1977), pp. 1–91; Charles B. Schmitt, "Reappraisals in Renaissance Science," *History of Science*, 14 (1978), pp. 200–18.

3 Paolo Rossi, "Hermeticism, Rationality and the Scientific Revolution," in *Reason, Experiment and Mysticism in the Scientific Revolution*, ed. M. L. Righini Bonelli and W. R. Shea (London, 1975), p. 257.

4 C. J. Scriba, "The Autobiography of John Wallis, F.R.S.," *Notes and Records of the Royal Society*, 23 (1970), pp. 17–46.

5 Giordano Bruno, *The Ash Wednesday Supper*, ed. and trans. E. A. Gosselin and L. S. Lerner (Hamden, Conn., 1977), pp. 186–7.

6 Frances A. Yates, "Giordano Bruno's Conflict with Oxford," *Journal of the Warburg and Courtauld Institutes*, 2 (1938–9), pp. 230–1; *Giordano Bruno and the Hermetic Tradition* (London, 1964), pp. 167–8, 205–11.

7 Allen G. Debus, *Science and Education in the Seventeenth Century: The Webster–Ward Debate* (London and New York, 1970), pp. 10–11; *The Chemical Philosophy, Paracelsian Science and Medicine in the Sixteenth and Seventeenth Centuries* (New York, 1977), I, 211–13; II, 382–5.

8 Peter J. French, *John Dee: The World of an Elizabethan Magus* (London, 1972), pp. 22–7.

9 Nicholas H. Clulee, "Astrology, Magic, and Optics: Facets of John Dee's Early Natural Philosophy," *Renaissance Quarterly*, 30 (1977), p. 637.

10 Terry Hawkes, "Ficino and Shakespeare," *Notes & Queries*, 203 (1958), p. 186.

11 Mordechai Feingold, *The Mathematicians' Apprenticeship: Science, Universities and Society in England, 1560–1640* (Cambridge, 1984).

12 Quoted by Robert McNulty, "Bruno at Oxford," *Renaissance News*, 13 (1960), pp. 302–3.

13 P. A. Welsby, *George Abbot, the Unwanted Archbishop, 1562–1633* (London, 1962); E. W. Gilbert, *British Pioneers in Geography* (Newton Abbot, 1972), pp. 43–5; *The Private Diary of Dr. John Dee . . .*, ed. J. O. Halliwell (-Phillips), Camden Society (London, 1842), p. 49; A. Cox-Johnson, "Lambeth Palace Library 1610–1669," *Transactions of the Cambridge Bibliographical Society*, 2 (1955), p. 109.

14 Quoted in Joan Rees, *Samuel Daniel: A Critical and Biographical Study* (Liverpool, 1964), p. 3.

15 Corpus Christi College, Oxford, MS. 303, fol. 208: Richard Hooker to John Rainolds, ca. 1591.

16 Bruno, pp. 192–3, nn. 56–8.

17 Yates, *Giordano Bruno*, p. 241.

18 *Francisci et Johanni Hotomanoreum epistolae* (Amsterdam, 1700), p. 333: Alberico Gentilis to John Hotoman, 8 November 1583. The English trans. is in Dorothea W. Singer, *Giordano Bruno: His Life and Thought* (New York, 1950), p. 43.

19 Giordano Bruno, *Dialoghi italiani* (Florence, 1957), p. 212; McNulty, p. 304.

20 Bodleian Library, Oxford, MSS. Savile 26–32.

21 The statutes make one thing clear: Although Aristotle was considered the prime authority on most subjects, Plato continued to be taught. The Edwardian statutes of 1549 assigned Plato, in addition to Aristotle, to all students of philosophy. The Elizabethan statutes of 1570 assigned Plato to the students of both philosophy and cosmography. Various college statutes also assigned Plato for their internal lectures. Trinity College, Oxford, prescribed Plato for the auditors of the philosophy lectures, while Clare

Hall, Cambridge, used Plato in conjunction with the Greek lectures. The
statutes of St. John's College, Cambridge (1545), specifically stipulated that
no lectures were to be delivered on philosophers other than Aristotle, with
the exception of Plato, and here the *Timaeus* was designated as the
appropriate text. Finally, the statutes of Trinity College, Oxford (1560),
prescribed that "primo anno Euclidis elementa, ita enim aditus fiet in
Platonis et Aristotelis . . . Secundo anno Platonem Latinum faciant et
philosophia Ciceronis Graeca." Even the Savilian statutes that initiated the
lectureships in geometry and astronomy and were unique in differentiating
between lawful and unlawful studies still regarded the study of Plato as
essential. Enumerating the qualifications of his intended professor, Savile
decreed that he be "very well versed in mathematics . . . after having in the
first instance drawn the purer philosophy from the fountain of Aristotle and
Plato" (*Collection of Statutes for the University and Colleges of Cambridge
. . .* [London, 1840], p. 7, appendix, p. 5; *Statutes of the Colleges of
Oxford*, IV, Trinity College, pp. 3–4).

22 C. L. Shadwell and H. E. Salter (eds.), *Oriel College Records*, Oxford
    Historical Society Pub. 85 (Oxford, 1926), p. 437.
23 Hastings Rashdall and Robert S. Rait, *New College* (London, 1901), p. 127.
24 Oxford University Archives, *Visitationes aularum*, N.E.P. supra 45, fol. 17
    (visitation of 1580–1), fol. 33 (visitation of 1583, when the relevant article
    carried the title "Blacke Science"). The 1613 visitation was published in the
    St. Edmund Hall Oxford Magazine, 2 (1929), p. 63.
25 *Oxford University Statutes*, I, 274.
26 Andrew Clark (ed.), *Register of the University of Oxford* (Oxford, 1888–9),
    II, i, 170–9.
27 Bodleian Library, Oxford MSS. Rawl. D. 272, 273, passim.
28 Cambridge University Library, MS. Add. 102, passim.
29 Queens' College, Cambridge, unclassified MS. notebook of Lawrence
    Bretton, ca. 1605; quoted by William T. Costello, *The Scholastic
    Curriculum at Early Seventeenth-Century Cambridge* (Cambridge, Mass.,
    1958), p. 30.
30 Francis Osborn, "Advice to a Son," *Miscellaneous Works* (London, 1722),
    I, 5.
31 *Historical Manuscripts Commission*, various collections (Westminster,
    1909), V, 246.
32 Feingold, chap. 2.
33 C. H. Josten, "A Translation of John Dee's 'Monas Hieroglyphica'
    (Antwerp, 1564), with an Introduction and Annotation," *Ambix*, 12 (1964),
    pp. 137, 86.
34 *Autobiographical Tracts of Dr. John Dee . . .* , ed. James Crossley,
    Chetham Society Pub. 24 (London, 1851), pp. 5–6.
35 Ibid., p. 56.
36 Robert Fludd, *Integrum morborum mysterium: sive medicinae catholicae,
    tomi primi, tractatus secundus* (Frankfurt, 1631), pp. 27–8; Frances A.
    Yates, *Theatre of the World* (Chicago, 1969), pp. 61, 43.
37 Quoted by French, p. 25.
38 For Cheke, who was one of the driving forces behind the introduction of
    mathematics to sixteenth-century Cambridge, see John Strype, *The Life of
    Sir John Cheke Kt., 1514–1557* (Oxford, 1821); Paul L. Rose, "Erasmians
    and Mathematicians at Cambridge in the Early Sixteenth Century,"
    Sixteenth Century Journal, 8, suppl. (1977), p. 58; John G. Nichols, "Some

Additions to the Biographies of Sir John Cheke and Sir Thomas Smith,''
Archaeologia, 38 (1860), pp. 98–127.

39 *Grace Book B*, pt. II, ed. M. Bateson (Cambridge, 1905), pp. 238, 241;
Lawrence V. Ryan, *Roger Ascham* (Stanford, 1963), p. 242 and passim.

40 Virginia F. Stern, *Gabriel Harvey: A Study of His Life, Marginalia, and
Library* (Oxford, 1979); British Library, MS. Add. 36,674, fols. 58–62ᵛ.

41 For Caius, see Vivian Nutton, ''John Caius and the Linacre Tradition,''
Medical History, 23 (1979), p. 373–91. A partial list of his library is given
by Philip Grierson, ''John Caius' Library,'' in *Biographical History of
Gonville and Caius College*, ed. M. J. Prichard (Cambridge, 1978), VIII,
509–25; R. S. Roberts, ''A Consideration of the Nature of the English
Sweating Sickness,'' *Medical History*, 9 (1965), p. 386; British Library, MS.
Add. 36,674, fols. 23–45.

42 In his copy of Luca Gaurico, *Tractatus astroglogicus* (Venice, 1552), which
he annotated in 1580. The book is in the Bodleian Library, shelfmark
C.60.e.13. The list is on sig. Yiʳ and is printed in Stern, p. 168.

43 J. Venn (ed.), *The Annals of Gonville and Caius College*, Cambridge
Antiquarian Society, Octavo Pub. 11 (Cambridge, 1904), pp. 232–3; ''An
Astrological Fellow,'' *The Caian*, 6 (1897), pp. 28–36; Gonville and Caius
College MS. 73, fols. 348–88.

44 John Aubrey, *Brief Lives*, ed. A. Clark (Oxford, 1898), I, 138–43; J. J.
Smith, *The Cambridge Portfolio* (London, 1840), II, 489–92; Cambridge
University Library, MS. Nn.IV.10; British Library, MS. Add. 36,674, fol.
45. Many of his medical notes can be found in British Library, MSS. Sloane
1087, 1602, 1664, 1991, 2077, 3329. See also Jeffrey Boss, ''William Butler
(1535–1618): Further Evidence on a Physician Between Two Ages,''
Medical History, 21 (1977), pp. 434–45; Vivian Nutton, ''Dr. Butler
Revisited,'' *Medical History*, 22 (1978), pp. 417–26; and the exchange
between Boss and Nutton, *Medical History*, 22 (1978), pp. 426–30.

45 John Geree, *Astrologo-Mastix* . . . (London, 1646), pp. 14–15.

46 Thomas Fuller, *Abel redevivus* (London, 1651), p. 432; W. P., *Foure Great
Lyers* . . . (London, 1585).

47 The relevant material is to be found in Richard Bauckham, ''Science and
Religion in the Writings of Dr William Fulke,'' *British Journal for the
History of Science*, 8 (1975), pp. 17–31.

48 Charles H. Cooper, *Athenae cantabrigienses* (Cambridge, 1861), II, 284–5;
DNB; British Library, MS. Sloane 2175, fols. 148–72; Bodleian Library,
MS. Ashmole 1421, fols, 165ᵛ–217.

49 Cooper, II, 530; British Library, MSS. Sloane 2193, 320, 3086; Charles
Webster, ''Alchemical and Paracelsian Medicine,'' in *Health, Medicine and
Mortality in the Sixteenth Century*, ed. Charles Webster (Cambridge, 1979),
p. 324.

50 W. M. Palmer, ''College Dons, Country Clergy and University Coachmen,''
Proceedings of the Cambridge Antiquarian Society Communications, 16
(1912), p. 188.

51 Bodleian Library, MS. Rawl. D. 253: ''A Treatise of Conjurations of
Angels, with Experiments in Crystal, and Directions for Fumigations''; H.
E. Salter (ed.), *Remarks and Collections of Thomas Hearne*, Oxford
Historical Society Pub. 68 (Oxford, 1915), X, 444–5.

52 Anthony à Wood, *Athenae oxonienses*, ed. P. Bliss (Oxford, 1813–20), I,
188.

53 Ibid., II, 541–4; Aubrey, I, 26–8; T. Rogers, *Leicester's Ghost*, ed. F. B.

Williams (Chicago, 1972), p. 14; Bodleian Library, MS. Selden supra 79, fol. 150^c; Andrew G. Watson, "Thomas Allen of Oxford and His Manuscripts," in *Medieval Scribes, Manuscripts and Libraries: Essays Presented to N. R. Ker*, ed. M. B. Parker and A. G. Watson (London, 1978), pp. 279–314; William D. Macray, *Catalogi codicum manuscriptorum bibliothecae bodleianae . . . codices . . . Kenelm Digby* (Oxford, 1883), passim; Michael Foster, "Thomas Allen (1540–1632), Gloucester Hall and the Survival of Catholicism in Post Reformation Oxford," *Oxoniensis*, 46 (1981), pp. 99–128.

54 Bodleian Library, MS. Corpus Christi College 260, fols. 82–4^v; MS. Wood F26; MS. Corpus Christi College 160 (given to Twyne by Allen); MS Selden supra 79, fols. 150^{a-c}.

55 Bodleian Library, MSS. University College 47–9.

56 *The Responsa Scholarum of the English College, Rome,* ed. Anthony Kenny, Catholic Record Society Pub. 54 (London, 1962), p. 91.

57 Bodleian Benefactors' Book, vol. I, fols. 14–21; *The Letters of John Chamberlain*, ed. Norman E. McClure, 2 vols. (Philadelphia, 1939), passim.

58 Clark, II, i, 287, 374; II, ii, 12; R. T. Gunther, *Early Science at Oxford* (Oxford, 1937), XI, 337.

59 Wood, I, 632–3; *Private Diary of John Dee*, p. 13; Bodleian Library, MSS. Ashmole 1415, fols. 33–40 ("Tractatus de lapide philosophico"); 1415, VI ("A Treatise . . . uppon ye philosophers stone"); 1408, fols 22^v–23 ("Lapis philosophicus est duarum materiarum").

60 Wood, II, 132–3.

61 W. H. Hart, "Observations on Some Documents Relating to Magic in the Reign of Queen Elizabeth," *Archaeologia*, 40 (1866), pp. 391–4; J. O. Halliwell (ed.), *A Collection of Letters Illustrative of the Progress of Science in England* (London, 1841), p. 34.

62 Hope Emily Allen, "'Dicing Fly' and 'The Alchemist,'" *Times Literary Supplement*, 27 June 1935, p. 416; *The Works of Ben Jonson*, ed. C. H. Herford, P. Simpson, and E. M. Simpson (London, 1925–52), X, 62–3.

63 B. Camm, *Forgotten Shrines*, 2nd ed. (London, 1936), pp. 387–8. In a similar manner, both Nathaniel Torporley and Thomas Harriot were charged with casting the nativity of King James I prior to the discovery of the Gunpowder Plot. For the interrogation of Torporley, see PRO SP 14.216, pt. II (Great Plot Book, no. 122).

64 Wood, I, 498; Thomas Heth, *A Manifest and Apparent Confutation of an Astrological Discourse, Lately Published . . .* (London, 1583), sig. B7^v.

65 *DNB*; Bodleian Library, MSS, Ashmole 209, fols. 131–203; 242, fols. 1–19; 576; British Library, MS. Sloane 1713, fols. 1–9; British Library, MSS. Royal 12.B.XI, 12.E.II; A Favaro, *Cartegio inedito di Ticone Brahe, Giovanni Keplero e di altri celebri astronomi e matematici dei secoli xvi e xvii, con Giovanni Antonio Magini* (Bologna, 1886), pp. 242–3, 246–8, 317–21, 335–6.

66 British Library, MS. Cotton Julius F.XI, fols. 70^v–71, 151^v–152^v; Bodleian Library, MS. Ashmole 1788, fols. 70–6; M. Aston, "The Fiery Trigon Conjunction: An Elizabethan Astrological Prediction," *Isis*, 61 (1970), p. 165 n. 15. The Copernicus volume is at Brown University, sig. *QB41 C76. See also F. J. Levy, "William Camden as a Historian," unpublished Ph.D. thesis, Harvard University, 1959, pp. 114–19. In a similar manner we find Camden's fellow antiquarian, Henry Spelman, also engaged in astrological studies during his abode at Cambridge in 1581–2: see British Library, MS.

Harl. 6360, fols 1–31, and C. H. Cooper, "An Early Autograph of Sir Henry Spelman," *Cambridge Antiquarian Society Communications*, 2 (1860–4), pp. 101–3.

67 Charles B. Schmitt, "John Case on Art and Nature," *Annals of Science*, 33 (1976), pp. 543–59.

68 McConica, pp. 302–9; Queen's College, Oxford, MS. 241, fols. 151–5; Bodleian Library, MS. Wood D.10; Mordechai Feingold, *An Elizabethan Intellectual: John Rainolds, the Man and His Library*, Oxford Bibliographical Society Occasional Pub. (forthcoming).

3

Analogy versus identity: the rejection of occult symbolism, 1580–1680

BRIAN VICKERS

It is my contention that the occult and the experimental scientific traditions can be differentiated in several ways: in terms of goals, methods, and assumptions. I do not maintain that they were exclusive opposites or that a Renaissance scientist's allegiance can be settled on
an either/or, or yes/no, basis. Rather, in many instances, especially in
the late sixteenth and seventeenth centuries, a spectrum of beliefs and
attitudes can be distinguished, a continuum from, say, absolutely magical to absolutely mechanistic poles, along which thinkers place themselves at various points depending on their attitudes to certain key
topics. One of these topics, not much discussed so far, is the relationship between language and reality. In the scientific tradition, I hold, a
clear distinction is made between words and things and between literal
and metaphorical language. The occult tradition does not recognize
this distinction: Words are treated as if they are equivalent to things
and can be substituted for them. Manipulate the one and you manipulate the other. Analogies, instead of being, as they are in the scientific
tradition, explanatory devices subordinate to argument and proof, or
heuristic tools to make models that can be tested, corrected, and abandoned if necessary, are, instead, modes of conceiving relationships in
the universe that reify, rigidify, and ultimately come to dominate
thought. One no longer uses analogies: One is used by them. They
become the only way in which one can think or experience the world.

The distinction I am making between two cognitive processes has
analogues. One of the main differentia in Robin Horton's juxtaposition
of traditional African thought with modern science[1] is that the traditional thinker, who knows one system only and has no concept of
alternatives, sees "a unique and intimate link between words and
things." Words seem to him to be "bound to reality in an absolute
fashion. There is no way . . . in which they can be seen as varying

independently of the segments of reality they stand for. Hence they appear so integrally involved with their referents that any manipulation of the one self-evidently affects the other'' (p. 156). The modern scientist dismisses this concept of ''the immediate, magical power of words over the things they stand for,'' because it would lead to an intolerable conception of reality. For if ideas and words actually shape and control reality, ''then a multiplicity of idea-systems means a multiplicity of realities, and a change of ideas means a change of things.'' This would have the further grave defect of suggesting that ''the world is in the last analysis dependent on human whim, that the search for order is a folly, and that human beings can expect to find no sort of anchor in reality'' (p. 157). Opposed to this conception, modern science has to believe that ''while ideas and words change, there must be some anchor, some constant reality. This faith leads to the modern view of words and reality as independent variables'' (p. 157). Words are no longer seen as acting magically upon reality, and a clear distinction is made between ''mental activities'' and ''material things.'' In traditional thought – and, I would argue, in the occult sciences – ''everything in the universe is underpinned by spiritual forces,'' words and things ''are both part of a single reality, neither material nor immaterial'' (p. 157).

The phenomenon I am approaching through this general anthropological context is the fusion of word and referent basic to many forms of magic. In his 1968 Malinowski lecture, ''The Magical Power of Words,''[2] S. J. Tambiah has described the widespread concept, in ritual, of ''sacred words,'' which are ''thought to possess a special kind of power not normally associated with ordinary language'' (p. 179). This clear-cut disjunction between sacred and profane language is not, in fact, ''necessarily linked to the need to embody sacred words in an exclusive language or in writing,'' but seems to derive rather from the widespread ''ancient belief in the creative power of the word'' (p. 182). The Vedic hymns ''asserted that the gods ruled the world through magical formulae''; the Parsi religion believed that ''it was through the spoken word that chaos was transformed into cosmos''; the ancient Egyptians, the Semites, and the Sumerians all believed that ''the world and its objects were created by the word of God; and the Greek doctrine of *logos* postulated that the soul or essence of things resided in their names'' (pp. 182–3). In the Bible we find passages making the word ''an entity which is able to act and produce effects in its own right,'' as in Isaiah 55:11: ''So shall my word be that goeth forth out of my mouth: it shall not return unto me void, but it shall accomplish that which I please.'' In Buddhism ''the Dhamma, the doctrines preached by the Buddha, and inscribed in the text are themselves holy objects in their own right, and can transmit virtue and dispel evil'' (p. 183). In all these instances the belief system has broken down the distinction between, on the one hand, objects in the physical world, and on the other, language as a system of concepts, signs, and sounds. The Trobriand islanders, with their view that "magical formulae, once voiced, acted and influenced the course of events," are just one example of many in primitive cultures of the belief that "language as such has an independent existence" with the power to shape reality (p. 184).

As Tambiah shows, several theorists of language have tried to account for the primitive's "magical attitude to words." Ogden and Richards described it as "the superstition that there was a direct, even causal relation between the word and the thing it referred to," what they called "the denotative fallacy" (p. 187). For Cassirer, this "hypostatisation of the word (which implied . . . that the name of a thing and its essence bear a necessary and internal relation to each other)" was a sign of "mythic thought," that stage before "theoretical discursive thought" in which the imagination tended toward "concentration, telescoping," positing "a relation of identity and substantial unity between name and thing." Setting aside Cassirer's "imaginary and speculative evolutionary scheme" of a global movement from mythical to logical thought (p. 187), the agreement among the linguists is striking: To put it in the language of Ferdinand de Saussure, in this type of thinking the *signifiant* is confused with the *signifié*. Saussure proposed the accompanying diagram as a model for the relationship between word and referent.[3]

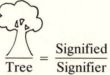

$$\frac{\text{Tree}}{} = \frac{\text{Signified}}{\text{Signifier}}$$

The linguistic sign is arbitrary; that is, any combination of letters and sounds may be used to designate a tree, depending on the established conventions of a language. As a linguist has observed, the crucial element in that formulation is the line separating the two realms, showing that they exist on different levels. In the occult and magical traditions the line is removed – or rather, it is never inserted; word and thing are not discriminated.

Here again the Renaissance occult tradition must be seen in a much wider historical context. The debate about words and things begins with Plato's *Cratylus*,[4] which discusses the origin of language and the appropriateness of names. Cratylus argues that names "are natural and not conventional" and can be assessed according to criteria of "truth or correctness" (383a), but Hermogenes denies "that there is any principle of correctness in names other than convention and agreement" and that "there is no name given to anything by nature; all is convention and habit of the users" (384d). Socrates first takes issue with Hermogenes, asking whether "the things differ as the names differ," that is, whether "things have a permanent essence of their own," uninfluenced by the appellation or changes in it, or – the alternative, obviously meant to be rejected – whether Protagoras is right in claiming that they are "relative to individuals," which would mean that there are no fixed scales of behavior and that wisdom and folly are indistinguishable (386a–c). This ethical relativism disposed of, Socrates can reassert that a reality exists independent of man: Things "must be supposed to have their own proper and permanent essence; they are not in relation to us, or influenced by us, fluctuating according to our fancy, but they are independent, and maintain to their own essence the relation prescribed by nature" (386d–e). Socrates' eagerness to establish this point at the outset of the dialogue confirms Robin Horton's analysis: the concept that reality is subservient to, and alterable by, words is as abhorrent to the Greek philosopher as to the modern scientist.

By cross questioning Hermogenes Socrates then establishes that "a name is an instrument" that we use to "give information to one another, and distinguish things according to their natures" (388a–b), an unexceptionable point, but one to which he adds the rather unconvincing rider that "not every man is able to give a name, but only a maker of names," the legislator, most skilled of artisans (389a). With this claim Socrates' argument, surprisingly enough, begins to veer toward the views of Cratylus. Having postulated that the giving of names was the work of a legislator, not of evolving social custom, he is unwilling to think that signification came about by usage or accident. The legislator "ought . . . to know how to put the true natural name of each thing into sounds and syllables" (389d), under the tutelage of a dialectician, of course (390d). Given this union of talents Socrates can conclude that no "light or chance persons" could have given names, but that "things have names by nature," thanks to the work of "an artificer of names . . . who looks to the name which each thing by nature has, and is able to express the true forms of things in letters and syllables" (390e). Asked by Hermogenes to explain this "natural fitness of names," Socrates embarks on a long etymological account (391–427) of how the names of Greek gods and heroes "express the nature" of these personages and how the same claim can be made for the names of the virtues and vices. Such names "are not given arbitrarily, but have a natural fitness" (397a), so that "the office and name of the god really correspond" (403b). This is an ingenious display – at times dazzling, at other times willful – and Socrates licenses his etymological speculation by claiming full rights of interpretation and "permutation" (400b), adding or subtracting letters (413e), attacking this practice (414d) but only as a way of legitimizing it, resorting to self-irony (416a) and other rhetorical devices to justify this exercise in linguistic essentialism.

This aspect of the *Cratylus* is well known, and it became a key text for Neoplatonists arguing for a realist view of language.[5] But the dialogue also contains a postlude, as it were, a section in which Socrates reverts to his original topic of the giving of names and argues against Cratylus.

First he asks whether the name that Hermogenes passes under "is a wrong name, or not his name at all?" Puzzled, Cratylus asks, "How can a man say that which is not?" (429c–d), initiating a sequence in which Socrates makes him "admit that the name is not the same with the thing named" and "further acknowledge that the name is an imitation of the thing" and "say that pictures are also imitations of things, but in another way" (430a–b). Here the stress has been shifted to the process of "assigning" words to things, pictures to objects, and we are back to the concept of language and reality existing on separate planes. Since "images are very far from having qualities which are the exact counterpart of the realities which they represent," then "how ridiculous would be the effect of names on things, if they were exactly the same with them! For they would be the doubles of them, and no one would be able to determine which were the names and which were the realities" (432d). If we collapse the line between *signifié* and *signifiant*, language and reality become indistinguishable, hopelessly confused. From this point Socrates restates Hermogenes' theory that "names are conventional, and have a meaning to those who have agreed about them" (433e). Cratylus still believes that "representation by likeness . . . is infinitely better than representation by any chance sign" (434a). Yet now Socrates shows that even on the basis of their preceding classification of letters as to their onomatopoeic quality, being "like" or "unlike" the activity they describe, nonetheless "the correctness of a name turns out to be convention, since letters which are unlike are indicative equally with those which are like, if they are sanctioned by custom and convention" (435a). The ultimate judge of meaning or appropriateness, then, is society.

Having argued for the conventional nature of language, Socrates returns to the relationship between language and reality. Cratylus states what he sees as "the simple truth . . . that he who knows names knows also the things which are expressed by them," for "as the name is, so also is the thing" (435d). The distinction he has failed to make is between discovery and instruction, believing that "in the discovery of [things] he who discovers the names discovers also the things." Socrates retorts, however, that "he who follows names in the search after things, and analyzes their meaning, is in great danger of being deceived," since "he who first gave names gave them according to his conception of the things which they signified," so that "if his conception was erroneous, and he gave names according to his conception," then those who follow him will also be deceived (436a–b). Between word and thing Plato has introduced a third term, conception, an important development in linguistic theory. Socrates uses it here to reinforce his view of the subordination of words to things, or the danger of confusing language with reality. It cannot be true that "things are only to be known through names" because things existed before names, and the givers of names had to invent them to express their conceptions of the yet-unnamed things (438b–c). If we are to judge whether the names are appropriate or not, we cannot appeal to other names, but must use "a standard which shows the truth of things" (438d) on the plane of reality. We can "learn things through the medium of names" if we wish, but this is to study the image rather than the truth: It follows that "the knowledge of things is not to be derived from names . . . they must be studied and investigated in themselves" (439a–b). It also follows that "no man of sense will like to put himself or the education of his mind in the power of names. Neither will he so far trust names or the givers of names as to be confident in any knowledge which condemns himself and other existences to an unhealthy state of unreality" (440c).

Far from supporting only a theory of natural language, Plato's *Cratylus*, catholic and puzzling work, also defends the concept of language as conventional and draws the clearest possible line between language and reality. Since the development of the experimental sciences in the Renaissance also placed *res* above *verba*, a science of reality above a philological science (especially above a science of *verba* presenting itself as a science of *res*), it is not surprising that the tradition inaugurated by Plato should have been echoed. In Galileo, for instance, we find that succinct statement that "names and attributes must be accommodated to the essence of things, and not the essence to the name, since things come first and names afterwards."[6] In truly Socratic vein, having asserted that "neither the satellites of Jupiter nor any other stars are spots or shadows, nor are the sunspots stars," he adds:

> It is indeed true that I am quibbling over names, while I know that anyone may impose them to suit himself. So long as a man does not think that by names he can confer inherent and essential properties on things, it would make little difference whether he calls these "stars." (p. 139)

Whether Galileo is alluding to the *Cratylus* is of no great importance: What matters is that the new science here distinguishes itself wholly from the occult, which indeed thought that "by names man can confer essential properties on things." Similarly, in *Il saggiatore* he tells Sarsi or his teacher that it is not enough to make a comet a quasi-planet merely by naming it. "If their opinions and their voices have the power of calling into existence the things they name, then I beg them to do me the favor of naming a lot of old hardware I have about my house, 'gold' " (p. 253). The climax of this distinction between language and reality is, of course, Galileo's attack on the peripatetic belief that "heat is a real phenomenon, or property, or quality, which actually resides in the material by which we feel ourselves warmed." In fact, "tastes, odors, colors, and so on are no more than mere names" that "reside only in the consciousness" and that "we have imposed upon" reality (p. 274). Many sensations "which are supposed to be qualities residing in external objects have no real existence save in us" and "when separated from living beings . . . are nothing more than names" (p. 277).

If the Platonic tradition could lend ammunition to the new sciences in their insistence on the difference between language and reality, so could another tradition much studied in Renaissance universities, that of Aristotelian logic. At the beginning of *De interpretatione*,[7] Aristotle offers the following definition:

> Spoken words are the symbols of mental experience and
> written words are the symbols of spoken words. Just as all
> men have not the same writing, so all men have not the
> same speech sounds, but the mental experiences, which
> these directly symbolize, are the same for all, as also are
> those things of which our experiences are the images. (16a
> 3–7)

Aristotle's emphasis is more toward individual psychology (he refers to his own treatise, *De anima*), but he recognizes, like Plato, the conceptual operation of language in his terms "symbols" and "images." Also like Plato, he stresses the social or conventional nature of language, linking it with *nomos* rather than *physis*:

> By a noun we mean a sound significant by convention . . .
> The limitation "by convention" was introduced because
> nothing is by nature a noun or name – it is only so when it
> becomes a symbol; inarticulate sounds, such as those which
> brutes produce, are significant, yet none of these constitutes
> a noun. (16a 19–29)

Language is a human, social activity in which meaning is assigned by general agreement and not "by nature." The prestige of Aristotle's philosophy ensured that this view of language had a lasting influence, and in many medieval philosophers – notably Saint Augustine – we find an explicit recognition of the notion of the linguistic sign, with the symbol or concept mediating between word and thing.[8] In Aquinas, as in medieval scholasticism in general, there is a careful emphasis on these discrepant levels. "The word is a sign of a thing, and this thing, in its turn, may be the sign or symbol of something different." Aquinas, for instance, refuses to identify man with the world, adding careful qualifiers: "Man has 'some' similarity with the world and, therefore, is called a microcosm; but he does not say that man is, strictly speaking, such a microcosm."[9]

The Aristotelian tradition, like the Platonic, was well known to the new sciences. Thus Francis Bacon, defining words, quoted this passage from the *De interpretatione*: "For the organ of tradition [communication], it is either Speech or Writing: for Aristotle saith well, *Words are the images of cogitations, and letters are the images of words*; but yet it is not of necessity that cogitations be expressed by the medium of words."[10] Bacon thought that the "real characters" of Chinese could represent "things and notions" immediately, without the intermediaries of letters and words, proof to him that in this area, where we "are handling the currency (so to speak) of things intellectual," then "as moneys may be made of other material besides gold and silver," so communication and exchange can also be performed by signs and symbols. As he had written earlier in the *Advancement of Learning*: "Words are but the current tokens or marks of Popular Notions of things" (III, 388). In the *Novum organum* Bacon pointed to the fragile nature of a deductive logic built on words, not things: "The syllogism consist of propositions, propositions consist of words, words are symbols of notions. Therefore if the notions themselves are confused and over-hastily abstracted from the facts, there can be no firmness in the superstructure" (IV,49). In the famous analysis that follows of the *idola*, the illusions or false appearances that afflict human knowledge, Bacon defined the "Idols of the Marketplace" as the "ill and unfit choice of words," which can produce confusion, controversy, and "idle fancies." Two kinds of illusions are created by words. The first occurs when *res* and *verba* do not correspond: "They are either names of things which do not exist (for as there are things left unnamed through lack of observation, so likewise are there names which result from fantastic suppositions and to which nothing in reality corresponds)," such as "Fortune," the "Prime Mover," and the "Element of Fire"; or "they are names of things which exist, but yet confused and ill-defined, and hastily and irregularly derived from realities." This kind of verbal illusion can be easily dispelled by rejecting the theories that produced it. The other kind is more insidious, since it "springs out of a faulty and unskilful abstraction." If we take the word "humid" and "see how far the several things which the word is used to signify agree with each other . . . we shall find the word *humid* to be nothing else than a mark loosely and confusedly applied to denote a variety of actions which will not bear to be reduced to any constant meaning" (IV, 61–2). Having listed nine different senses and shown that flame, air, fine dust, and glass can all be said to be "humid," Bacon, like

Galileo, concludes that the greatest distortion is caused by the names for qualities, which suggest that they exist. The propagandists for the new sciences had two enemies, Aristotelian philological science and the occult: Both tended to manipulate language as if that could effectively describe or control reality. The only remedy is a science built not on words but on the observation of reality, developing a proper scientific method.

Bacon's influence on the seventeenth century in this, as in so many areas, was great. Hobbes was at one time Bacon's amanuensis, and Baconian ideas, even Baconian metaphors, are frequently found in *Leviathan*. To Hobbes words are also conventional signs, representations of reality that are not to be confused with reality: "For words are wise men's counters, they do but reckon by them; but they are the money of fools."[11] Elsewhere Hobbes displayed his Aristotelian inheritance. In *De corpore* (ca. 1642) he defines communication as taking place through signs, some of which are natural, such as thick clouds portending rain, "others are *arbitrary*, namely those we make choice of at our own pleasure," such as a boundary stone.[12] "Words so connected as that they become signs of our thoughts, are called SPEECH, of which every part is a *name*," and a name is *"a word taken at pleasure to serve for a mark"* to recall a thought (pp. 15–16). Hobbes finds it "unquestionable" that the original of names was "arbitrary," and he might be taking issue with the Cratylean phase of Socrates' argument:

> For considering that new names are daily made, and old ones laid aside; that diverse nations use different names, and how impossible it is either to observe similitude, or make any comparison betwixt a name and a thing, how can any man imagine that the names of things were imposed from their natures? For though some names of living creatures and other things, which our first parents used, were taught by God himself; yet they were by him arbitrarily imposed [and have since been forgotten, replaced by] others, invented and received by men at pleasure. (p.16)

In addition to stressing the arbitrary nature of the linguistic sign, Hobbes insists on the line separating signified and signifier. As the section title has it,

> *Names are signs not of things, but of our cogitations.* But seeing names ordered in speech (as is defined) are signs of our conceptions, it is manifest they are not signs of the things themselves; for that the sound of this word *stone* should be the sign of a stone, cannot be understood in any sense but this, that he that hears it collects that he that pronounces it thinks of a stone. (p.17)

Disputes "whether names signify the matter or form" are maintained only by "erring men, and such as understand not the words they dispute about" (p.17). In the late treatise *De homine* (1658) Hobbes returns to the attack, adding a new element to his theory, the role of "the human will" in the act of signification: "Speech or language is the connexion of names constituted by the will of men to stand for the series of conceptions of the things about which we think."[13] The origin of language as described in Genesis is endorsed: "The first man by his own will imposed names on just a few animals, namely, the ones that God led before him to look at; then on other things." These names, "having been accepted, were handed down from fathers to their sons, who also devised others" (p. 38). Given this sequence, "speech could not have had a natural origin except by the will of man himself." This is made "even clearer by the confusion of languages at Babel," and since that time "the origins of language are diverse." Once again Hobbes rejects theories of a natural language:

> What others say, however – that names have been imposed on single things according to the nature of those things – is childish. For who could have it so when the nature of things is everywhere the same while languages are diverse? And what relationship hath a *call* (that is, a sound) with an *animal* (that is, a body)? (p. 39)

One of the forms of confusion discussed in *Human Nature* is that of men "deceiving themselves, by taking the universal, or general appellation, for the thing it signifieth."[14]

From the Platonic-Aristotelian tradition of language as conventional and arbitrary, Renaissance philosophers, including those connected with the new sciences, derived a series of clear distinctions between language and reality. To cite one more expression of this awareness, Kenelm Digby wrote at the beginning of his treatise on the soul and the body, that

> it is true, words serve to express things, but if you observe the matter well, you will perceive they do so, onely according to the pictures we make of them in our own thoughts, and not according as the things are in their proper natures. Which is very reasonable it should be so, since the soul, that giveth the names, hath nothing of the things in her but these notions: and . . . therefore cannot give other names but such as must signifie the things by mediation of these notions.[15]

Digby adds the corollary that the most dangerous of all confusions is when men "confound the true and reall natures of things, with the conceptions they frame of them in their own minds. By which fundamentall miscarriage of their reasoning, they fall into great errours and absurdities" and can produce nothing but "uselesse cobwebs or prodigious Chymeras" (p.2). Digby goes on to warn against the "mistaken subtilties, which arise out of our unwary conceiting that things are in their own natures after the same fashion as we consider them in our understanding" (p. 3.). The way in which Digby develops this argument owes more to the Aristotelian categories of substance and accident than to Galileo's distinction between primary and secondary qualities, but the end result is very similar to the attitudes of the new sciences, as in this caveat – which could be paralleled in Kepler, Bacon, or Galileo – to

> take heed, lest reflecting upon the notions we have in our mind, we afterwards pin those aiery superstructures upon the materiall things themselves that begot them; or frame a new conception of the nature of any thing by the negotiations of our understanding upon those impressions which it self maketh in us. (p. 5)

Yet, in addition to what I have called the Platonic-Aristotelian tradition of conventional language, there existed a rival school of thought in the Renaissance, deriving from several traditions, according to which words not only expressed but embodied the nature of things, somehow containing their very essence. Neoplatonists were wont to invoke Plato as their authority for this concept, basing their claim on the middle part of the *Cratylus*, ignoring the opening and closing sections where Socrates insists that names are conventional and that words are not directly interchangeable with things. Despite this, Neoplatonists such as Iamblichus believed that divine languages existed in which words expressed the essence of things, and the Jewish Neoplatonist Philo of Alexandria held that " 'with Moses the names assigned are manifest images of the things, so that name and thing are inevitably the same.' "[16] The innate power of the word is an important concept in many mystical traditions and in some related theurgical and magical practices.[17] In the texts ascribed to Hermes Trismegistus words and letters are granted creative power, and in Jewish mysticism the elements of the word are said to create the elements in the world.[18] If the rediscovery of hermetic texts in the Renaissance aided the revival of a concept of natural language, so did the spreading interest in the cabala.[19]

The result of this synthesis of magical and mystical traditions in Renaissance Neoplatonism, beginning with Ficino and Pico, was a revival of the belief in a natural language. As Allison Coudert has shown, Ficino follows the Neoplatonists by drawing on the Cratylean phase of Plato's dialogue: " 'For indeed, a name, as the Platonists say, is nothing else than a certain power of the thing itself, first conceived in the mind, so to speak, then expressed by the voice and finally indicated by letters. Moreover, divine things by necessity contain divine power.

For this reason Plato . . . commands that the names of God be venerated since they contain a far greater power than shrines and divine statues.' ''[20] Pico drew on the cabalist idea that Hebrew letters embodied different powers, as did the German Neoplatonist and cabalist, Johannes Reuchlin in his *De verbo mirifico* (1494), a discourse on the "wonder-working word" which Charles Zika[21] has placed in the late fifteenth-century debate over *vis verborum*, the possibility of performing magical operations by using words and names. Reuchlin believed that words, especially divine names, could perform marvelous deeds, granting Hebrew this power (especially the tetragrammaton), and evolved a new science of wonders called *soliloquia*, a ritual act of prayer in which the correct invocation of the divine name could perform miracles (Zika, p. 117). Pico's similar cabalist arguments had been attacked by Pedro Garsias, who drew on the passages in *De anima* where Aristotle stated that sound and smell, being mere accidents, can have no power over substances (pp. 126, 131), but Reuchlin replied with what Rika has called "mystical philology": " 'When the Word descended into flesh, then the letters passed into voice' " (p. 132). Reuchlin claims that the pentagrammaton, IHSUH, is the word through which man achieves knowledge and can perform miraculous deeds (p. 133).

Throughout this mystical philology the word is reified, turned into a concrete object with magical powers. In Cornelius Agrippa's *De occulta philosophia*, for instance, the letters of the Hebrew alphabet, which he takes to be the holiest language, represent the actual structure of the universe, so that "manipulations of them have intrinsic power." To rearrange language is to rearrange reality. The characters of the alphabet "are like secrets or sacraments, and are vehicles, as it were, of their material referenda and of the 'essences' and powers these contain."[22] Magical operations through language have transitive effects on the world:

> Words therefore are the fittest medium betwixt the speaker
> and the hearer, carrying with them not only the conception
> of the mind, but also the vertue of the speaker with a certain
> efficacy unto the hearers, and this oftentimes with so great a
> power, that oftentimes they change not only the hearers, but
> also other bodies, and things that have no life.[23]

In the Paracelsian doctrine of "signatures" we find the related idea that essences can be expressed in tangible form, confused, as Alexandre Koyré pointed out, with the notion of "correspondences" and simile magic.[24] (As with Agrippa on the Hebrew letters and the structure of the universe, in occult thought ideas on language and ideas of cosmology are closely related.)

In Jacob Boehme the concept of a natural language leads not to a belief in magic but to a mystical philology in which signified and signifier become fused in that "union of opposites" so typical of mystical thought.[25] The word or the "sound" of an object expresses its inner qualities directly (Koyré, p. 144), a concrete or corporeal view of language that cannot sustain abstract thought or clear distinctions: "La distinction entre l'expression et l'exprimé est trop souvent confondue avec celle de la 'matière' première et des objets formés de cette matière" (p. 154). Boehme identifies God's word with God's person: "Dieu, par son Verbe, engendre sa nature, il parle à sa nature et celle-ci, reproduisant cette parole créatrice, produit *elle-même* l'être de ce monde . . . les objets matériels, les plantes et les animaux" (pp. 275–6). The divine word expresses God's spirit in letters and syllables that have a creative dynamism. This "divine alphabet" produces a world of forces and essences intermediate between God and the real world, whereas human words constitute "les choses du monde elles-mêmes, les êtres individuels et séparés" (pp. 398–9). This view of language expresses a view of reality in which the world is simultaneously distinguished from God yet incarnates God's word in a temporal plane: "Les êtres sont des 'signatures' de Dieu, ce sont des mots de la parole divine" (pp. 447f.). Microcosmic man, sharing in all things, is capable of expressing all things, and spiritualizes the universe through language, reexpressing the word of God. As Boehme says in *Mysterium magnum*: "Dan dass der Mensch redet und verstehet, das komt . . . dem Menschen aus dem eingeleibten, geformten Worte Gottes her, es ist der Name Gottes" (p. 460).[26] Because language is the corporeal expression of meaning, there must be a natural connection between thought and verbal expression, which reaches its highest intensity in such divine revelations as those accorded to Boehme: "Und was noch grösser, ist mir die NATUR-SPRACHE eröfnet worden, dass ich kann im meiner Mutter-Sprache die allergrösten Geheimnisse verstehen." This natural or *sensual* language is that of paradise and is the language spoken by the angels to each other, as man will discover when he is reborn into the next world (p. 457). Similar ideas were expressed in the Rosicrucian pamphlets (p. 457).

Boehme's belief that the natural or Adamic language would be restored in the afterlife is found in slightly different form in the millenarian versions of the schemes for a universal language, which were intended to undo the confusion of tongues that beset humans after the Fall and restore us to that happy state before the Tower of Babel.[27] In his *Prodromus pansophiae* (1657) the Protestant educational reformer Comenius proposed a panglottia, a universal language in which words would express the essence of things, to be realized with the millennium (Koyré, p. 458). Like other Neoplatonists, Comenius believed in the existence of a sum of " 'universal Notions, original and innate, not yet perverted by monstrous conceptions, the divinely laid foundations of our reason,' " common to all mankind before perverted by the ambiguity and equivocation of words.[28] Boehme was extremely well known in mid-seventeenth-century England, since all his works had been translated by the order of parliament, and Comenius's schemes and works had also been widely disseminated. The occult tradition, with its essentialist, corporeal-mystical view of language, is found in parliamentary, "puritan," and Baconian contexts. John Webster, for instance, who represents the fusion of all four elements, in his *Academiarum examen* (1653),[29] attacking traditional university education, invokes many occult writers, including Della Porta, Trithemius, Agrippa, in favor of hieroglyphical and cryptographical languages (p. 24), and links Boehme and the Rosicrucians in his panegyric to the "wonderful secret . . . of the language of nature" (pp. 26–32). In a mystical-religious rhapsody that outdoes Boehme, Webster praises "this *Angelical*" language of nature, as spoken by "the *Protoplast Adam*," the "*ineffable words*" of "the *Paradisical* language of the outflown word which *Adam* understood while he was unfaln in *Eden*." Warming to his theme, Webster celebrates this language for being able to breathe forth

> those central mysteries that lay hid in the heavenly *magick*,
> which was in that ineffable word . . . wrapped up in the bo-
> some of the eternal essence, wherein were hidden and in-
> volved in the way of a wonderful and inscrutable mystery,
> all the treasury of those *ideal*-signatures, which were mani-
> fest and brought to light by the . . . outflowing *fiat* of God.
> (p. 27)

It follows that when Adam named the animals he understood "both their internal and external signatures," finding an "absolute congruency" between their names and natures. Just as Adam "understood by his intrinsick and innate light" what Eve was, so animals understand each other by the language of nature. Only man does not comprehend the "immediate sounds of the soul" and their relation to "the internal notions impressed," and he has therefore "imposed others that do not altogether concord and agree to the innate notions."

In accepting the existence of innate notions Webster resembles Comenius and other Neoplatonists, while his belief in a natural language links him with Ficino, Pico, Reuchlin, Boehme, and many more. The occult tradition is eclectic and syncretist, of course, but homogeneous on many issues. The equation of words and things, the reification of the word, identifying it with the "nature" of its referent, is found in others drawing on the occult tradition, such as John Dee and George Dalgarno.[30] That this whole tradition was antipathetic to the new sci-

ences will be readily conceded even on the brief outline I have given
of their views on the relationship between language and reality. In fact,
a reaction against the occultists' views soon established itself. It is
visible, I believe, in the later works of Hobbes, quoted above. It is
directly, if succinctly, expressed by Seth Ward, who joined with John
Wilkins to write *Vindiciae academiarum* (1654), refuting Webster point
by point. The use of symbols instead of words or quantities is perfectly
correct in mathematics, but it has been inappropriately applied, Ward
says,

> to the *nature of things*, by the Pythagorean Philosophers,
> and diverse of the Cabalists, and to the *Art of Speaking*, by
> diverse both Jewes and others: and this Symbolicall art is
> that *Ars Combinatoria*, from which *Picus Mirandula* & oth-
> ers, make such large undertakings. The Pythagoreans did
> make Symbols of numbers, designing (*ex Arbitrio*) the parts
> of nature (as the supreme mind, the first matter &c.) by
> them. (p.19)

While Ward will consider the possibility that a universal character may
become a natural language (pp. 20–2), he rejects the idea that a sign
can incarnate the essence of a thing, reiterating instead the Platonic-
Aristotelian concept of "notions" intermediate between words and
things. Since words either signify simple notions or are "resolvible into
simple notions, it is manifest that if all the sorts of simple notions be
found out, and have Symboles assigned to them," then they will "represent to the very eye all the elements of their composition, & so deliver
the natures of things," that is, by an intermediate classification of concepts (p. 21). Further, the names should "be made up of the definitions
of things, or a complexion of all those notions whereof a Complexe is
compounded," and in this sense, "where every word were a definition
and contain'd the nature of the thing," it might be not unjustly called
a natural language (p. 22). As for Webster's rhapsody, Ward simply
denies "that ever there was any such Language of Nature" (p. 22).
When Wilkins came to produce his *Essay Towards a Real Character
and a Philosophical Language*, he, too, briefly discussed the possibility
of words having in their sound "some Analogy of their natures," but
dismissed it as impossible, falling back, with Plato and Aristotle, on
the concept of meaning being assigned by human will and convention.[31]
The antioccult movement found a vociferous exponent in Samuel Parker, who reiterated the orthodox view that

> the use of Words is not to explaine the Natures of Things,
> but only to stand as marks and signes in their stead . . . it
> has been an ancient and creditable Opinion of the *Platonists*,
> that Names have in them a natural resemblance and suita-
> bleness to things . . . But words . . . can have no likeness to

anything but sounds . . . And I therefore conclude that the
office of Definitions is not to explain the Natures of things,
but to fix and circumscribe the signification of Words.[32]

The culmination of this critical response is represented by John
Locke, and it is thanks to the work of Hans Aarsleff[33] that Locke's
discussion of language can be seen as in part a reply to the theories of
natural language revived by occultists and mystics. Locke's relationship to the Royal Society has been stressed by other recent studies,[34]
and we can now accept that the *Essay Concerning Human Understanding* (1690) was in some sense "intended as a manual in the epistemology of the Royal Society" (Aarsleff, p. 178). In the Preface,
Locke alludes to the "Master-Builders, whose mighty Designs in advancing the Sciences, will leave lasting Monuments to the Admiration
of Posterity" (Boyle, Sydenham, Huygens, and Newton). Locke sees
himself – in a very Baconian metaphor – "employed as an Under-
Labourer in clearing Ground a little, and removing some of the Rub-
bish, that lies in the way to Knowledge." Philosophy, he writes,
"which is nothing but the true knowledge of Things," has been ob-
structed by "vague and insignificant Forms of Speech, and Abuse of
Language," a complaint (so reminiscent of the famous passage in Book
I of the *Advancement of Learning*) to which Locke devotes the whole
of Book III.[35] Although Locke claims that this section was an after-
thought,[36] Aarsleff has shown that in the 1671 drafts Locke already
announces that in discussing "humane Intellect I could not avoid saying
a great deale concerning words because soe apt and usuall to be mistaken for things"; and that in his 1677 journal he had noted that "words
are, in their own nature, so doubtful and obscure . . . that if, in our
meditations, our thoughts busy themselves about words, and stick at
the names of things," they are bound to be confused (Aarsleff, p. 167
n. 7).

Locke's *Essay* includes a refutation of natural language theories,
beginning with Book I, Chapters 2–4, which show that there are no
innate principles in the mind, no primary notions "stamped" or "im-
printed" upon "the Mind of man," but that ideas are acquired through
our upbringing, education, and culture: "We by degrees get *Ideas* and
Names, and learn their appropriated connexion one with another" (p.
60). Moral principles "lie not open as natural Characters ingraven on
the Mind," but "require Reasoning and Discourse" (p. 66) and are
subject to questioning and disagreement (pp. 76–8). They may seem
innate because we cannot remember when we first learned them, but
all mental development is gradual, controlled by the relationship between our knowledge of language and our conceptions. "For Words
being but empty sounds, any farther than they are signs of our *Ideas*,"
we can assent to words only so far "as they correspond to those *Ideas*

we have" (p. 61). As Durkheim was to bring out so strongly, we are born into a culture and its language. Men are "furnished with Words, by the common Language of their own Countries" and inevitably acquire "some kind of *Ideas* of those things whose Names" they frequently use (p. 89). One bad effect of the doctrine of innate principles is to discourage further inquiry and turn the opinions of other men, especially leaders of sects, into unquestioned authority. Whereas, Locke argues, we can make progress in knowledge only if we seek it "in the Fountain, *in the consideration of Things themselves*," not in other men's thoughts. "In the Sciences, every one has so much as he really knows and comprehends." The authority of Aristotle is nothing if it means accepting "another's Principles without examining them" (p. 101).

Locke's emphasis on knowing for yourself, testing propositions by reason and judgment, is obviously in line with the goals and methods of the new sciences. That he should devote about a fifth of his work to a consideration of language shows the force of his realization that "the Extent and Certainty of our Knowledge" depends so much on language that "though it terminated in Things, yet it was for the most part so much by the intervention of Words," which interpose themselves "between our Understandings and the Truth which it would contemplate and apprehend," like some distorting medium (III, ix, 21; p. 488). This sharp distinction between language and reality recalls Hobbes, Bacon, and the long tradition back to Aristotle and Plato, all of whom are echoed in this sequence of argument. Words are "*articulate sounds*," which man uses "*as Signs of internal Conceptions*," standing as "marks for the *Ideas* within his own Mind" (III, i, 1–2; p. 402). The association between sound, sign, and meaning, Locke states, is conventional: So it is that

> *Words* . . . come to be made use of by Men, as *the Signs of*
> their *Ideas*; not by any natural connexion that there is be-
> tween particular articulate Sounds and certain *Ideas*, for
> then there would be but one Language amongst all Men; but
> by a voluntary Imposition, whereby such a Word is made
> arbitrarily the Mark of such an *Idea*. (III, ii, 1; p. 405)

The "signification of Sounds is not natural, but only imposed and arbitrary" (III, iv, 11; p. 425). Then, "words being voluntary Signs" (ibid.), which "*signify* only Men's peculiar *Ideas*, and that *by a perfectly arbitrary Imposition*" (III, ii,8; p. 408), it is erroneous for men to "*suppose their Words to stand also for the reality of Things*" (III, ii, 5; p. 407). The whole thrust of Locke's argument is toward clarity, separation not fusion, favoring abstract thinking and perception of relationships, at a diametric opposite to the corporeal "holism" of a Boehme. Since "the signification and use of Words" depend on "that connexion which the Mind makes between its *Ideas* and the Sounds it uses as Signs of them, it is necessary, in the Application of Names to things, that the Mind should have distinct *Ideas* of the Things, and retain also the particular'' and appropriate name (III, iii, 2; p. 409): Clear and distinct ideas are not what we find in hermeticism. Again, anyone who has reflected on the proliferation of synonyms in alchemy, say, will appreciate the gap that lies between that thought world and Locke's, in which *"a Definition is* nothing else but *the shewing the meaning of one Word by several other not synonymous Terms"* (III, iv, 6; p. 422). In Paracelsus and Boehme we find the related phenomenon of the same term meaning several different things in different contexts. To Locke this *"Inconstancy"* is a "great abuse of Words," for "Words being intended for signs of my *Ideas*, to make them known to others not by any natural signification but by a voluntary imposition, 'tis plain cheat and abuse when I make them stand sometimes for one thing and sometimes for another; the wilful doing whereof can be imputed to nothing but great Folly, or greater dishonesty" (III, x, 5; pp. 492–3), as if a man at the market were to sell "several Things under the same Name" (III, x, 28; pp. 505–6).

One major leitmotif of Locke's work that opposes it sharply to the occult tradition is his attack on the confusion of language and reality made, for instance, "by those who look upon *Essences* and *Species* as real established Things in Nature" (III, v, 10; p. 435). The truth is that the *"Genera* and *Species* of Things . . . depend on such Collections of *Ideas* as men have made; and not on the real Nature of Things" (III, vi, 1; p. 439). As for essences, Locke distinguishes *"real Essence . . .* that real constitution of any Thing," the "foundation of all those Properties" combined in it, from *"nominal Essence,"* the "abstract *Idea"* that we form (III, vi, 61; p. 442). When we classify substances, we do so by their nominal, not by *"their real Essences,* because we know them not" (III, vi, 7–9; pp. 443f.). Since the real essences are unknowable, names were originally given to things not by reference to any "internal real Constitutions," but by *"their obvious appearances"* and "sensible Qualities." This operation was performed not by "philosophers or Logicians" (so much for the "legislators" praised in the *Cratylus*), but by "ignorant and illiterate People" (III, vi, 25; pp. 452–3). Since these *"nominal Essences"* are variously defined by men, it is *"evident they are made by the Mind,* and not by Nature" (III, vi, 26; p. 453), a consideration that weakens further any theory proposing that words represent the inner nature of things. Another thrust in this attack is to restate the social, conventional nature of language. Since the gift of language is to enable us to communicate with each other, when men "speak of Things really existing they must, in some degree, conform their *Ideas* to the Things they would speak of" or else risk "being intelligible only to" themselves (III, vi, 28; p. 456). Otherwise we are all at liberty to conceive of complex ideas and to give them whatever name we will, just as Adam was. Far from being uniquely privileged, Adam had the same freedom, and the same constraints, that we have, the only difference being that since "common Use" has already "appropriated known names to certain *Ideas*, an affected misapplication of them cannot but be very ridiculous" (III, vi, 51; pp. 470–1). By this criterion the coinages of Paracelsus would seem antisocial, an instance of *"affected Obscurity*, by either applying old Words to new and unusual Significations; or introducing new and ambiguous Terms without defining either" (III, x, 6; p. 493).

Locke's most scathing attack on what can only seem to him the misuse of language by the occult tradition comes in the last three chapters of Book III, discussing the "Imperfection" and "Abuse" of words, and some "Remedies." Among the imperfections of language Locke singles out the uncertainty of meaning caused when it is not clear to which idea a word refers. In this case, "since Sounds have no natural connexion with our *Ideas*, but have all their signification from the arbitrary imposition of Men," the weakness lies not in the "incapacity" of one sound more than another "to signify any *Idea*: For in that regard they are all equally perfect" (III, ix, 4; p. 477). Having rejected the argument from onomatopoeia often used by proponents of natural language, Locke reiterates the relational view, by which words communicate only if they "excite in the Hearer exactly the same *Idea* they stand for in the Mind of the Speaker" (III, ix, 6; p. 478). The relationship between sound, sign, and idea being the foundation of all language, it follows that the greatest abuse of words is to use them "without clear and distinct *Ideas*; or, which is worse, [as] signs without any thing signified" (III, x, 2; p. 490); that is, lacking any distinct meaning. Another *"abuse of Words is the taking them for Things*," an abuse to which "those Men are most subject who confine their Thoughts to any one System," since "there is scarce any Sect in Philosophy has not a distinct set of Terms that others understand not" (III, x, 14; p. 497). Closed systems have closed languages. Worse still is to apply words to things that they cannot possibly signify, as when we substitute "their names for the real Essences of *Species*," a "preposterous and absurd" step, to "make our names stand for *Ideas* we have not, or (which is all one) Essences that we know not, it being in effect to make our Words the signs of nothing" (III, x, 17–21; pp. 499–502). The man who has

> imagined to himself Substances such as have never been, and fill'd his Head with *Ideas* which have not any correspondence with the real Nature of Things, to which yet he gives settled and defined Names, may fill his Discourse and,

perhaps, another Man's Head, with the fantastical Imaginations of his own Brain; but will be very far from advancing
thereby one jot in real and true Knowledge. (III, x, 30; p.
506)

In that conclusive dismissal we hear the very tones of the new sciences, from Kepler and Bacon to Galileo, Hobbes, and Descartes. All
who shared the ethos of that movement would have endorsed Locke's
belief in the importance of establishing a truly *"Philosophical Use* of
Words,'' that is, "such an use of them as may serve to convey the
precise Notions of Things, and to express in general Propositions certain and undoubted Truths, which the Mind may rest upon and be
satisfied with in its search after true Knowledge'' (III, ix, 3; p. 476).
All the leaders or spokesmen of the Royal Society shared Locke's
conception of language, even if they could not have expressed it in
such a powerful and coherent system. Hans Aarsleff has cited passages
in Boyle that similarly recognize that qualities have never been properly distinguished, but that certain "species of bodies . . . have had
the luck to have distinct names found out for them, though perhaps
divers of them differ much less from one another than other bodies,
which (because they have been huddled up under one name) have been
looked upon as but one sort of bodies'' (p. 177). The inaccurate use
of language can falsify reality. Boyle, like Galileo, would prefer to alter
the words so that "they may better fit the nature of things, than to
affix a wrong nature to things that they may be accommodated to forms
or words'' (p. 178).

In *The Sceptical Chymist* (1661)[37] Boyle delivered his own attack on
the language of the occult tradition, as well as on its science. In addition
to scathing comments on the Paracelsian alchemists' "obscure, ambiguous . . . aenigmatical way of expressing what they pretend to
teach'' (p. 3), their "ambiguous or obscure terms'' (p. 6), their "enig-
matical obscurity'' (pp. 22, 99), at the beginning of Part IV Boyle made
a sustained exposure of their misuse of language that anticipates Locke
on several points. The chymists take the "unreasonable liberty . . . of
playing with names at pleasure,'' so abusing "the termes they employ,
that as they will now and then give divers things one name; so they
will oftentimes give one thing many names,'' some of which properly
refer to quite distinct bodies (p. 113). Their "equivocal expressions''
may be intended to conceal arcane knowledge, but it is more likely
that their confusion is due to the fact that, "not having clear and distinct
notions'' of their three principles, "they cannot write otherwise than
confusedly of what they but confusedly apprehend'' (p. 114). Their
obscurity may be excusable if they are dealing with arcana that they
do not wish to divulge, but when they pretend to write natural philosophy, "where the naked knowledge of the truth is the thing principally

aimed at,'' the use of mystical terms "darkens what [they] should clear up" (p. 115). The goal of natural philosophy is quite the opposite, for it believes "that principles ought to be like diamonds, as well very clear as perfectly solid" (p. 23).

It may seem self-evident that Locke or Boyle, writing between the 1660s and the 1680s, should develop a theory of scientific language totally distinct from the practice or theory of the occult. Yet, as I have already shown, many of their ideas on the relationship among word, sign, and thing, insisting on a clear separation between language and reality, had already been expressed by Hobbes, Bacon, and Galileo, and can be traced back to Aristotle and Plato. I shall now argue that a parallel separation between the occult tradition and the new sciences in their attitude to language can be found a hundred years before Locke's *Essay*, in the context of the use of analogies and symbols. In the Renaissance scientific tradition, whose debts to classical rhetoric are only just beginning to be recognized, metaphor and simile, whether granted a heuristic or a merely illustrative role, are subordinate devices, clearly distinguished from the normal level of discourse, which is nonmetaphorical. In the *Poetics*[38] Aristotle writes that "a 'metaphorical term' involves the transferred use of a term that properly belongs to something else" (1457b 7). Of the conditions governing its use he writes that using such "strange expressions" – everything "out of the ordinary" or "over and above standard words" – confers distinction on our language, but must not be overdone: "If anyone made an entire poem like this, it would be . . . a riddle if it were entirely metaphorical . . . For it is the nature of a riddle that one states facts by linking impossibilities together (of course, one cannot do this by putting the actual words for things together, but one can if one uses metaphor) . . . So there ought to be a sort of admixture of these,'' a judicious blend of metaphor and "standard terms" (1458a 24–35).

Metaphor is a departure from normal usage, and its mode of action – I add, bringing out something implied by Aristotle, if not explicitly stated – is mental, not physical or corporeal. In the writer "it is a sign of natural genius, as to be good at metaphor is to perceive resemblances" (1459a 5ff.). When these resemblances are transmitted to us we equally perceive them as mental events, not as actual occurrences in the real world. As Aristotle writes in the *Rhetoric*,[39] metaphors must be drawn "from things that are related to the original thing, and yet not obviously so related – just as in philosophy also an acute mind will perceive resemblances even in things far apart" (1412a 9–11). Metaphors convey liveliness, but have "the further power of surprising the hearer; because the hearer expected something different, his acquisition of the new idea impresses him all the more. His mind seems to say, 'Yes, to be sure; I never thought of that' '' (1412a 17–21). Between the main subject and the thing to which it is compared there exists not a real, but a mental, channel: The gap is sparked by a mental act, but the two poles of the metaphor do not fuse or coincide in reality. Just as with Saussure's line separating signified and signifier, so in the theory of metaphor one must distinguish the two separate levels, as in the familiar terminology of I. A. Richards, "tenor" and "vehicle."[40] In traditional theory, metaphor is an abbreviated form of simile. Instead of saying "A resembles B," we say "A is B" or "A has B attributes." Utterances of the "A is B" type are taken to imply resemblance, not identity: They actually mean "A resembles B in one or more ways, but not in all." One can distinguish a positive area of analogy, where there are true resemblances, and a negative one, where there are differences. This remains a constant feature of the rhetorical and grammatical tradition from Aristotle to the early nineteenth century, at least.

Metaphor is important not only in verse: "In the language of prose, besides the regular and proper terms for things, metaphorical terms only can be used with advantage" (1404b 31), giving clarity and force, provided that they are "fitting," that is "fairly correspond to the thing signified" (1405a 6–10). They are useful in all forms of discourse, since "the arts of language" are important "whatever it is we have to expound to others: the way in which a thing is said does affect its intelligibility" (1404a 6–9). However, Aristotle adds, in the dismissive tone so often used by scientists and philosophers when they talk about language, "all such arts are fanciful and meant to charm the hearer. Nobody uses fine language when teaching geometry" (1404a 9–11). That restriction (which was not lost on the new sciences)[41] typifies the general caution of both Plato and Aristotle in regard to metaphor in philosophy and science, as Geoffrey Lloyd has shown.[42] In the *Phaedo* (92c–d) and *Theaetetus* (162e–f) Plato distinguishes between proofs and merely probable arguments, which are often based on images or likenesses, and in the *Sophist* we are told that the " 'careful person should always be on his guard against resemblances above all, for they are a most slippery tribe' '' (231a). The distinction is between the argument from analogy for heuristic or didactic purposes – both preached and practiced by Plato – and the confusion of metaphorical arguments with proofs. Using the same distinction, Aristotle dismisses Empedocles' description of the sea as the sweat of the earth as being " 'adequate perhaps, for poetic purposes,' but 'inadequate for the purposes of understanding the nature of the thing.' '' Similarly, in the *Metaphysics* (991a 20ff., 1079b 24ff.) he attacks the theory of Forms, since " 'to say that they [the Forms] are models ($\pi\alpha\rho\alpha\delta\epsilon\iota\gamma\mu\alpha\tau\alpha$) and other things [particulars] share in them is to speak nonsense and to use poetic metaphors.' '' In the realm of logic or scientific reasoning Aristotle

consistently warns against ambiguity. In the *Posterior Analytics* (97b
37f.) he says that "if metaphors should not be used in reasoning it is
clear that one should not use metaphors in giving definitions" – since
metaphors are by definition two-leveled or equivocal; while in the *Topics* (139b 32ff.) he criticizes definitions that contain metaphors, including Plato's definition of the earth as a "nurse" (*Timaeus*, 40b).

Nothing in this account will surprise the modern reader with some
knowledge of literary criticism, since, despite the loss of favor suffered
by rhetoric in the nineteenth century, our theory of metaphor derives
from the tradition inaugurated by Aristotle. Yet, just as with theories
of language, so with the use of analogy: The occult tradition in the
Renaissance showed a consistent desire to break with this tradition.
Given their tendency to treat words as things and essences, to believe
in innate notions, to collapse the concept of a linguistic sign, it is not
surprising that the occult use of language should also not recognize the
distinction between tenor and vehicle. Nor is it surprising that those
who held to the main linguistic and rhetorical tradition should draw
attention to the occult's subversion of it. This process of deviation and
restoration can be traced in several of the occult sciences: namely,
magic, alchemy, and medicine in its occult form.

Once again Neoplatonism, with its emanations and hypostatizations,
the successive overflows by which spirit gradually extends itself into
matter, seems to be the crucial influence. This whole process might
be described as a progressive reification of the immaterial, whether
that be described as mental (word, idea) or spiritual (soul, spirit). Any
discussion of this issue must begin with Ficino and with three authoritative modern accounts.

Consider, first, D. P. Walker's placing of Ficino in the Neoplatonist
magical tradition.[43] Ficino's whole philosophy, at least as seen in *De
triplici vita* (1489), is based on the materialization of spirit. To Ficino
the cosmic spirit, *spiritus mundi*, flows through "the whole of the sensible universe . . . thus providing a channel of influence between the
heavenly bodies and the sublunar world" (p. 12). Ficino defines this
spirit as " 'a very subtle body; as it were not body and almost soul.
Or again, as it were not soul and almost body . . . . It vivifies everything
everywhere and is the immediate cause of all generation and motion' "
(p. 13). Although not quite body, this cosmic spirit – which resembles
the alchemists' – is still sufficiently material for man to benefit by
consuming "things which contain an abundance of pure cosmic spirit,
such as wine, very white sugar, gold, the scent of cinnamon or roses"
(p. 13). Music plays an important role in Ficino's thought, for it shares
the same medium as spirit, namely, the air (p. 7), and "reaches the air
through . . . spherical motion," most suitable to the soul (p. 9). Further,
it can have additional power if linked up with astrology, for by using

" 'tones chosen by the rule of the stars, and then combined in accordance with the stars' mutual correspondences . . . a certain celestial virtue will arise' " (p. 16). What is involved here is a whole series of correspondences between music and the planet-gods, correspondences that, as so often in the occult tradition, are invoked not as analogies but as identities. Instead of "A is like B," we have "A is B." Thus (in Walker's summary) Ficino argues that "since the planets have the moral character of the gods whose names they bear, this character can be imitated in music; by performing such music we can make ourselves, especially our spirit, more Jovial, Solarian, Venereal, etc." The identity, or reification, involved is brought out clearly by Walker: "Such mimetic music *is* a living spirit and the heavens also *are* musical spirit" (p. 16). The occult tradition, typically, moves from analogy to identity, from suggestion to assertion. Thus Ficino claimed that his own practice of spiritual music had actually cured someone,[44] and he frequently asserted that his magical amulets would make astral influences materialize in the body of their wearer.

Walker's admirable study of the Neoplatonists will provide many more instances of the reification of spirit or even the fusion of matter and spirit in this branch of the occult tradition.[45] One I would like to pick out concerns Lodovico Lazarelli, who is linked with Ficino through a common interest in the *Hermetica*. Lazarelli draws on the *Kabbalah* for the belief that "God created the universe through the 22 letters of the Hebrew alphabet" and shows in a mystical hymn the typically occult magical theory of language: "He believes that words have a real, not conventional connexion with things and can exert power over them" (p. 69). Walker's interpretation of this extremely obscure hymn sees it as

> a magical operation by which the master provided his disciple with a good demon. The operation consisted mainly of words sung in some special manner. These sounds themselves became the demon; it is easy to understand how, *if we take literally Ficino's probably metaphorical description* of the matter of song: "warm air, even breathing, and in a measure living, made up of articulated limbs, like an animal, not only bearing movement and emotion, but even signification, like a mind, so that it can be said to be, as it were, a kind of aerial and rational animal" . . . Lazarelli was not summoning demons; he was making them. (pp. 70–1; my italics)

While wishing that Walker had been more clear about whether he is merely summarizing Lazarelli's views or endorsing them ("he was making" demons), I agree that one of the crucial points in this instance of breaking down the distinction between the immaterial and the ma-

terial is Lazarelli's taking Ficino's metaphorical account literally.
Walker has proposed a suggestive "general theory of natural magic"
(pp. 75–84), in which he distinguishes two levels of utterance, an A
and a B level, the A being nonmagical (words refer to, or denote, things
or ideas) whereas in the B, or magical, level words are essences of
things, as in incantations or in their existence as figures or characters
on talismans. In this theory the connection between words and things
is real, not conventional. "Moreover the word is not merely like a
quality of the thing it designates, such as its colour or weight; it is, or
exactly represents, its essence or substance" (pp. 80f.). What is involved here, it seems to me, is a further characteristic of the occult,
namely, substitution. If both levels are of equal status, then either may
be applied for the other. "A formula of words, therefore, may not only
be an adequate substitute for the things denoted, but may even be more
powerful. Instead of collecting together groups of planetary objects,
we can, by naming them correctly by their real, ancient names, obtain
an even greater celestial force" (p. 81). The occultist, in fact, often
prefers the more remote of the two levels, as if to demonstrate his
control over occult qualities.

   The identification of, and mediation between, discrete levels of existence are two of the main features of Renaissance Neoplatonism, and
the chapter, "Hierarchy of Being," in Kristeller's study of Ficino gives
an admirably clear and full account of their working.[46] Having separated, the Neoplatonists reconnected, and Kristeller distinguishes three
modes of connection: "'symbolism', 'continuity', and 'affinity'" (p.
92). In his use of symbol and metaphor Ficino derives much from Plato
and Plotinus; yet, Kristeller finds, in his images there is "a strange
mixture of ridigity and delicacy that is quite distinct" from his classical
models.

> More important than the impression, however, is the func-
> tion of the metaphor, the relation between image and idea.
> For Plato, as well as for Plotinus, the metaphor's primary
> task serves as a means of making abstract ideas evident to
> intuition, and since the relation of the image to the idea is
> produced by an arbitrary act of thinking, the metaphor can
> claim validity only for our thought, without stating anything
> definite about real entities. For Ficino, on the contrary, the
> relation of image to idea is not merely suggested by thinking
> but also corresponds to a real relationship existing among
> objects. (p. 93)

(I note in passing that Kristeller's analysis paid exemplary attention
to Ficino's debt to the classical philosophical tradition, yet gave less
space to his use of occult sources. It seems to me that Ficino's reification of metaphor is implicit in the whole occult development of the

concepts of macrocosm and microcosm, "correspondences," and sympathy.)

The difference can be explained, according to Kristeller, by Ficino's addition of "a new, ontological element" to metaphor, implying that "underneath the external connection of concepts is hidden an internal symbolism of things" (p. 94). Kristeller's discussion of symbolism is based, unfortunately, on a rather too materialist definition of symbol as "an object which by virtue of a similarity of character indicates another object," where one misses the concept of a sign. But we can readily assent to his description of the reification involved when a metaphor is "freed from its connection with thinking, transferred into reality, so to speak, 'substantiated'" (p. 94). When metaphors become treated like objects or essences in the occult tradition, this is not merely an issue of style or good taste: What is involved is another conception of reality, which posits that all the elements are interlinked according to theories of correspondence. I have argued that in such a concept we are dealing with identities, not analogies, and this view gains some support from Kristeller's observation that when "the relation of image and idea is transformed into a real relation between real objects, there appears a hidden connection between the individual objects in the world. The manner of thinking symbolically, therefore, seems to take metaphorical elements for immediate attributes of things" (p. 94) – thus fracturing the distinction between image and essence. The symbolism of light, for instance, so important in Plato, Plotinus, and Dionysius the Areopagite, in Ficino "loses the form of a metaphor" and becomes *identified* with "the divine truth and goodness," the "supercelestial spirits," the "splendor of the heaven," and God himself (p. 95).

If an object or concept symbolizes a higher reality, then one can "know an existing thing not only through its direct attributes but also through its relation to that higher reality" (p. 95): The two levels are interchangeable. More important is the contrary procedure; namely, "the definition of the originals themselves by means of their symbols. Because of the internal unity between the original and the symbol, the essence and attributes of a symbol can also be predicated, in a metaphorical sense, to its corresponding original" (p. 96). This technique of "transference," Kristeller writes, "sometimes surrounds Ficino's abstract considerations with the veil of an obscure and shadowlike concreteness" (p. 96). In other words, since a binary category is involved, a process of substitution or equation takes place, object can become symbol, or symbol object. Thus Ficino's demonstration of the incorporeality of the soul depends on the concept of food being "transferred to the soul in a symbolic sense," but in his handling of this Platonic idea "the metaphorical food of the soul and the proper food

of the body are almost united in a kind of genus, in such a way that
the attitude of the one may be derived from that of the other by a direct
analogy" (p. 97). This fusion of metaphorical and literal may, as Kristeller argues, "enrich immediate ontological knowledge by means of
symbolism" (p. 96), but we may also conclude that the reification of
symbols is by no means an aid to abstract thought.

Further light on the Neoplatonist treatment of symbolism can be
gained from the art historians. In his study, "Icones symbolicae,"[47]
E. H. Gombrich begins with the point that "to primitive mentality
distinction between representation and symbol is no doubt a very difficult one. Warburg described as '*Denkraumverlust*' this tendency of
the human mind to confuse the sign with the thing signified, the name
and its bearers, the literal and metaphorical, the image and its prototype." To Gombrich the "fusion between the image and its model" is
the sign of "more primitive states" (p. 125); but if so, then "primitive"
is not to be taken in a chronological or evolutionary sense, for wholesale reification is, as he acknowledges, the mark of the late Renaissance. Dionysius the Areopagite, indeed, commenting on the celestial
hierarchies, warned of the "danger" involved in the symbolic language
used by Revelation, representing "spiritual entities by way of analogy
through such dignified concepts as Logos or Nous or through the image
of Light," for it "may lead to the very confusions the religious mind
must avoid. The reader of the Scriptures might take it literally and
think that the heavenly beings are really 'god-like men, radiant figures
. . . clad in shining robes'" (p. 151). It is to avoid this confusion, Dionysius writes, that the authors of Revelation used deliberately inappropriate symbols "so that we should not cling to the undignified literal
meaning," but are led on to some higher truth. The "analogical
symbol," then, "has its dangers if it leads the mind to take the reflection
for the reality," and the Latin church stressed these dangers "lest the
fusion between image and prototype leads to idolatry" or image worship (p. 151).

With Ficino's use of symbols such caveats seem appropriate. Where
the image of the serpent biting its tail is explained by Horapollo, in his
Hieroglyphica, as a symbol of the universe, with its cycle of decay and
rejuvenation, Ficino interprets it as a symbol for time and its circularity
(p. 159). This raises one of the problems of symbolism: the variable,
arbitrary, and in the last resort personal nature of interpretation. Gombrich comments that "where symbols are believed not to be conventional but essential, their interpretation in itself must be left to inspiration and intuition . . . The symbol that presents to us a revelation
cannot be said to have one identifiable meaning assigned to its distinctive features" (p. 159). Where the linguistic sign is seen as conventional, signification is essentially a social, historical process. Where

the sign is held to be an essence, signification can become purely personal. It is typical, too, of this essentialist attitude that, as Gombrich puts it, "the distinction between the representational and the symbolizing function of the image becomes blurred. Ficino does not accept the image of the serpent as a mere sign which 'stands for' an abstract concept. To him the essence of time is somehow 'embodied' in the mysterious shape" (p. 160). Here is one major difference between the occult and the experimental sciences in their attitude to the use of symbol and analogy. Gombrich quotes from Goethe the passage where Faust opens the mysterious book of Nostradamus and sees the universe in and through the "sign of Macrocosmus":

> Wie alles sich zum Ganzen webt!
> Eins in dem Andern wirkt und lebt!
> Wie Himmelskräfte auf und nieder steigen . . .
> Harmonisch all' das All durchklingen!
> Welch Schauspiel! aber ach! ein Schauspiel nur.

There Goethe juxtaposes an "esoteric conception of the visual symbol" with "rational categories of representation and symbolization." Here – as in the occult tradition in general, we might add –

> the magic sign "represents" in the literal sense of the word.
> Like the name it gives not only insight but power. The Neo-
> Platonic theory has indeed accepted this consequence. For if
> the visual symbol is not a conventional sign but linked
> through the network of correspondence and sympathies with
> the supracelestial essence which it embodies, it is only con-
> sistent to expect it to partake not only of the "meaning"
> and "effect" of what it represents but to become inter-
> changeable with it. (p. 172)

The network of correspondences, although conventionally described as analogical, does not in fact work by analogy, which posits an imaginative or imaginary connection between discrete entities. Rather, as I have shown elsewhere,[48] it posits a real connection, an inter-equation or identity of elements on the corresponding levels of classification. But, I have argued, the identity is merely the juxtaposition of separate and preformed categories, not of innate likeness.

The occult sciences' practice of substitution or interchangeability of concepts depends fundamentally on the reification process, the breakdown of the line between literal and figurative. Commenting on Ficino's concept of "the virtue of the visual image," Gombrich describes it as a "most extreme position," one in which "not only the distinction between symbolization and representation is removed but which threatens even the distinction between the symbol and what it symbolizes" (p. 172). Although he hedged his bets occasionally, Ficino evidently believed in the magic potency of images and used the argument from analogy to explain how they work. As Gombrich summarizes it: "Just as one lute resounds by itself when the strings of another are plucked, the likeness between the heavenly bodies and the image on the amulet may make the image absorb the rays from the stars to which it is thus attuned" (p. 173). Gombrich invokes both Saxl's account of the consequence of a belief in the magic efficacy of images (namely, that the "rational division between 'form' and 'content'" is elided) and Warburg's description of the Astrologers' "Schlitterlogik" (shifting, slipping, arbitrary logic). The point being that

> rationally there is of course no likeness whatever between the image Ficino bids us to engrave and the star as a "heavenly body." What he means is the image of astrological tradition, of Saturn with his falx or Mars with his sword. These images, then, are not to be regarded as mere symbols of the planets nor are they simply representations of demonic beings. They represent the essence of the power embodied in the star. (p. 173)

As such they both embody and express this essence. This idea is simultaneously abstract and concrete, metaphysical and rhetorical or psychological. Ficino thought that "the numbers and proportions of a thing preserved in the image reflect the idea in the divine intellect and therefore impart to the image something of the power of the spiritual essence which it embodies." According to Neoplatonic magic, the images affect the mind, reproducing in it the powers of the divine essence, cosmic harmony, or whatever is being invoked. "In other words," Gombrich continues,

> the Neo-Platonic conception favoured not only a removal of the distinction between the representational and the symbolizing functions of the image, but also the confusion of these two levels with what we have called the expressive function. All the three together are not only seen as various forms of signification but rather as potential magic. (p. 174)

The sign is the thing it represents, and as such it works in us, and we can use it to work on the world. The reification is not accidental, but functional, performative. The collapse of analogy into identity enables a substitution process, by which manipulation of one item can affect the related items. The lute strings affect each other, the star's image affects us; by wearing a magic amulet we can tap the health-giving forces in the invisible world. Analogy leads to identity and to actual connections between things.

In my second category of reification in the occult sciences, I move to an area where symbols were peculiarly important, but problematic: alchemy. Maurice Crosland, in his valuable study of chemical symbolism,[49] has drawn attention to the tendency toward reification in the alchemists' use of symbols. This begins with the extremely ancient habit of using the names of the planets to describe metals and their derivatives, "due to a supposed analogy between the seven planets and the seven metals," which not only weakened alchemy's status as an independent discipline but "also gave rise to confusion in so far as the names of the planets were those of the gods of mythology" (p. 5). In some alchemical works, as in the following excerpt from the *Turba philosophorum,* it is hard to know if the text refers to alchemy or astronomy: The envious have said "that the splendour of Saturn does not appear unless it perchance be dark when it ascends in the air, that Mercury is hidden by the rays of the sun, that quicksilver (*argentum vivum*) vivifies the body by its fiery strength, and thus the work is accomplished" (cited in Crosland, p. 6). Obviously we are dealing here with the substitution process in the occult sciences, by which one category stands for another and may be invoked in place of it. Yet it was clear to some seventeenth-century writers that the substitution process involved a movement between literal and metaphorical levels. Jean Brouault, in a work called *Abrégé de l'astronomie inférieure* (1644) – a term going back to the Middle Ages, "inferior astronomy" acknowledging the transplantation of astrology into alchemy – "argued that many documents which appeared to be concerned with astronomy were really about alchemy, because if they were taken literally, many of the statements of the ancients would be absurd. Rather than agree to this he suggests a metaphorical interpretation" (Crosland, p. 6).

The substitution of, say, Saturn for lead or Venus for tin (or Mercury or Jupiter, depending whose system one uses) "perpetuated ancient superstitions of a real connection between the two categories," and it was not until Lemery's famous textbook, *Cours de chymie* (1675), which ridiculed any associations between the planets and the metals, that the correspondence theory was called in question (Crosland, p. 80). A similar movement from analogy to identity can be traced with alchemical symbols. Whereas to a modern scientist symbols are arbitrary signs, chemists as late as the eighteenth century analyzed symbols "in search of a rational justification for every line, cross or circle" (p. 233). Like the medieval concept of signatures, where each plant was signed, supposedly representing by its shape, color, or texture its relevance for medical treatment, chemists took a cross to denote "anything sharp and corrosive, whereas perfection was indicated by a circle. It was therefore appropriate that a half circle should stand for silver (since it was 'half Gold')," and the symbol for copper, ♀, was seen as denoting a metal consisting partly of gold but "with crude, sharp and corrosive matter joined with it." These interpretations come from no less a figure than Boerhaave and show the persistence of occult thought habits. It is not just that he treated arbitrary signs as if they were symbolical signs, but that he confused sign and substance. So, from the conventional sign for iron, ♂, Boerhaave deduced that "this too is intimately Gold; but that it has with it a great deal of the sharp and corrosive; though with but half the degree of Acrimony as the former, as you see that it has but half the sign that expresses that quality." Therefore, he reasoned, the philosophers' gold lies concealed in iron, and thus "here therefore we must seek for metalline Medicines" (p. 233). The use of symbols for substances "often depended on the literal interpretation of chemical names," as shown in the accompanying illustration (p. 237).

Regulus of antimony

Plume alum

Even though we might feel that the iconic representation of the substance was unmistakable, the fact is that considerable confusion existed in the seventeenth and eighteenth centuries due to the inherently arbitrary nature of symbolism. The same substance "was often denoted by a variety of different symbols"; for tartar one lexicon gives thirty-two symbols, and another gives thirty-nine symbols for mercury (pp. 235f). The substitution process does not necessarily lead to consistency.

Another concept in alchemy that was subjected to reification was fire. Rosaleen Love, studying Boerhaave's concept of fire as an "all-pervasive fluid medium" (p. 157),[50] suggests that the ultimate source is Dionysius the Areopagite, who, in *The Celestial Hierarchy*, used fire as a recurring metaphor for the divine. It became "the particular sensible representation of God's universal power and presence" (p. 160). This sensible fire is in everything, yet passes through everything unchanged; it gives light, but is concealed; it is "both incomprehensible and invisible"; "comprehending, incomprehended"; and so on through a series of the paradoxes used by so many mystics to describe the ineffable nature of the divine, transcending or reconciling contradictions. There would be "many characteristics of fire," Dionysius writes, "appropriate to display the Divine Energy, as it were, in sensible images." But, he adds, this fire is only a metaphor, nothing more, giving "the warning that 'those who diligently contemplate the Divine imagery should not rest in the types as though they were true'" (p. 160).[51] Here is the orthodox caveat from the rational tradition of philosophy and rhetoric, stressing that the types or symbols are not to be confused or identified with reality. Such caveats were ignored by the occult tradition, and it makes a significant link with E. H. Gombrich's analysis that the figure whom Dr. Love describes as "pivotal" in the "reification process" by which the metaphorical concept of fire was taken more literally "was undoubtedly Marsilio Ficino" (p. 161). As she says, Ficino attempts "to link the Cabalist and Neoplatonic cosmologies in such a way that the relationship of symbol to object, e.g. light to God, implied not only a meaningful likeness, but a secret identity." Where Dionysius "had been clear about the non-literal relation of symbol to object, Ficino blurred the distinction" and effected in his philosophy "an internal unity between the object and its symbol such that the word and the thing, merging together, became interchangeable in argument" (p. 161). While Ficino preserved a degree of caution, his followers were less tentative. Henry Cornelius Agrippa, who drew much from Ficino, was the first who "ascribed the metaphorical attributes literally to the fire," paraphrasing Dionysius in such a way as to transfer the divine qualities wholly to the material, physical fire (p. 162). Some occult writers, such as the alchemist Sendivogius, followed Agrippa in this reification of fire symbolism; others, such as Pico, applied it to the *Kabbalah*'s division of the universe into three worlds: the terrestrial, celestial, and supercelestial. These included the numerologist Francesco Giorgi and the Paracelsian chemists Joseph Du Chesne and William Davisson (pp. 162f.). With the Paracelsians, indeed, ethereal fire was conceived of as a force circulating "from the heavens throughout terrestrial things, conveying its life-enhancing characteristics to plants and animals by means of the air" and becoming "the medical counterpart to the philosopher's stone," the universal healing force (p. 165). The final reification of fire in Paracelsian alchemy was a rather incongruous taming of Dionysius's sublime spirit, for the Paracelsians identified it with saltpeter.[52] But this reduction of symbol to object was wholly in line with occult thought habits.

The last and most remarkable instances of reification in the occult tradition that I wish to examine here are those made by Paracelsus himself. In his work, where alchemy, astrology, numerology, magic, and medicine unite, where Neoplatonist and Gnostic influences merge with Aristotelian cosmology and Galenic medicine (despite his professions to the contrary),[53] the processes of reification, substitution, fusion of levels, identification of opposites, become habitual, constitutional.[54] It is generally recognized that the whole of Paracelsus's system is based on the distinction between macrocosm and microcosm. Yet where many thinkers treated the relationship analogically, Paracelsus collapsed the two poles into one. Man does not merely resemble the macrocosm, he *is* the microcosm. The move from analogy to identity is total. In his account of the Creation Paracelsus states that God made man out of *limus terrae,* fusing the four elements into one, the quintessence, so that man contains in himself all the minerals, plants, beasts – indeed, the whole of creation:

> aus diesem limo hat der schöpfer der welt die kleine welt gemacht, den microcosmum, das ist den menschen. also ist der mensch die kleine welt, das ist, alle eigenschaft der welt hat der mensch in ime. darumb ist er microcosmus, darum ist er das fünft wesen der element und des gestirns oder firmaments in der obern sphaera und in der undern globul. (*Astronomia magna*; cited in Fischer, p. 281)

For Paracelsus, man, formed out of *limus terrae* and *limus coelorum,* is "den centrum aller ding. der centrum ist der mensch und er ist der punkt himels und erden" (p. 282). But this anthropocentric metaphor coexists with other irreconcilable metaphors, such as the microcosm being the "spigelbilt" of the macrocosm, each possessing a soul (Fischer, p. 304), mind, spirit, and body (King, pp. 104f.). Indeed, although he claims that man is the center of both heaven and earth, Paracelsus prefers to look in the mirror, examine the macrocosm for information about the microcosm. Since man is only a mirror image and shadow of the macrocosm, the doctor can glean no knowledge from a study merely of the human body. "The firmament, the great world, the macrocosm offers the true picture, from whose powers the doctor must win the insights needed to understand the 'spigelbilt' and cure its illnesses" (Fischer, p. 287).[55] The macrocosm is also identified with nature, itself the cause, apparently, and also the cure of illness: "Die natur die krankheit selbs ist, darum weiss sie allein, was die krankheit ist; sie ist allein die arznei . . . aus dem arzt kompt kein krankheit, aus im kompt auch kein arznei" (p. 288). Because the doctor is not the source of illness, he cannot cure it by merely human insight. This means that there is no point in examining the human body or cutting it up, for this kind of seeing is merely that of a peasant who looks at a psalter but cannot even understand the alphabet (p. 291). With this typically dogmatic, assertive invocation of the Platonic and mystic distinction between outer and inner nature, Paracelsus rejects anatomy of the human body in favor of anatomy of the astral body. In practice this turns out to mean the traditions of iatromathematics, astrological medicine, with the supposed correspondences between metals, herbs, stars, parts of the human body, and so forth (e.g., Fischer, pp. 290–2).

The occult sciences' double process of reification and substitution, formulating ideas as essences, then making them identical and exchangeable, inevitably broke down the distinction between metaphorical and literal. Man has often been described as an image of the universe, usually by reference to some common principle, such as order or harmony. But Paracelsus takes the image literally. Each individual, he writes elsewhere, is "an offspring of a particular part of the earth and so are each of his organs and limbs," so the doctor is urged to study cosmography: "Look at *anatomia terrae*, find in what order its hands and feet are distributed and . . . its fingers" (Pagel, *Paracelsus,* p. 138). Such animism is typical of the Neoplatonic tradition, of course, if seldom carried to such lengths. At every stage Paracelsus converts the nonmaterial into the material. For him "every object is . . . but coagulated smoke" (ibid., p. 95); the assimilation of macrocosmic nourishment is performed by the "archeus," or spirit of digestion personified (Fischer, p. 300). The magus can make an image, devoid of flesh and blood, "to act as a comet" (Pagel, *Paracelsus,* p. 63). Imagination is a spiritual power that "acts through magnetic attraction on an object in the outside world" (ibid., p. 122). Paracelsus's concept of the plague is one by which human sin infects heaven, the human passions arising in the form of a "body" to the relevant and appropriate star, where they rest as seeds until the wrath of God shoots them back down to earth. This is not so much an "anthropocentric" view, as Walter Pagel calls it, as a moralized, theodicy-supporting view, where man is the cause of his own disease. But the remarkable aspect is the reification of human passions, the conversion of them into arsenical (or corrosive) substances, to be trapped in a coagulate (tartar), this whole process being reechoed in human contagion (ibid., pp. 181f.). As Pagel says, this theory "finally leads to the concept of a psychic element in bodies, and vice versa, and thus to an abolition of strict dualism. The noncorporeal spirit begets corporeal matter" (p. 181).

In such sequences fundamental distinctions are collapsed in the process of fusion so typical in occult thought. Kurt Goldammer classifies Paracelsus's system as "a 'vitalistic monism,'" and Pagel says that his ideas of God, the world, nature, and man are all based on "the unity of spirit and nature." A persistent trend in Paracelsus is "to dissolve the body and to trace in it the all-pervading spirit. The latter, in turn, is not regarded as alien to matter, but as a substance of finest corporeality" (p. 208). The basic conception is familiar from Ficino; what again distinguishes Paracelsus is that he takes it to its utmost extension. His way of acquiring total knowledge was by a kind of union with the object brought about by communication between man's astral body and the "super-elementary world of the 'astra,'" *astrum* here denoting, as Pagel glosses it, "not only a celestial body but the 'virtue' or activity essential to any object" (pp. 50f.), a remarkably all-embracing concept. The quest is for totality. Paracelsus seeks knowledge "through union of the object with something alike in the observer" (p. 52) and urges the physician to make himself a part of the phenomenon he is investigating: "By virtue of his union with his objects – the patient, the disease and the cure – the physician indeed acts like an Archeus" (p. 110). As Koyré puts it: "Connaître, n'est-ce pas s'assimiler, n'est-ce pas devenir en quelque sort identique à l'objet ou à la personne que l'on veut connaître?" Hence, we might add, the justification for the union of knower and known in a universe of interlinked essences: "Pas de connaissance sans sympathie, et pas de sympathie sans similitude. C'est le semblable qui connait son semblable" (p. 52) – preferably by becoming like him. In Paracelsus's ideal system, there is no way of telling the dancer from the dance, the observer from the observed, a fusion of categories[56] that looks back to Nicholas of Cusa and on to Jacob Boehme.

The materialism of Paracelsus is part of the general tendency in the occult sciences to shun abstractions, to think in wholly concrete terms. Although Paracelsus conceives of each part of the universe reflecting every other part, every solid body containing a soul, these bodies are not unreal containers for souls: "Le monde et le corps ne sont pas de purs 'symboles,' ne sont pas des images. Paracelse n'est pas Idéaliste" (Koyré, p. 54). The conception of an immaterial, incorporeal spirit would have seemed absurd to him: no soul without body (p. 55). When the imagination produces an image, according to Paracelsus, it is something real, "a natural organic product of the soul's astral body . . . a 'body' which 'incarnates' thought. – This expression must be taken literally. The image is a body in which are incarnated the soul's thoughts and wishes" (p. 60). To "conceive" a thought in Paracelsus's vocabulary is literally to give birth. The *mysterium magnum* of life, the stuff from which the universe is formed, materializes itself in degrees, "and we have only to condense this impalpable matter progressively to obtain, in more and more material coagulations, the astral matter, the firmament, and finally the matter from which our bodies are made" (p. 63). This whole process resembles the alchemists' techniques of distillation and precipitation, a particularly appropriate analogy because Renaissance alchemy believed that the changes in the external world moved in parallel with those in the soul. That is, in alchemy, as throughout the occult sciences – cosmology, psychology, astrology, numerology – a continuous two-level model is used.

> The alchemical books always speak in symbols . . . and always talk of two things simultaneously, of nature and man, of the world and of God. The philosopher's stone is the Christ of nature, and Christ is the philosopher's stone of the spirit. Mercury, being intermediary between the sun and the moon . . . *is* Christ in the world of matter, in the same way that Christ, mediator between God and the world, is the spiritual mercury of the universe. (p. 70)

We note again how the basis of occult analogy rests on identity:

> This is more than a simple allegory or comparison. The
> analogy is more profound. The same symbols apply both to
> material and spiritual processes because, at bottom, the two
> are identical. The identity of the symbols is explained by the
> identity of the processes. (p. 70)

To Koyré such an analogy is "more profound": I would question
whether it remains an analogy at all. In the continual movement in
Paracelsus's thought that Koyré distinguishes, between high and low,
creator and creation, ascent and descent, the organicist conception of
the world can hardly be adapted to logical categories: At the level of
thought, as Koyré admits, it constitutes a vicious circle (p. 71). Instead
of the circle I would prefer Paracelsus's own image of a "spigelbilt":
Man projects his own ideas and categories onto the macrocosm, reflects
them in the microcosm, and reads off an anthropomorphism that simply
confirms itself in an endless self-reflection.

Whether we use the metaphor of circular or of mirrorlike movement
in Paracelsus's thought, it seems clear that the process is so dynamic as to break down distinctions of kind. As F. R. Jevons has
said, placing Paracelsus in the tradition of mystics from Plotinus to
Hildegard of Bingen, the introspection of the mystic "led to a generally
unifying type of thought, a tendency to 'lump' rather than to 'split,'"
an urge toward unity, integration (p. 155). These two-way movements
in Paracelsus's thought are reflected, naturally enough, in his language,
and we can see that for him it will be perfectly natural, indeed unavoidable, to turn metaphor into reality. As Jevons has said: "Paracelsus
had trouble in keeping his similes for the non-corporeal both invisible
and intangible" (p. 142 n.). In his discussion of quintessences "the
distinction between the elemental and the astral becomes hazy," while
"the borderline is unequivocally violated by references to the stars of
each of the four elements" (p. 145). This typical fusion of categories
suggests that Paracelsus's thought processes, for all their fertility, did
not include the facilities of distinction and differentiation: like a calculating machine that can multiply but not divide. Virtually every commentator on Paracelsus notes his confusions and contradictions.[57] One
cause of confusion is his use of the same term for two things linked
by the correspondence theory of the macro–microcosm, as if their
natures were not just analogous but identical, so much so that the same
term can cover both of them. Thus Jevons points out that when Paracelsus discussed the elements, he often "seems to have been referring
not to the constituents of matter but to regions of the cosmos," perhaps
because he associated "the same generative force or 'mother' (matrix)
with a region of the cosmos and with certain sets of properties in earthly
bodies" (p. 145 n.). And he makes the acute point that "these situations
exemplify the pervasive use of analogy in Paracelsus's explanations

and, correspondingly, of metaphor in his terminology, the difficulty of which is largely due to the way in which the metaphors are not only mixed but also superimposed, so that the meanings of words double and redouble" (p. 145 n.). If, to use a very simple model, metaphor works by comparing the familiar with the "foreign," in a single movement linking two concepts, Paracelsus moves from the foreign to the foreign without reverting to the familiar. Leaving the fixed point, he builds variable on variable.

The problem with such an associative movement is that the writer risks losing contact with his reader as he moves into more personal thought patterns. The associative process and the preference for concrete over abstract turn analogies into substitutes for argument. The effect this has on Paracelsus's writing has been well described by Owsei Temkin, commenting on passages in which Paracelsus interpreted dropsy as resulting from an invisible rain or equated an epileptic attack with a thunderstorm. Instead of analyzing causes, Paracelsus "presents us with pictures which he expects us to see just as he sees them," without accompanying explanation or argument. "The picture may have something compelling, but it remains a picture. There is no necessity for its choice" (p. 210). Paracelsus, of course, rejected logical proof and even human reason, resorting instead to fables or visions. "To make his reader see the truth of his interpretation, Paracelsus has no other means[58] but to lead him as near as possible through examples." Hence his style "is marked by a series of statements connected by analogies or by open or hidden biblical references" (p. 211). Analogies merge into statements: Dropsy *is* a hidden rain. The description of the heavens, the elements, nature itself, as books, may seem at first a manner of speech, a "mere simile," as Temkin calls it (p. 213). But the metaphor is used so often and in such a literal way that the distinction between tenor and vehicle collapses. Countries are leaves of the book of nature, plants and stars are read, the physician makes a gloss on the text only. Yet the way in which the glossator works is confusing, since it involves rejecting the terminology of both ancients and moderns and inventing a whole set of new terms, the notorious, barbaric neologisms – iliaster, cagaster, and so on. The names are personal, and their connotations often so. Yet in one place Paracelsus claims that he was forced to coin the name of tartaric diseases because the old names, stone and gravel, were inadequate. They are unsatisfactory, he says, because "the concretions found in the bladder, kidneys, and elsewhere are not real stones. The old names are metaphors" (p. 216). In his objection Paracelsus talks like the Aristotle of the *Posterior Analytics*: "'And this I declare, because it is a lack of skill to use *metaphora* in medicine and nothing but an error to give names metaphorically'" (p. 216). Brave words! His replacement, "tartaric," derives from his theory that the matter formed in such diseases was identical with the concretions found in wine casks. Now it was clearly not possible for Paracelsus to prove or disprove his theory by chemical analysis. But in fact the identity he makes rests only on an analogy, and instead of justifying his new term by rational explanation, he gives this series of tautologies and assertions:

> Moreover, mark my words, the name I give to this disease
> is *tartara*, that is *aegritudo tartari* or *tartareus morbus*,
> taken from *tartaro* which is called *tartarum* by its inborn
> name; *tartarum* because it yields an oil, a water, a tincture,
> a salt which inflames and burns the sick like a hellish fire,
> for *tartarum* is the hell. (p. 216)

With its repetitions and its avoidance of giving a definition, that passage is uncomfortably close to some of Shakespeare's inarticulates[59] and far removed from Locke's account of definition as "shewing the meaning of one Word by several other not synonymous Terms." The significant detail is the phrase "inborn name," which suggests the familiar occult preference for a natural language, where the connection between things and words is not arbitrary and fixed by history and social convention, but real and fixed – all too often, alas – by the individual's own fiat. Temkin sees a similar point: "Since Paracelsus offers this etymology [*sic*!] instead of the alleged metaphors, we may assume that to him the connection between the substance that causes hellish pain and the name of the disease that means hell is a real one. In short, the magic glossator is not satisfied with referring to a thing, but wishes to express the thing itself" (p. 216). In collapsing the distinctions between signifier and signified, in confusing literal and metaphorical, the occultist ultimately produces a private language that no one can follow. This is the exact reverse of the goals of the new sciences.

I have based this brief account of Paracelsus's habits of thought and speech with regard to analogy on the work of some of his modern commentators in order to show how, largely independently of each other, readers in our age have reached very similar conclusions about his practice of reification and fusion. Yet the whole analysis, and the negative conclusions, can be found in his original critics, from Erastus to Van Helmont. These opponents of Paracelsus objected precisely to his breaking down fundamental distinctions. Thus Thomas Erastus's extremely thorough attack of 1572, his *Disputationes de medicina nova Paracelsi*, took issue with his Neoplatonist unification of spiritual and corporeal. As Walter Pagel puts it: "In contrast to Paracelsus's monism and pluralism, Erastus' position is that of dualism. The strict separation of the spiritual and corporeal is in his case associated with a disbelief in and abhorrence of all that is 'occult'" (*Paracelsus*, p. 331). Andreas Libavius, author of the first systematic chemical textbook, attacked

Paracelsian alchemy and its claim to be able to manipulate divine powers in nature through the agency of the alchemist. As Owen Hannaway
has said, the Paracelsian world view "was predicated on the belief of
the immanence of divine powers in man and nature, which broke down
all barriers between the natural, human, and divine. Libavius resists
all such conflation of knowledge and power: his is still a hierarchical
world in which nature, man, and God have their own appropriate
spheres and modes of operation."[60] The Paracelsians, then, confused
the power of man and the powers of God. Similar criticisms were made
by Francis Bacon, in an early work, the *Temporis partus masculus* (the
Masculine Birth of Time). In this polemic against previous philosophic
systems preliminary to his own reforms, Bacon arraigned Paracelsus
on "graver charges":

> By mixing the divine with the natural, the profane with the
> sacred, heresies with mythology, you have corrupted, o you
> sacrilegious impostor, both human and religious truth. The
> light of nature, whose holy name is ever on your lips, you
> have not merely hidden, like the Sophists, but extin-
> guished.[61]

On several occasions Bacon attacks Paracelsus for confusing the book
of nature with the Bible, for pretending "to find the truth of all natural
philosophy in the Scriptures," a claim that does not "honour . . . the
Scriptures, as they suppose, but much imbase them. For to seek heaven
and earth in the word of God . . . is to seek temporary things amongst
eternal, and . . . to seek the dead amongst the living."[62] Daniel Sennert,
writing in 1619, similarly attacked practitioners of Paracelsus's "false
Chymistry" with its "peculiar Religion," for they "proceed to Divinity
and mix prophane and holy things together."[63]

The Paracelsians' thoroughgoing collapse of the distinction between
the immaterial and the material was also singled out for attack. Bacon
took issue with their treatment of "Fascination," that is, "the power
and act of imagination, intensive upon other bodies than the body of
the imaginant . . . wherein the school of Paracelsus and the disciples
of pretended Natural Magic have been so intemperate, as they have
exalted the power of the imagination to be much one with the power
of miracle-working faith" (III, 381; also II, 640f.; IV, 400). Bacon also
attacks Paracelsus's belief that dew is an exudation from the stars (I,
356; IV, 239) and his literalist theory of nutrition, by which "Archaeus,
the internal artist, educes out of food by separation and rejection the
several members and parts of our body" (I, 339; IV, 224). Bacon's
sharpest criticisms, however, are reserved for Paracelsus's elaboration
of correspondences, especially the microcosm–macrocosm analogy:
"You have a passion for taking your idols in pairs and dreaming up
mutual imitations, correspondences, parallelisms, between the products

of your elements. As for man, you have made him into a pantomime,"[64] that is, a mere mimic of the processes in the macrocosm. The basis of the analogy is weak, Bacon points out, since it is not built on any profound relationship, but depends on the "obvious and superficial qualities" of things; by using it Paracelsus forced his own a priori scheme onto nature: "The evidence drawn from things is like a mask cloaking reality and needs careful sifting. You subjected it to a pre-ordained scheme of interpretation" (p. 66).[65] In the *Advancement of Learning* Bacon went on to attack the overelaboration of the analogy into the realm of literal, particulate correspondence:

> The ancient opinion that man was Microcosmus, an abstract or model of the world, hath been fantastically strained by Paracelsus and the alchemists, as if there were to be found in man's body certain correspondences and parallels, which should have respect to all varieties of things, as stars, planets, minerals, which are extant in the great world. (III, 370; also IV, 379–80)

Bacon's criticism of the collapsing of analogy into identity is made with more force in his book of fables reinterpreted to show his main scientific ideas, the *De sapientia veterum* (1609): "The Alchemists, when they maintain that there is to be found in man every mineral, every vege-table, &c., or something corresponding to them, take the word *micro-cosm* in a sense too gross and literal, and have so spoiled the elegance and distorted the meaning of it" (VI, 747; see also II, 640f.).

Bacon's remarks are typical of the two main points made against Paracelsus's use of the macro–microcosm analogy in the period be-tween the 1570s and the 1640s. The first is that if taken literally, the analogy implies an absolute one-to-one correspondence that assimilates man to the universe and destroys his humanity. This claim can be rejected either on purely scientific grounds, denying that the human body is mineral-like or insect-like, or on theological grounds, asserting that man was created in God's image. The second line of attack was directed not against the content but against the form of Paracelsus's thought, especially the way in which it reduced analogy to identity.

Thomas Erastus led the charge, describing Paracelsus's concept of the microcosm as a trope, but not a reality, "a pleasant allegory," but no more. Taking it literally would have had the dangerous conse-quence of eliminating completely "the differences between plants, an-imals, and man." If man's body really contained the virtues and ma-terials of the rest of creation, then why could he not fly or lay eggs or live in the sea?[66] Andreas Libavius attacked Paracelsus's "similitudes, analogies and harmonies," especially the doctrine of sympathetic in-teraction between the astra in the world of man and those in the great world.[67] The Paracelsians "have reduced all knowledge to uncertainty in their search for correspondences in the great and little world. Indeed, no science is possible by this means because true knowledge resides in an understanding of the specific, inherent, and immovable causes of things, which causes are comprehended by scientific definitions and principles arrived at by reason and experience and confirmed by the judgment and experience of scholars."[68] Libavius is opposing to occult procedures something that might best be called protoexperimentalism, and like many thinkers of this period did not succeed in wholly freeing himself from the system he attacked. Indeed, on several occasions he defended alchemy.

Yet the dispute between himself and Oswald Croll, Paracelsian alchemist, highlights one of the crucial issues separating the occult and experimentalist traditions: the nature of scientific language. The followers of Paracelsus took over his mystical-magical concept of language, the so-called language of nature, where correspondences were held to link not tropes but real things. As Hannaway's analysis has shown, "similitude and analogy were not for Croll figures of speech which illuminated the essentially incomparable; they were the very fabric and glue of the universe and the means by which it spoke" (p. 107). The human, social language of the protoexperimentalist Libavius works in a quite different way: "It defines, divides, distinguishes, and establishes criteria for judgment – a judgment which separates things. It seeks to discriminate knowledge, whereas Croll's language sought to reveal . . . the resemblances of things" (p. 108). As we have already noted, the difference between the occultists and the experimentalists is the difference between the Neoplatonists and Saussure: on the one hand, a natural language in which words embody things in a real equation of signified and signifier to form a magic object; and on the other, a language where the connection between signified and signifier is arbitrary, socially given. It is symptomatic of the incompatibility of these two attitudes that Libavius should attack Croll's use of a cabalistic "archetypical language of signs":

> The Cabala is a falsehood and a deceit. For it presents
> things, not as they are, but as they are compared with other
> things in an indeterminately external fashion. Thus we are
> not able to know what constitutes a thing, for the gateways
> [to knowledge] are surrounded by deceiving images.[69]

This reaction is typical of opponents of the occult sciences, from Pico onward,[70] who complained that the occultists tended to heap analogy onto analogy, spiraling off into the void. Libavius's reaction is to insist on fixed points, holding to the difference between literal and figurative language, thus keeping analogy in its subordinate position. He attacked the magia and cabala for elevating the use of rhetorical figures (*tropologia*) to the point where it could transform God into man, man into

God, in metamorphoses more marvelous and more dangerous than anything in Ovid. He did not want to exclude analogy from philosophy and science, but rather to restrict it to a form of argumentation, proportionality, used to differentiate elements of a proposition, so that "analogy could not possibly enjoy primary epistemological status" (pp. 109f.). This marking of the distinction between literal and figurative was carried out in practice, too, for although he retained traditional terminology in his *Alchemia* (1595), avoiding the neologisms and obscurity of the Paracelsians (p. 119), he points out in several places that names such as "quintessences, arcana, mysteria" have "only a tropological or analogical significance. The whole endeavour of the *Alchemia* is to define and distinguish definitively the species represented by these names from one another and thereby to identify them uniquely" (p. 148).

One could sum up the difference between the occult and the experimental scientific traditions at this point in the form that where the experimentalist will say "this is not reality, but only a trope," the occultist will say "this is not just a trope, but reality." Typical of the occult tradition is Joseph Du Chesne, expounding the Paracelsian concept of the three principles (salt, sulfur, and mercury) with the comment that "les susdites qualitez virtuelles & sensibles se trouvent en ces trois principes hypostatiques non par imagination, analogie & conjecture mais reellement & d'effect."[71] In the same way John Webster argued that the doctrine of signatures is no trope but a reality: "Many do superficially and by way of *Analogy* (as they term it) acknowledge the Macrocosm to be the great unsealed book of God," whereas in the Adamic language it is literally legible (p. 28), and Sir Thomas Browne wrote in 1635: "To call ourselves a Microcosm, or little World, I thought it only a pleasant trope of Rhetorick, till my neer judgment and second thoughts told me there was a real truth therein."[72] The experimentalist or antioccult tradition rejoins by stressing the difference between trope and reality. The two most sustained critiques of Paracelsian analogies come from the pens of Daniel Sennert[73] and Johann Baptista Van Helmont. Sennert's criticism is partly linguistic, alleging that the Paracelsian sect tries "to deceive by Names and Titles, and to get the opinion of being wise thereby," inventing strange words from no known language and using the same term in different senses. Anticipating Boyle and Locke in criticizing the Paracelsians' confusion, shifting terminology, and inconsistency,[74] Sennert invokes the traditional rhetorical concept of language as the source of clarity and mutual help in human communication:

> But the goodness of every thing consists in that for whose
> cause it was made. Speech is a great gift of God given to
> men, that one might declare their meaning to another, and

that which doth not so, is not worthy the name of a speech;
for the knowledg of things follows the knowledg of their
Names. (p. 22)

If, as Galen says, "*speech is the character of the mind,*" then "*a
monstrous speech is the sign of a monstrous mind*" (p. 23). So far men
have held "reason and experience" to be the basis of all knowledge,
but

> *Paracelsus* and his followers propound all their opinions
> without foundation, and begin a new way of knowing, of
> which *Crollius* speaks at large. A *Physitian* (saith he) *must
> have the light of Nature and grace, from the internal and
> visible Man, an internal Angel, and light of Nature.* And if
> you ask what this light of Nature is, he saith, *It is the Fir-
> mament that gives man all things naturally.* (p. 23)

Pressed further, he will take refuge in "the light of grace," which he
cannot or will not explain. Sennert, however, like Galileo, Kepler,
Bacon, and Locke, believes in the primacy of *res* over *verba*, holding
that knowledge "is so far true as it agrees with things, for things measure our knowledge, but not on the contrary; nor are things so, because
we think them so" (p. 24). Knowledge is derived from God by revelation, or from man by an empirical investigation of reality: "This being
not naturally in us, must come from without." Since the Fall no man
has had illumination; we must all work on reality with the use of our
external senses, our imagination, and our reason (pp. 24 f.).

Sennert expresses the criteria of the experimentalist traditions: an
appeal for proof, testing, verification, clear distinctions between body
and soul, words and things. He frequently invokes reason, denies that
the Paracelsians offer valid proofs, and shows that their invocation of
an invisible realm is antiempirical.[75] Like all the critics of the occult,
Sennert draws attention to the way in which that tradition blurs fundamental differences, breaking down the distinction between matter
and spirit, animate and inanimate. Paracelsus, he says, "wrote not only
absurd but wicked things, showing how a little man may be made by
Chymistry without a Father or Mother, and saith it is not a great secret"
(p. 18). In reply Sennert defends the uniqueness of the Creation (pp.
18, 68).[76] He also attacks the Paracelsians' concept of invisible elements. They dismiss the visible elements – fire, air, water, earth –
"which they scarce think worthy of the names of Elements, but call
them *dead bodies without secrets*" and claim that their elements are
invisible and are "the essence, life and act of all beings. *Paracelsus*
calls them the *Matrices* that bear and nourish fruit . . . So his Element
is whatsoever produceth and nourisheth fruit, or any created species"
(p. 31). Sennert's comment is that "many things here cannot be
proved." The chymists "foolishly say [the elements] are dead, because

they never were alive . . . As for their invisible Elements, we shal believe it when they prove any such hidden under these: for their affirmations only cannot create new beings" (pp. 32f). Sennert goes on to attack the whole basis of vitalism:

> The modern Chymists abuse the Name of Life, and extend it too large, when they give life to the stars, and say they have vital seed, when indeed nothing can truly be said to live, in which there appears no operation of any soul, at lest of a vegetative or growing soul. They say, *Metals, Minerals, Gemms,* and *Stones* do live, but life in them is nothing but an Energy or operation, which is in all things. (p. 42)

Sennert will accept that concept of "natural heat, mentioned by *Galen* and the Peripateticks" (p. 44), but not the occultists' innate life.

Sennert's sharpest attack on the occult sciences for blurring fundamental distinctions comes, not surprisingly, in connection with magic. Paracelsus requires that the physician learn magic, "*the Art of Arts, and the Inventor of all hidden things,*" together with astrology, pyromancy, chyromancy, and other occult arts. His work, *De morbis invisibilis*, states that the effects of magic depend upon the heavens or on spirits, good and evil; that heaven and the spirits "*are subordinate to man, and the force of the Heavens and Stars may be brought into Characters*"; and that the combination of "*words and wax*" can effect miracles (pp. 76f). What Paracelsus calls "*Characteral*" magic gives occult powers to words, claiming that "*whatsoever a Physitian can do by Medicines, may be done by words*" (p. 78). In his system *verba* have replaced *res*. Another of the Paracelsian types of magic is cabalistic, which "shows the way how wonders are wrought by Characters, Seals, Figures, and Words. By this a voice may be heard from beyond *Sea* . . . And Trithemius fetch his Supper out of *France* or *Italy*, saying this word, *Affer*, that is, *Bring to me*" (pp. 78f.). Other Paracelsians believed the same, Croll claiming that "whatsoever we see in the greater World, may be produced in the imaginary world; so all herbs and things that grow, and Metals, may be produced by imagination, and the true Cabal," by the operation of the "internal Heavenly or starry Man, who by the affinity of Magnetick vertues, can attract to himself all the strength of the Stars" (p. 79). Croll says that these gifts depend on prayer, faith, and the power of the imagination. In reply Sennert agrees with Libavius in condemning this magic as "ungodly and blasphemous" and warns that natural magic can easily lead to diabolic magic. As for the "natural faith" he speaks of, to say that it is given equally to all men "is a meer lye. For true miracles onely belong to the Church" (pp. 80f.).

Sennert is well aware of the tradition behind Paracelsian magic, Neoplatonism as revived by Ficino (p. 92), but dismisses Ficino's claims that imagination can unite the soul of man with angels and spirits, and can make spirits do man service, as being "the work and invention of Conjurers" (p. 92). He rejects Croll's account of the power of the imagination as "a mere fable" (p. 81). Imagination cannot act on bodies directly: It "is only a knowing power" and "doth nothing effectually but know" (p. 83). The effect of imagination is only through intermediaries: Fancy "moves the desiring faculty or appetite, and by the passions of the mind affects the body," causing gladness or sorrow (p. 84). The occultists are mistaken in thinking that "the fancy can affect strange bodies"; indeed, they "disagree among themselves" as to its causes (p. 91). Subsequently, in the chapter on semiotics in medicine Sennert asks "whether there be any force in Words and Character in Physick? *Paracelsus* caused it, when he said Characters would cure diseases otherwise incurable, and he saith it is lawful to fetch remedies from the Devil, if they will cure a man" (p. 134). Sennert's reply once again shows the difference between the occultist tradition, which believes in a natural language, an innate union of signified and signifier, and the experimentalist one, which holds that the linguistic sign is conventional, its meaning given by society.

> We answer as for words they signifie from a compact and
> convention of men. For thoughts are the same in all men,
> but the words or notes by which they are expressed, are di-
> vers, and the same words signifie divers things in divers Na-
> tions. Therefore words do only declare the sense of the
> mind, and work no further, for all principle of operation by
> which bodies are changed is a quality and a natural power,
> and things have their efficacy by their qualities. *Paracelsus*
> saith that words have an hidden force and vertue as Roots
> and Plants, but because he proves it not we ought not to be-
> lieve it. (pp. 135f.)

Obviously echoing Aristotle, that seems to me a coherent and decisive statement of the differences between the occult and the experimental traditions as to the nature of language.

Turning to seals and the characters, words, or signs engraved on them, Sennert says that they were invented by astrologers and magicians, as Pliny already recorded (*Natural History*, Book 30, chapter 1), and that much of the lore of the – now thought to be spurious – Paracelsian *Archidoxes of Magic* derives from Galen's chapter on magic stones (p. 135). If the antiquity of the doctrine is typical of the occult, not less so is the fact that none of the exponents of magic seals agree as to the causes of their effects. Sennert's rejection of this aspect of the occult again insists on clear distinctions:

> Two things are in Seals, the *Matter* and the *Character*, to
> neither of which can this force be ascribed: not to the *Mat-*

> *ter* which is from nature, nor hath it that strength as they
> confess; and if it had, it would have it without a figure or
> Character, as a Loadstone under what figure soever, hath
> power to draw iron without a Character. The *Characters* are
> from the Artificer, and from the Idea in his mind, which
> cannot work upon external things, therefore cannot have
> force from themselves or from the Artificer; of themselves
> they are nothing but figures, but a figure is not active being
> but a quality of the quantity: nor do artificial things act upon
> natural, and change them, or affect them, as being such, but
> they act upon them as they have natural matter . . . There-
> fore images or names *graven upon Matter*, can do nothing of
> themselves. (pp. 136f.)

Those clear statements of the differences between natural and artificially induced forces, between mind and matter, look back to Galileo
and on to Descartes.

It would be wrong to present Sennert as a wholly empirical scientist,
however. He still believes in sympathy, signatures, and a certain formal
correspondence between macrocosm and microcosm.[77] Yet he will not
accept that innate natural forces can be controlled by man or manip-
ulated by means of magic; nor will he accept the Paracelsians' handling
of the concept of analogy between macro- and microcosm. Like other
critics of the occult, he objects to their claiming a total identity between
upper and lower worlds. Chapter 6 of *Chymistrie Made Easie* is called
"Of the *Analogie of the great and Little World*" and begins with a
summary of the occult position:

> The whole Philosophy of *Paracelsus* is built upon the Analo-
> gie of the great and little World which they extend largely.
> And they of *Marpurg*[78] write the opinion of the Chymists
> thus. *The Chymists call Man a little World, not from a su-
> perficial likeness, but because he comprehends indeed, and
> according to the species (though invisibly) all things in him-
> self that are contained in a visible form, in the three King-
> doms, Vegetal, Animal, and Mineral, and in the whole
> World.* This is gathered from divers places in *Paracelsus* his
> works. (p. 25; also p. 98)

So much for the conversion of resemblance into identity; now for the
claim for absolute equivalence, item to item.

> *Crollius* in his Preface writes thus. *A man is a circle that
> contains in it all creatures. Man carries all things about
> him, the whole Firmament, and all the Stars, and Planets.
> Man hath the parts of all the world, and there is nothing in
> it that is not really in him.* (p. 25)

The doctrine further implies that man has two bodies, one physical, tangible, "which the first man had from the earth. Another invisible, insensible from the Stars" (p. 26).

Sennert himself is an Aristotelian and a Galenist. He notes that the concept of microcosm is very ancient, and he is willing to accept it, provided that some clear distinctions are made. Man, he writes,

> hath a visible Elementary body, a Heavenly soul, that hath power to grow and nourish as in Plants, sensible as in Brutes, and the mind is Angelical. Also he is like the World in the position of his members, and rise of them. For there are three parts of the great World, *The Elementary, Coelestial,* and *Supercoelestial.* To these three, man answers by the head, breast, and belly.

The last corresponds to religion; the breast and heart to the sun, therefore to the celestial realm; as for the head, where the mind and senses are placed, man "is not only like the Coelestial and Angelical world, but in that particular the image of God" (p. 26). Although this comparison may seem too far-fetched to some, Sennert has been careful to frame it as a comparison: Man "is like the world" in one aspect, "like the Coelestial" world in another. Thus he can fairly accuse the Paracelsians of having removed the word "like" and reduced the relationship to a bald "is," claiming not just a likeness but an identification:

> Hence we may gather that the Analogie of the great and little World is extended too large by the Chymists, because they make not an Analogie, but an identity, or the same thing. For *Paracelsus* requires in a true Physitian that he say this is a Saphire in man, this is Quicksilver, this Cypress, this a Walflower; but no Paracelsian ever shewed this. (pp. 26f.)

Not only does this way of using the trope collapse the analogy into identity, it also reduces man from the image of God, made "in our image and likeness" as Genesis puts it, the glory of Creation, to the level of the rest of the universe. But "there is more in man, who is the end of Natural things, than in other Naturals, and what man hath in him, he hath as a man" (pp. 27f.). The logical conclusion of Paracelsus's reduction of analogy to identity was his construction of retorts for the analysis of urine in the shape of life-size models of the human body, which Sennert dismisses as "foolish":

> What doth a furnace as high as a man concern the constitution of the urin? Why should the bigness of the vessels in a certain part answer to the just stature of a man? All men are not of one stature, and therefore this proportion will not fit

all, and you must make other furnaces and glasses for others: but these are trifles. (p. 121)

This is to wholly misinterpret the concept of signs or semiotic in medicine.

Sennert is concerned as much with the process of such correlative
thinking as with its resulting valuations or practical techniques. In an
extremely acute passage he points out that the overextended argument
from analogy moves not from the known to the unknown, but from the
unknown to the unknowable:

> Therefore the soul that loves truth is not satisfied with simi-
> litudes onely, but desires solid demonstrations; and volves
> things from their own, not from the principles of another.
> And as they who think they have demonstrative arguments,
> are often deceived, much more may they who use only com-
> parisons. There is nothing so like, but in some part it is un-
> like. Moreover, the Chymists know not the great World all
> over, how then can they bring us to the knowledg of the lit-
> tle World thereby? If they know it perfectly, let them exam-
> ine themselves, if they can arrogate that unto themselves
> truly. (p. 28)

This is a most pertinent comment on the occult tradition's tendency
to see similarities and avoid or elide differences, and on its way of
slighting an empirical study of the human body in favor of deductions
from the stars to plants, metals, and other of their interrelated categories. The Paracelsians argue rigidly from macrocosm to microcosm,
and Sennert, as a Galenist, is naturally incensed that this move should
result in the abandonment of the theory of humors: *"There is no flegm,
choler or melancholy in the great World, therefore not in Man"* (p.
27). But to "toss the humors" and dismiss them as "bare words" is
to confuse demonstrative proof with analogy and to displace attention
from the immediate area of knowledge to a remote and nonempirically
knowable one. Sennert questions this procedure:

> And why should we prove the humours from the Analogy
> between the great and little World? It is foolish, without
> sense or experience to flie to such Analogical proofs. For as
> in other creatures, so in man there is blood which nourish-
> eth: now sense teacheth that blood is made of meats re-
> ceived, but not Salt, Sulphur, or Mercury. (p. 96)

The "names of the humours are not insignificant without essence and
properties," nor does Paracelsus have the right "to give names to
things" (p. 97). Paracelsus cannot set himself up as another Adam.

Whatever we think of Sennert's conservatism as a doctor, on the
level of language he is clearly aware of this crucial issue and distinguishes literal and figurative levels quite coherently, saying that while

some writers deny that stones grow, others "allow a seed to metals and Minerals, if not univocal or proper, yet Analogical or like" (p. 44), similarly qualifying a later discussion with the words, "let this seed be called Analogical, if not Univocal" (p. 70). Like Aristotle, he stresses the importance of avoiding metaphor and ambiguity in definitions, complaining that the "chymists" use the concept of tartar without explaining it and that when Paracelsus does define it he has to resort to metaphor:

> *Paracelsus* rails against the Galenists, because they call tartarous diseases sand or the stone, because it is a Metaphorical appellation; but in Physick we must speak properly, and things must be denominated from their Nature, which he doth not observe. He saith the cause of this appellation is because an oyl, and a water, and a tincture are made of it, which burns the sick as the Tartar of Hell; therefore if the name be from the likeness of Hell fire, it is taken from a similitude, and is not proper. (p. 109)

And since they also call various things by the name of tartar, to which it does not apply, then he can fairly conclude that "the Chymists have no clear definition of *Tartar*" (p. 110). Sennert, we may recall, is writing fifty years before Boyle and Locke.

Sennert's appeal for clarity of definition, consistency of terminology, and avoidance of confusion of discrete levels of reality all sound remarkably modern; indeed, his criticism (in 1619) of the metaphorical and unstable base of Paracelsus's concept of tartar was repeated, quite independently, by Owsei Temkin in 1952. While belonging to the same critical, rational tradition as Galileo, Mersenne, and Descartes, Sennert also belongs to the humanist tradition, with its training in rhetoric and its clear distinction between figurative and nonfigurative language. We find the same critical spirit, and the same awareness of the difference between scientific discourse and rhetorical or persuasive discourse, in a writer and scientist who was much closer to Paracelsianism than the Aristotelian-Galenist Sennert, namely, J. B. van Helmont.[79] Van Helmont shows what might seem to us a surprising knowledge of rhetoric, referring easily to such technical terms as *hysteron proteron* (p. 222), identifying an "*antonomasia*, or taking one name for another" (p. 666), noting an "improper metaphor, or *hyperbole* or excessiveness" (p. 169), invoking "the license of Paradox" to apply the term "gas" (p. 69). While being totally familiar with rhetoric, Van Helmont – like Aristotle, Bacon, Sennert – believes that tropes should not form part of scientific reasoning. He writes that where he himself has sometimes attributed to water the three principles of Paracelsus, "that was spoken Analogically, or by way of suitable resemblance" (p. 410). Attacking the alchemists' indiscriminate use of the term "essence" he concludes that "the name of Essence is plainly Metaphorical. Wherefore very many things have not an Essence," so that "essence" is "an improper Name, and a [Fifth] Essence is an unsavoury Epithite" (p. 414). The scholastics built their theory of heat on a metaphor of fire, but they "have been forced to confesse that fire not to be fiery, yet devouring; but they have said, *It is sufficient for them to have described the Fewel or Torch, or Beginning of heat Metaphorically*; As if," he adds contemptuously, "nature should admit of Metaphors" (p. 178). His antipathy is to the misplacing of metaphor in scientific argument: "Surely," he writes, "I have hated Metaphorical Speeches in serious matters" (p. 719). Metaphors are constructs of the human imagination, not objects in the physical world.

Given such an insistence on separating the levels of literal and figurative speech, it is no surprise that Van Helmont, for all his debts to Paracelsus, should have rejected his master for basing his whole system on analogical reasoning. He attacks the too-literal insistence on the universal presence of salt, mercury, and sulfur:

> Surely I have hated the proportional resemblance [analogy] of the principles of Paracelsus brought back into the three principles of nature: because they are those things which are neither in bodies actually, nor are they present, nor are seperated, unless by changing them first as it were by the fire

– that is, by chemical processes. To assert that they are literally present is to confuse analogy with scientific statement:

> For truly, I do willingly behold a naked naturall Phylosophy every where; surely I do not apply [rhetorical] figures or moving forces [rhetorical persuasion, or *movere*] in Mathematicall demonstration unto nature: I shun proportionable resemblance, as also metaphorical speeches as much as I can. (p. 112)

Van Helmont objects to Paracelsus's treating analogies as if they were logical arguments: "He will have us bring back the Microcosme or little World, unto the Rule" (p. 405). Yet his analogies are too inconsistent to be deemed logical: He "doth oft-times define a Feaver to be an Earth-quake of the Microcosm; which trembling of the earth, he sometimes defines to be our Falling-sickness." But elsewhere he says that it is caused by "burnt or smoaking Mercury" and defines fever as "a Disease of Sulphur and Nitre" (p. 406). Analogical reasoning is arbitrary, unsystematic.

Van Helmont devotes long sections of his book to attacks on Paracelsus's misuse of the argument through analogy. Thus he summarizes Paracelsus's theory of "tartar," the coagulation of matter in the bloodstream, digestive, and excretory systems, into "a brief tract" (p. 230), showing that it began with the allegory of the Fall (Genesis 3:17–18).

"Nature being at first a beautiful Virgin, was defiled" by the sin of ungrateful man, and in punishment whereof God "appointed that the Earth should hence-forward bring forth Thistles and Thorns: under the allegory whereof," according to Paracelsus, "the curse and rise of *Tartarers* are designed unto us; To wit, their matter which should exceedingly sharply prick us," so that diseases "should at length be incorporated in us." This is a religious or moral allegory, ascribing the cause of disease to human sinfulness, and Van Helmont's own theory of disease is very similar.[80] What he objects to is Paracelsus's subsequent application of this metaphorical concept to the deposits inside "Wine-Hogsheads," which are "on every side incrusted with a Stony bark." Paracelsus equates this tartar with the divine-punishment tartar "by a Microcosmical Law" (p. 232). In Van Helmont's eyes the identification by analogy is "altogether impertinently taken according to the likeness of coagulated things in us" (p. 233), and the move from analogy to identity is illicit. Since "*tartar* is not an excrement of wine," then Paracelsus "doth badly accommodate or fit the Tartar of Wine by the identity of Being, and framing, with diseasie *Tartarers*, which he calls an excrement." The analogy in no way explains the "cause of Diseases" since the "cause of coagulations" in wine and in the human system "do not any way agree" in matter or manner. "Therefore the whole metaphorical transumption[81] of name and property is frivolous, and a bold rashness of altering" (p. 234). Van Helmont's appeal is to observation and experiment. Whereas Paracelsus thinks that the deposits in the human system are "hardened out of meats and drinks, by a co-like curdling" to the process in the wine vats, to call those encrustations a "stone" is false: It is only "a Metaphorical Stone I say, because resolveable in waters" (p. 236). The wine deposits are dissoluble in water; the human stone not. The analogy breaks down when each side is examined empirically.

Van Helmont is no more a fully-fledged empiricist than Sennert, of course; indeed, his medical theory is full of mystical and occult attitudes. But his attack on Paracelsus's misuse of analogy derives from a clear sense of the different levels of language and the "transfer," "translation," or "transumption" that the occult tradition so frequently makes. In his awareness of this "improper" use of language, Van Helmont stands outside occult science, nowhere more so than in his sustained attack on Paracelsus's use of the microcosm–macrocosm analogy. Section 15 of Chapter 31 has the title, "*He was deceived by the* Metaphor *of a* Microcosm *or little World*," and Section 18 has as its title, "*That the* Metaphor *of a* Microcosme *differs from the truth*" (p. 235). In the first Van Helmont alleges that Paracelsus fled "unto another the last Anchor of his hope" in the final stage of his theory.

> To wit, he translated the Metaphor of a *Microcosme* into the
> truth it self; Willing, that we should express every way and
> fully, the whole Universe exactly . . . to contain in it all the
> differences of Earths, Mountains, Fountains, Stones, Mines,
> Plants, Fishes, Birds, four-footed Beasts, creeping things,
> also of the Stars, with all the properties, motions, Tempests,
> Diseases, Defects, and interchangeable courses of the same:
> Asserting, that unless we do . . . believe this . . . in every
> created thing, we are unfit for to exercise Phylosophy, to
> practise Medicine, or to dispute against their suppositions.
> (p. 237)

Van Helmont's answer to this categorical assertion is in part theological, invoking the biblical account of Creation: "Let eternal prayse and
glory be to my Lord in all Benediction, who hath formed us not after
the Image of the most impure World, but after the figure of his own
divine Image." This alchemical concept of creation destroys God's
scheme for man. The "condition of that similitude" of the microcosm
would cause much grief, if mankind "before sin . . . should onely be
the engravement of so abjected a thing: as if the World had been framed
for itself, but not for us as the ultimate end; but we for the World,
whose Images indeed onely we should be." Furthermore, if the image
is taken literally then we would not only resemble but *be* the macrocosm, "to wit, we ought to be made stony, that we may represent
Stones and Rocks: And so we should all of right, be altogether stony,
leprous, &c." (p. 237). If the parallels are taken literally on the side
of the animal creation, then we too

> ought to fly; Seeing it is more rational, for us sooner to
> shew our selves Birds than great Stones, or storms of the
> Air, or water. Therefore let allegorical and moral senses de-
> part out of nature. Nature throughly handles Beings as they
> do in very deed and act . . . neither doth it admit of any
> other interpretation, than by being made, and being in es-
> sence, from ordained causes. (pp. 237–8)

The basis of Van Helmont's rejection is partly empirical, claiming that
nature should be known direct, at first hand, "in very deed and act,"
and partly religious. As he adds, dismissing this "fable" or "fiction of
Tartars," "for I being a Christian, could not admit of Microcosmical
Dreams, as they have been delivered by *Paracelsus*! That is, by literally, and not metaphorically understanding them, which sense or
meaning doth alwayes banish it self from the History of natural things"
(p. 239).

Van Helmont's insistence that we distinguish literal from metaphorical meaning, and that science concern itself only with the literal
level, to be studied directly and not by transference, emerges again in

Chapter 41, "The Scabs and Ulcers of the Schools," a discussion of leprosy which is primarily directed against Galen and his followers, but again criticizes Paracelsus. Section 32 itemizes "*The trifles of* Paracelsus *concerning the Microcosmical birth of wounds*," and Section 33 delivers a formal indictment: "Paracelsus *is urged with an actual and true Identity of the microcosm or little world*," that fusion of idea and essence so widespread in the occult. Where Van Helmont would work directly, from observable causes in the human organism, Paracelsus, he claims, worked by translation to the analogical plane. Those "ulcers I refer unto a seminal, and poysonous Ferment, *Paracelsus* after his manner, hath transferred on the minie and saltish minerals of the microcosme or little World" (p. 322). Thus Paracelsus claims

> that man (whom elsewhere by an Etymologie or Zodiack, he boasts to be a drawn Epitome of the whole Universe, and feigneth that he is more glorious by the dignity of that extraction, than by the Image of the Creator) is a most miserable monster, every way formed by minerals alone. (p. 322)

But if the theory of analogical correspondences believes "the *Macrocosme* or great World, to consist no lesse of Stars and Planets, than of minerals," then it is an "absurd thing" to hold "that it should resolve itself rather into Salts, than into Plants and four-footed Beasts" (p. 322). The analogy, used in one direction only, is inconsistent and reductive: "*Paracelsus* reducing all things into an under-earth off-spring . . . grew mad a while" (p. 322), indulged "his own Fiction" (p. 405), and "endeavoured to bind nature under his own idiotism" (p. 418). In this way he "heaped up great Fables" by "sporting with the Zodiack or compass of the microcosm at [his] own pleasure" (p. 322). It was a personal, arbitrary, and irrational activity.

Van Helmont flatly rejects the theory of minerals as the basis of life:

> Away with the trifles: For we have no fountains of Salt, no reducements of venal bloud into feigned and lurking mettals. Neither are there minerals in us . . . Neither also are there microcosmical Lawes in us, any more than the humors of four Elements mutually agreeing in us, and the fights or grudges of these: For with *Nazianzen*, I cannot tie up man unto the sporting Rules of a Microcosme: For I had infinitely rather to be the Image of God, than the Image of the corruptible and torturing World. (pp. 322–3)

The fact that some processes are shared is no basis for a total identification. Man grows and increases, as do beasts and plants, "yet Beasts shall not therefore be the Image of Plants." Man feels and perceives, as do beasts, "yet nothing speaks but a man" (p. 323). The theory of correspondences would in effect erode all the crucial distinguishing features established by God's creation. Man does not contain hail, snow, rocks, stones in all their variety, and the "stone" that humans suffer from may share its name with the mineral stone but not its properties:

> For a Peare is indeed changed into the flesh of a Cow
> sooner than the stone in a man can decline into a Mineral
> Rock or stone. The name therefore of Microcosm or little
> World is Poetical, Heathenish, and metaphorical, but not
> natural, or true. It is likewise a phantastical, hypochondria-
> cal, and mad thing to have brought all the properties and
> species of the Universe into man, and the art of healing: But
> the life of man is too serious, and also the medicine thereof,
> that they should play their own part of a Parable or Simili-
> tude, and metaphor with us. (p. 323)

Rhetorical tropes, one might paraphrase, are suitable for works of the imagination, but not for research on which human life depends.

Paracelsus's use of the microcosm analogy is not only reductive and opportunistic, but it forces the human organism to conform to some a priori model derived from the macrocosm. Another rejection of a priori science is made in a section called "*A Modern Pharmacopolium and Dispensatory*," where Van Helmont attacks the belief that provided the foundation of astrological medicine, the doctrine of signatures:

> I believe that God doth give the knowledge of Simples, to
> whom he will, from a supernatural grace: but not by the
> signes of nature! For what Palmestrical affinity hath the
> Boars tooth, the Goats blood, the peisle of a Bull, the dung
> of a Horse, or the Herbe Daysie, with a Pleurisie? or what
> signature have those Simples with each other? (p. 458)

Rejecting the traditional subordination of the earth to planetary influences, Van Helmont affirms that organic growth is a self-contained process:

> The earth hath of itself a seminal virtue of producing
> Herbes, the which, therefore, it doth not beg from the Heav-
> ens. For the whole property of Herbes is from their Seed,
> and the seminative power is drawn from the earth, according
> to the holy Scriptures: but not from the faces of the lights of
> Heaven.

Once the macrocosm–microcosm analogy has been rejected, astrology, the signatures, and correspondences – the whole occult system – is seen to be futile. One can manipulate a few limited categories, the twelve zodiacal signs, or the thirty decans,

> But in what sort could so few Stars contain the essences,
> seeds, faces, and properties perhaps of five hundred plants,
> differing in their species and internal properties? Moreover,
> besides a thousand vain attributions of so many things, as

well humane as politick? Away with these trifles! The properties of Herbes are in the Seeds, but not in the Heaven or
Stars. The powers of the Stars are grown out of date, the
which by an old Fable have stood feigned unto heats, colds,
and complexions. For the Stars, in whatsoever manner they
are taken, do differ from Plants much more than Herbs do
from mists and frosts, or fishes from precious stones. Let it
therefore be a faulty argument, to have attributed effects to
causes which do contain nothing at all like a cause in them.
(p. 458)

Some of the key attitudes of seventeenth-century experimental science show up there: the rejection of the symbolic dimension added by
the occult tradition; the need to begin observation or classification
direct from nature, and not by correlation with some preexisting matrix
or category; the assertion of differences, as between stars and plants,
animals and stones; the refusal to link them all in one grid as demanded
by the system of correspondences. Van Helmont's critique of Paracelsus, thorough and devastating in terms of methodology, has many
points of contact with the analyses of the occult sciences made, as it
would seem, independently of each other by Sennert, Libavius, and
Bacon. In these critiques we see a shift of attitude that defines the
emergent new sciences.

The critics are not wholly experimental scientists, of course, since
they retain some parts of the occult mentality. Yet they share what
might be called a relative distancing from the occult. They have moved
farther away from the mystical-magical tradition than have Paracelsus
or Fludd, and in some areas they are perfectly clear as to the gulf that
separates them. This sense of irreconcilable difference is especially
sharp as concerns language, metaphor, and symbolism.

One scientist who was constantly alert to the existence of various
levels of language was Kepler, to whom analogy was a heuristic or
explanatory device necessary to science, but not to be confused with
scientific discourse itself. He distinguished a "popular style of speech"
from that needed by "the precision balance of natural science," invoking Copernicus as proof of the distinction that "laymen control
language and express what they see in familiar speech, [whereas] the
philosopher seeks the truth which lies behind the apparent forms of
phenomena."[82] To penetrate to this deeper truth one must use "the
thread of analogy" by which man can make his way through "the
labyrinth of the mysteries of nature."[83] Kepler used analogy freely and
consciously in both his mathematical and astronomical works. In the
Ad Vitellionem paralipomena (1604), an epoch-making work on optics,
he grades the five types of conic sections "analogicè magis quam Geo-
metricè loquendo,"[84] meaning that he is using "analogy rather than

rigorous proof in the manner of Euclid.'' Kepler superimposes the four figures (circle, ellipse, parabola, and hyperbola) that, together with the straight line, account for the five types of sections, and calls one of their centric points the "focus," by analogy with light, naming the vertical axis the "chorda" or chord, the horizontal one "sagitta" or arrow, then calculating the proportions of "chord" to "arrow" in the various sections. He consciously extends the analogy by ascribing two foci to the straight line ("we speak in this manner contrary to normal usage, but only to give a content to the analogy") and justifies his rather "improper" procedure in these terms:

> But for us the terms in Geometry should serve the analogy
> (for I especially love analogies, my most faithful masters,
> acquainted with all the secrets of nature) and one should
> make great use of them in geometry, where – despite the in-
> congruous terminology – they bring the solution of an infin-
> ity of cases lying between the extreme and the mean, and
> where they clearly present to our eyes the whole essence of
> the question.
>
> Further, analogy has been of great help to me in drawing
> the sections.[85]

As he gives examples of how to construct these sections, he notes the interplay between the two procedures: "Analogy has shown, and Geometry confirms.''

In geometry, then, especially in non-Euclidean procedures, analogy is a heuristic tool, a different way of thinking, unusual but helpful. Analogy is more usual in the astronomical and physical sciences, yet while it was to be valued as formulating explanatory or predictive models, Kepler warned that a similarity should not be taken as an identity. This is particularly clear in his *Astronomia nova*[86] of 1609. Here, as so often, Kepler insists that the whole business of the astronomer is with reality, not with human, verbal categories: "The divine voice, which commands men to learn astronomy, expresses itself in the world, not in words and syllables, but through things themselves and through the agreement of the human intellect and senses with the entirety of celestial bodies and phenomena.''[87] When he came to analyze the motive force causing the planets to rotate around the sun, in order to provide a causal and structural explanation of this force, which is not directly accessible, he could only proceed, as Koyré puts it, "by analogy with other forces and other more usual, better known, emanations." By analogy, then, he showed that the motive force is "of an intangible nature, closely related to light and magnetic force.''[88] Kepler develops this similarity at some length and in a conscious way, creating a model (partly derived from geometrical optics) which he describes as "a comparison" used to "render the force of my argument all the more obvious.''[89] The analogy is subordinate, and the boundary line between it and the scientific argument is clearly marked with such signposts as "to continue, reasoning by analogy shows" and "to pursue the analogical argument further.''[90]

Kepler uses the magnet "as an example," drawing on William Gilbert's proof that the earth is a large magnet to suggest that "because the Earth moves the Moon by its *species*, and is a magnetic body; and because the Sun moves the planets in a similar manner by the *species* which it emits, therefore, the Sun, too is a magnetic body."[91] Yet, as Koyré records, Kepler was extremely well aware "that analogy is not the same as identity; immaterial properties, or species, though similar in some respects, are nevertheless different in others." This is the clearly distinguished heuristic model of experimental science, which recognizes both positive and negative analogies. "For example, light is generally stopped by a screen, but magnetic force is never, or hardly ever, stopped. Now, the analogy between motive force and magnetic force is even more valid than the analogy with light, but it still remains an analogy." In reply to the question whether an occultation (or conjunction) of planets would lead to an absence of motion as well as an absence of light, Kepler wrote:

> One should reply in the first place that the analogy between light and motive force should not be falsified by a rash confusion of their properties. Light is stopped by anything opaque; it is not stopped by the body as such, simply because it is light, and does not act on the body itself but on its surface, as it were.[92]

Kepler goes on to remove "the obstacle which the absence of any effect from the occultation of one planet by another seemed to raise against the similarity of the *species motrix* to light and magnetic force," the objection being, as Koyré points out, "only in respect of identity." Kepler then investigates "whether or not this similarity involves consequences which are incompatible with the very data that [his] theory was intended to explain."[93] This is the procedure of the experimental scientific tradition, which uses analogies (and is not used by them) to provide models that can be tested for their explanatory and predictive usefulness.[94] The successful use of such models depends on the clear awareness of the difference between analogy and identity. As Kepler wrote to Maestlin on 5 March 1605: "Every planetary body must be regarded as being magnetic, or *quasi*-magnetic; in fact, I suggest a similarity, and do not declare an identity."[95] The word "quasi" is important in both these quotations: We find it again in his remark that in the motion of the universe "a kind of *quasi*-power" is found.[96]

Kepler uses analogy as a heuristic device and is in no danger of fusing the two levels of comparison into an identity. Similarly, he regards symbols as existing on a verbal or conceptual plane distinct from reality. He consistently rejected numerology as an arbitrary symbol system that divorced mathematics from quantities and relations existing in the real world.[97] These attitudes are clearly expressed in what Kepler considered to be his chief work, the *Harmonice mundi* (1619),[98] and involve, as ever, a definite concept of reality. In the Preface Kepler acknowledges Proclus as the outstanding theoretical philosopher in mathematics, praising his distinction between the finite, as representing the form of geometrical bodies, and the infinite, as providing the matter. Quantity in geometry depends on figuration, which is determined by fixed points or boundaries, and proportion, the relationship between bodies. The human mind can grasp the finite and circumscribed, but not the infinite and indeterminate (p. 15). In geometry, to know means to measure according to a known measure, and the knowable is that which is either immediately mensurable or whose measurement can be deduced (p. 20). For Kepler, analogy, in turn, depends on the mathematical computation of bodies or relationships existing in the real world, and for this reason Pythagorean numerology is suspect, depending as it does on symbols, which can be interpreted variously, and on numbers that are discrete and do not represent continuous quantities (pp. 94–101). For, as Aristotle had shown in refuting the Pythagoreans, numbers nowhere exist separated from sense objects in the real world (pp. 217, 222). A true and solid science depends on establishing exact boundaries and discriminations; not on identifying things that are merely similar, but on preserving fundamental differences (p. 234).

The demand that analogies be based on the finite and knowable in the real world is one of several points made against Bodin, to whose analogy between the three types of state and the three forms of proportion (democracy: arithmetical; aristocracy: geometric; monarchy: harmonic) Kepler devotes a long Political Appendix to Book III (pp. 175–205). In addition to making erroneous analogies with numbers and confusing geometrical with arithmetical proportions, Bodin goes so far as to define harmonic relationships within the soul: But, Kepler comments, "these are only symbolic, not visibly expressed in connection to some solid body, as mathematics desires" (p. 204). Any such resemblance is qualitative or figurative, not truly analogical, in which the related parts have a body or quantity. Kepler can accept symbols if they have some reference to the real world which can be expressed in terms of quantity and proportion. He praises quantity as something wonderful, since it can express both the human and the divine in the same symbolic terms (p. 224). From this basis, in the Appendix to Book V, Kepler criticizes both Ptolemy and Robert Fludd. Having studied and translated Ptolemy's *Harmonics*, together with Porphyry's commentary, he is clear about the difference between his own legitimate mathematical demonstrations and the weakness and imperfection of Ptolemy's symbolisms, which, moreover, rest on an astronomy whose principles are false (p. 369). If the conception of reality is false, analogies based on it will be false. In such passages Kepler seems to be using the symbolic and the real as antithetical and exclusive reference points. Ptolemy's error was to search for the principles of harmony in abstract numbers, to which Kepler would deny any meaning (p. 370), preferring to deal with the geometrical bodies to which numbers are subject.[99] Lacking this base in reality, Ptolemy's symbolisms are neither necessary nor compelling, neither causal nor natural, but resemble those used by poets and orators (p. 371). Summarizing a section of the *Harmonics* that proposes correlations between planetary motions and keys in music, Kepler comments: "I have shown that Ptolemy luxuriates in using comparisons in a poetical or rhetorical way, since the things that he compares are not real things in the heavens" (p. 372). Kepler's reformed astronomy, by contrast, which has excluded apparent planetary movements derived from the false testimony of the eyes, has shown that "all harmonic proportions appear in the heavens according to a true and genuine, quantitative and mensurable cause, but not according to mere trivial symbolisms" (p. 372).

The line Kepler draws between his use of analogy and Ptolemy's symbolisms is drawn even more sharply to differentiate himself, and all mathematicians, from Robert Fludd and the methods of the alchemists, hermeticists, and Paracelsians (p. 373). Fludd goes around with an idiosyncratic picture of the world in his head, which he sets out in the many pictures in his book, while Kepler puts down mathematical diagrams derived from reality (pp. 374–6). As for the harmonies Fludd teaches, they are "pure symbolisms, of which I say, as I did of the symbolisms of Ptolemy, that they are poetical and rhetorical, rather than philosophical or mathematical" (p. 374). Fludd divides the world into three parts and attempts to apply to them the Hermetic belief that "what is below is like that which is above . . . But in order to make this analogy fit all cases he often has to drag in his comparisons by the hair" (p. 375). Kepler refers back to the Appendix on Bodin for a fuller account of his views on analogy, merely noting that with geometrical figures analogies based on harmonic proportions are not just formal but material. "For while harmonic proportions define a fixed quantity, analogies, on the contrary, are apt to extend themselves to infinity" (p. 375) and so to take on the nature of the unknowable, which is antithetical to the concerns of science. Further, Fludd bases his harmonies on the Pythagoreans' abstract numbers, finding whatever numerical concordance he can, whereas Kepler never seeks to find harmonies "when the things between which the harmonies exist cannot be measured by one and the same scale of magnitude," as with the proportions that can be gauged between strings under the same tension with respect to their length (p. 375).

Kepler's final criticisms concern the intersecting triangles or "pyramids" which play such a large part in Fludd's symbolism. Fludd used two equal, intersecting cones (in the illustrations they look like isosceles triangles) to represent the two fundamental principles in his conception of the universe, that is, form or light descending from above, and matter or darkness ascending from the earth. This symbolic opposition became incorporated or reified in his thought in a typically occultist fashion. As Wolfgang Pauli said: "Fludd never distinguishes clearly between a real, material process and a symbolical representation."[100] Kepler's complaints are, first, that Fludd divides his cones into three, "as if he really had equal units," even though he knows that the parts (the elemental, ethereal, and empyrean realms) are not of equal dimensions, because he wishes to represent them pictorially (p. 375). Then he makes these two imaginary cones intersect and derives musical proportions from their mixture, a procedure Kepler dismisses as illicit. "For he compares light (the dispenser of form and soul) with matter, two things which are completely different, and whose quantities can never be measured on the same scale." Kepler, in absolute contradistinction, admits as "components of a harmonic proportion to be discovered in the world only those things whose quantities can be measured on the same scale, such as the daily movements of Mars and Jupiter" (p. 376). Fludd's units of proportion "are again arbitrary," while Kepler's are drawn from nature. Fludd's harmonies derive from his own conception of the world; Kepler's from the world itself, according to the principles of a reformed astronomy based on observation and measurement.

The *Harmonice mundi* is a remarkable book, developing all manner of analogies among geometry, music, and astronomy that result, inter alia, in the discovery of the relation between period and orbital radius, which we call his third law. It includes many speculative elements and preserves several characteristics of the occult tradition – a belief in a world soul, the correlating of preformed categories – which make it a unique hybrid rather than an ideal exemplar of the new sciences. Yet on several issues it maintains an absolutely sharp distinction from the occult sciences, notably in its rejection of numerology, of idiosyncratic and imaginary world views, and of symbols being taken for realities. What I have described as the mainstream philosophical-rhetorical view on the necessary distinction between words and things, between the metaphorical and the literal levels of language, seems to have been a constant element in Kepler's thought. Max Caspar and D. P. Walker have drawn attention to an exchange between Kepler and Joachim Tanckius in 1608, concerning symbolism in music.[101] Kepler himself was fond of using analogies from gender to describe and differentiate musical intervals (major thirds are male, minor thirds are female), and he linked them with geometrical figures. But he still felt clear about the distinction between using analogy heuristically, to move from the known to the unknown, as opposed to using symbols, which can only relate things already known:

> I too play with symbols, and have planned a little work, Geometric Cabala, which is about the Ideas of natural things in geometry; but I play in such a way that I do not forget that I am playing. For nothing is proved by symbols, nothing hidden is discovered in natural philosophy through geometric symbols; things already known are merely fitted [to them]; unless by sure reasons it can be demonstrated that they are not merely symbolic but are descriptions of the ways in which the two things are connected and of the causes of this connexion.

Again we note the expulsion of symbolism from the domain of scientific argument, the appeal for proof, the demand for a rational explanation of causes. The new sciences depended in part on the establishing of such critical attitudes toward language and its relationship to reality.

All these attitudes are reaffirmed in the work of Marin Mersenne, especially in his attack on the cabala. As Robert Lenoble has shown, Mersenne drew attention to the arbitrariness of interpretation in cabalistic symbolism, whereby the correlation of letters and numbers could be interpreted differently by each practitioner. By juxtaposing two cabalistic alphabets, Mersenne showed how totally individual and variable the process of interpretation was.[102] Like the other critics of the occult, forerunners of the new sciences, Mersenne has not freed himself from all traces of the system he attacks: He, too, uses the macrocosm–microcosm analogy.[103] However, like Kepler, he sees these analogies as literary ornaments, not scientific proofs, protesting energetically "lorsqu'on veut faire du jeu de mots le principe de la recherche scientifique" (p. 107). Like Sennert or Libavius or Van Helmont, he demands that theories be proved, attacking Fludd's cabalistic astrology for being purely arbitrary: "Il avance tout ce qu'il dit de cette harmonie sans aucune demonstration" (p. 108). Mersenne similarly attacks the cabalists for not demonstrating anything; indeed, he calls their dreams worse than ignorance because they prevent us from observing nature correctly. Their antiempirical attitude, detached from reality, is accompanied, as so often in the occult sciences, by an essentialist concept of language: "They deceive us as to the nature of language. For them the word signified the essence of things," not an agreed relationship between signified and signifier. In order to "destroy the secular prestige of onomancy," as Lenoble puts it, Mersenne discusses the issue at length and produces the decisive dismissal: "The word is merely a *flatus vocis*, a purely conventional sign, an agitation of the air, whose nature depends on acoustics and physiology" (p. 108). (This again is fifty years before Locke.) In Mersenne's eyes only true science can deliver us from false science. The danger of the cabala to Mersenne is that the universal correspondences it proposes make human destiny become absorbed in cosmic history. The new sciences separated these realms, as they separated and distinguished the various levels of language. The difference between the two traditions emerges in many forms, not least in this awareness that science cannot be built on figures of speech. To return to Francis Bacon and to his designation of the "first distemper of learning" as being "when men study words and not matter":

> It seems to me that Pygmalion's frenzy is a good emblem or portraiture of this vanity: for words are but images of matter; and except they have life of reason and invention, to fall in love with them is all one as to fall in love with a picture. (III, 284)

Notes

1 Robin Horton, "African Traditional Thought and Modern Science," *Africa*, 37 (1967), pp. 50–71, 155–87; repr. (slightly abridged) in, and quoted from, *Rationality*, ed. B. R. Wilson (Oxford, 1970), pp. 131–71.

2 S. J. Tambiah, "The Magical Power of Words," *Man*, n.s. 3 (1968), pp. 175–208.

3 Ferdinand de Saussure, *Course in General Linguistics*, ed. C. Bally, A. Sechehaye, and A. Riedlinger, trans. W. Baskin (New York, 1959), pp. 65–70.

4 Quoted in the Benjamin Jowett translation from *The Collected Dialogues of Plato*, ed. E. Hamilton and H. Cairns (New York, 1963), pp. 421–74.

5 Cf. Allison Coudert, "Some Theories of a Natural Language from the Renaissance to the Seventeenth Century," in *Magia Naturalis und die Entstehung der modernen Naturwissenschaften: Studia Leibnitiana*, Sonderheft 7 (Wiesbaden, 1978), pp. 56–114, at p. 65, and R. T. Wallis, *Neo-Platonism* (London, 1972), p. 19, on the later Neoplatonists, for whom the *Cratylus*, with its "account of divine names – concerning which the school displayed its usual blindness to Plato's irony – was vital to theurgy." On the persistence of this idea in Stoic linguistic theories, see R. Pfeiffer, *History of Classical Scholarship* (Oxford, 1968), and A. A. Long, *Hellenistic Philosophy* (London, 1974).

6 *Discoveries and Opinions of Galileo*, trans. Stillman Drake (New York, 1957), p. 92: from the first letter on sunspots. Subsequent quotations are from this useful anthology. Cf. also other remarks, such as this on Apelles' error in placing Mercury after the moon, followed by Venus: "To get the cart before the horse in this way would not matter much so far as the words are concerned, if only he had kept the things arranged correctly" (p. 96).

7 Quoted from the translation by E. M. Edghill in the *Works of Aristotle*, trans. ed. W. D. Ross, 12 vols. (Oxford, 1908–52), I. On the influence of

Aristotle's view of language as conventional, see G. A. Padley, *Grammatical Theory in Western Europe 1500–1700* (Cambridge, 1970), pp. 11–13, with the references there.

8 See Marcia Colish, *The Mirror of Language: A Study in the Medieval Theory of Knowledge* (New Haven, 1968), pp. ix, 12, 54, and passim. See also R. A. Markus, "St. Augustine on Signs," in *Augustine,* ed. R. A. Markus (New York, 1972), pp. 61–91; R. H. Robins, *Ancient and Medieval Grammatical Theory in Europe* (London, 1951), pp. 21, 26f.

9 Rudolph Allers, "Microcosmus: From Anaximander to Paracelsus," *Traditio,* 2 (1944), pp. 319–407, at pp. 341, 384.

10 Francis Bacon, *The Works of Francis Bacon,* ed. J. Spedding, R. L. Ellis, and D. D. Heath, 14 vols. (London, 1857–74), III, 399 (*Advancement of Learning*), and IV, 439 (*De augmentis*). Unfortunately this point seems to have escaped both G. A. Padley, who dismisses Bacon as a nominalist antipathetic to language ("The Seventeenth Century: Words versus Things," in *Grammatical Theory,* pp. 132–53, at pp. 136ff.), and James Knowlson, *Universal Language Schemes in England and France, 1600–1800* (Toronto, 1975), pp. 36f., even though he has quoted earlier Bacon's definition of words as "the tokens current and accepted for conceits" (p. 16).

11 Thomas Hobbes, *Leviathan,* ed. M. Oakeshott (Oxford, 1946), p. 22.

12 Thomas Hobbes, *English Works,* ed. W. Molesworth (London, 1839–45), I, 14.

13 Thomas Hobbes, *De homine,* trans. C. T. Wodd, T. S. K. Scott-Craig and B. Gert, in *Man and Citizen,* ed. Bernard Gert (New York, 1972), p. 37. I discuss Hobbes at this length because his renovation of the Aristotelian tradition is important and because some recent studies (e.g., Padley, pp. 141–3) ignore Hobbes's concept of "sign."

14 Hobbes, *English Works,* IV, 22.

15 Kenelm Digby, *Two Treatises* (London, 1645), p. 2.

16 Coudert, pp. 65f., citing Beryl Smalley, *The Study of the Bible in the Middle Ages* (Oxford, 1952), p. 6.

17 See, for example, Franz Dornseiff, *Das Alphabet in Mystik und Magie* (Berlin, 1925; Leipzig, 1977).

18 Ibid., pp. 118–22.

19 See, for example, Gerson Scholem, *Major Trends in Jewish Mysticism* (London, 1941; rev. ed. 1955), and *Kabbalah* (New York, 1974).

20 Coudert, p. 75.

21 Charles Zika, "Reuchlin's *De Verbo Mirifico* and the Magic Debate of the Late Fifteenth Century," *Journal of the Warburg and Courtauld Institutes,* 39 (1976), pp. 104–38.

22 Wayne Shumaker, *The Occult Sciences in the Renaissance* (Berkeley and Los Angeles, 1972), pp. 135–7.

23 Cornelius Agrippa, *Three Books of Occult Philosophy,* trans. J. F. (London, 1651), p. 152; cited in Coudert, p. 82.

24 Alexandre Koyré, *La Philosophie de Jacob Boehme* (Paris, 1929), p. 21; also Coudert, pp. 87–91, and K. Quecke, "Die Signaturenlehre im Schrifttum des Paracelsus," *Beiträge zur Geschichte der Pharmazie und ihrer Nachbargebiete,* 2 (1955), pp. 41–52.

25 On Boehme's "union of opposites," see Koyré, pp. 74, 93, 109, 125, 131, 169, 226, 255, 262, 287, 360, 362f., 368f., 384, 386f., 394f., 396, 397, 405, 455, 490, 506. The belief in onomatopoeia as a proof of natural language

extends in Boehme to the point that the phoneme is elevated to the status of lexeme, according to S. A. Konopacki, *The Descent into Words: Jakob Böhme's Transcendental Lingustics* (Ann Arbor, 1979).

26 Cf. also this passage from Boehme's *Von dreyfachen Leben* (1620) in the English trans. by J. Sparrow (1650): "*Now* as the spirit of the eternity hath formed and framed *all* things, so also the spirit of man formeth *them* in his word, for all ariseth from his centre: for the *human* spirit is a form, figure, and similitude of the Number Three of the Deity; Whatsoever God is in his nature, that the spirit of man is in itself: and therefore he giveth every thing its *name,* according to the spirit and form of every thing, for the inward speaketh forth the outward" (cited in Coudert, p. 89). See other similar passages from Boehme quoted by Hans Aarsleff, "Leibniz on Locke on Language," *American Philosophical Quarterly,* 1 (1964), p. 180 and n. 51.

27 See Paul Cornelius, *Languages in Seventeenth- and Early Eighteenth-Century Imaginary Voyages* (Geneva, 1965).

28 Cited in Coudert, p. 96. On Comenius, see Charles Webster, *The Great Instauration* (London, 1975), and Vivian Salmon, *The Study of Language in 17th-Century England* (Amsterdam, 1979).

29 Quotations from John Webster are from the facsimile edition included in A. G. Debus, ed., *Science and Education in the Seventeenth Century: The Webster–Ward Debate* (London, 1970), which also includes the reply by Wilkins and Ward. I retain the original pagination.

30 See Wayne Shumaker, *Renaissance Curiosa* (Binghamton, N.Y., 1982), pp. 48, 134.

31 Cf. Coudert, p. 104.

32 Samuel Parker, *A Free and Impartial Censure of the Platonick Philosophie* (Oxford, 1666), pp. 61–3; cited in Padley, pp. 139–40.

33 Aarsleff's first published essay on this topic, "Leibniz on Locke on Language" (cited in n. 26 above and repr. in Aarsleff's collection, *From Locke to Saussure* [London, 1982], pp. 42–83), remains important, despite being overdocumented and awkwardly structured. A more straightforward – if less finished – account was given by Aarsleff in some lectures delivered at Princeton in 1964, deposited in the library of the Warburg Institute, London, under the title "Language, Man and Knowledge in the Sixteenth and Seventeenth Centuries"; chap. 2, "Adamic Language and Mysticism," discusses Boehme and Kuhlmann; chap. 5, "The Royal Society," discusses Hooke, Ray, Boyle, and Locke.

34 See, for example, S. K. Land, *From Signs to Propositions: The Concept of Form in Eighteenth-Century Semantic Theory* (London, 1974), pp. 6–20, and the further references given at p. 6 n. 1.

35 Quotations are from John Locke, *An Essay Concerning Human Understanding,* ed. Peter H. Nidditch (Oxford, 1975).

36 See ibid., II, xxxiii, 19 (p. 401), and III, ix, 21 (p. 488).

37 Robert Boyle, *The Sceptical Chymist.* Quotations are from the Everyman Library edition, introduced by E. A. Moelwyn-Hughes (London, 1964). For other comments on the obscurity, contradictions, and deceptions in the language of the alchemists, see pp. 95, 116, 130, 143, 146, 166, etc. For the new sciences' commitment to "solid" knowledge and "truth," see pp. 2, 11, 164, 165, etc.

38 Aristotle, *Poetics,* trans. M. E. Hubbard, in *Ancient Literary Criticism: The Principal Texts in New Translations,* ed. D. A. Russell and M. Winterbottom (Oxford, 1972), pp. 85–132.

39 Aristotle, *Rhetoric,* trans. W. Rhys Roberts, in *The Works of Aristotle,* ed. W. D. Ross (Oxford, 1924), XI. For further discussion of metaphor in the classical rhetorical tradition, see Brian Vickers, *Francis Bacon and Renaissance Prose* (Cambridge, 1968), pp. 141–54, 288–90.
40 I. A. Richards, *The Philosophy of Rhetoric* (New York, 1936), pp. 96ff.
41 Cf. Bacon's *Parasceve,* in the "Aphorisms on the Composition of the Primary History," on the need to reject superfluous matters, such as the irrelevant citation of authorities or merely verbal controversies: "And for all that concerns ornaments of speech, similitudes, treasury of eloquence, and such like emptinesses, let it be utterly dismissed. Also let all those things which are admitted be themselves set down briefly and concisely, so that they may be nothing less than words. For no man who is collecting and storing up materials for ship-building or the like, thinks of arranging them elegantly, as in a shop, and displaying them so as to please the eye; all his care is that they be sound and good, and that they be so arranged as to take up as little room as possible in the warehouse" (*Works,* IV, 254–5). Needless to say, Bacon's remarks apply to the amassing of observations, not to the formation of theories or to general communication, in both of which he gave metaphor and analogy an important role.
42 Geoffrey Lloyd, *Polarity and Analogy: Two Types of Argumentation in Early Greek Thought* (Cambridge, 1966), pp. 394ff.; subsequent quotations in the text are from pp. 363f., 403, 404f.
43 D. P. Walker, *Spiritual and Demonic Magic from Ficino to Campanella* (London, 1958).
44 Full of admiration as I am for the work of D. P. Walker, I must nevertheless record a feeling of unease at the extent to which he seems willing to accept Ficino's claims to have achieved *practical* magic: see ibid., pp. 45f., 63, 70f., 72, 84, 89, 94, 120, 124, 126, 137, 150f., 207, 209, 210, 230, 233f., 236. A few more caveats would have been in order.
45 See, for example, ibid., pp. 32f., 40f., 45, 55f., 108 (Pomponazzi: an amazing instance), 113, 115, 131, 134f. (Paolini thinks that the invention of clocks was due to the help of the *anima mundi*), 142, 189f., 211 etc.
46 Paul Oskar Kristeller, *The Philosophy of Marsilio Ficino,* trans. V. Conant (New York, 1943), pp. 74–91.
47 E. H. Gombrich, "Icones Symbolicae," *Journal of the Warburg and Courtauld Institutes,* 11 (1948); enlarged version in *Symbolic Images: Studies in the Art of the Renaissance* (London, 1972), pp. 123–95, 228–35.
48 Brian Vickers, "On the Functions of Analogy in the Occult," *Renaissance Tradition,* ed. Allen G. Debus and Ingrid Merkel (Associated University Presses, forthcoming).
49 Maurice Crosland, *Historical Studies in the Language of Chemistry* (London, 1962), esp. chap. 1, "Allegory and Analogy in Alchemical Literature," pp. 3–24.
50 Rosaleen Love, "Some Sources of Herman Boerhaave's Concept of Fire," *Ambix,* 19 (1972), pp. 157–74.
51 Dionysius, *The Celestial and Ecclesiastical Hierarchy,* trans. J. Parker (London, 1894), pp. 44–5, 20–1.
52 See Allen Debus, "The Paracelsian Aerial Niter," *Isis,* 55 (1964), pp. 43–61.
53 See Walter Pagel, *Paracelsus: An Introduction to Philosophical Medicine in the Era of the Renaissance* (Basel, 1958); "Paracelsus: Traditionalism and Medieval Sources," in *Medicine, Science, and Culture,* ed. L. G.

Stevenson and R. P. Multhauf (Baltimore, 1968), pp. 50–75; "Religious
Motives in Medical Biology," *Bulletin of the Institute of the History of
Medicine,* 3 (1935), pp. 97–128, 213–31, 265–312.

54 The following works are cited: A. Koyré, "Paracelse," in his *Mystiques,
spirituels, alchimistes du XVIe siècle allemand* (Paris, 1955); F. R. Jevons,
"Paracelsus's Two-Way Astrology," *British Journal for the History of
Science,* 2 (1964), pp. 139–55; Pagel, *Paracelsus,* "Traditionalism," and
"Religious Motives"; O. Temkin, "The Elusiveness of Paracelsus,"
Bulletin of the History of Medicine, 26 (1952), pp. 201–17; L. S. King, *The
Growth of Medical Thought* (Chicago, 1963); H. Fischer, "Die
kosmologische Anthropologie des Paracelsus als Grundlage seiner
Medizin," *Verhandlungen der naturforschenden Gesellschaft in Basel,* 52
(1940–1), pp. 267–317.

55 Cf. also Koyré, "Paracelse," pp. 51 and n. 2, 64f.

56 For other examples of reification and the collapse or fusion of categories,
see Fischer, pp. 289, 290, 301f. (both the *anima vegetativa* and the *spiritus
vitae* conceived in material terms); King, pp. 96, 103f, 107; Pagel,
"Traditionalism," pp. 57f., 63, 71, 74; Koyré, "Paracelse," p. 59; Jevons,
pp. 140, 142, 144, 149, 151.

57 See, for example, Pagel, *Paracelsus,* pp. 83, 105, 152; King, pp. 125f., 134;
Temkin, pp. 206, 209 n. 37, 215.

58 "Paracelse ne peut penser autrement que par des analogies psychologiques
ou organiques" (Koyré, "Paracelse," p. 61).

59 Such as the following exchange from 2 *Henry IV,* 3.2.66ff.:
   Bardolph: "Sir, pardon, a soldier is better accommodated than with a
   wife."
   Shallow: "It is well said in faith sir, and it is well said indeed too. Better
   accommodated. It is good, yea indeed is it; good phrases are surely, and
   ever were, very commendable; Accommodated – it comes of accommodo;
   very good, a good phrase."
   Bardolph: ". . . Accommodated, that is, when a man is, as they say,
   accommodated, or when a man is, being whereby 'a may be thought to be
   accommodated, which is an excellent thing."
   Shallow: "It is very just."

60 Owen Hannaway, *The Chemists and the Word: The Didactic Origins of
Chemistry* (Baltimore, 1975), p. 87.

61 Quoted from Benjamin Farrington's translation of this and two other early
works, *The Philosophy of Francis Bacon: An Essay on Its Development
from 1603 to 1609 with New Translations of Fundamental Texts* (Liverpool,
1964), p. 66.

62 Francis Bacon, *Advancement of Learning* (1605), in *Works,* III, 485f.; see
also *Works,* I, 835 (*De augmentis,* 1623), English trans. at V, 117.

63 Daniel Sennert, *De chymicorum cum Aristotelicis et Galenicis consensu ac
dissensu* (1619), trans. Nicholas Culpeper and Abdiah Cole as *Chymistry
Made Easie and Useful: Or, the Agreement and Disagreement of the
Chymists and Galenists* (London, 1662), p. 24.

64 Farrington, p. 65; see ibid., p. 122, for a similar point in the *Redargutio
philosophorum.*

65 As Bacon remarks in his *History of the Winds,* since Paracelsus proclaimed
three principles, he could only recognize three winds, so the east wind had
to be dropped (*Works,* V, 154).

66 Pagel, *Paracelsus,* pp. 323f.

67 Hannaway, pp. 99, 101.
68 Ibid., p. 103.
69 Trans. in ibid., p. 108.
70 Cf. Shumaker, *Occult Sciences*, p. 22, translating Pico's *Disputationes adversus astrologiam divinatricem*: "In this way anything can easily be proved, since nothing exists which it is impossible to imagine by an argument of this kind to have some similarity and dissimilarity with something else."
71 Joseph Du Chesne, *Traicté de la matiere, preparation et excellente vertu de la medecine balsamique des anciens philosophes: auquel sont adioustez deux traictez, l'un des signatures externes des choses, l'autre des internes & specifiques, conformément à la doctrine & pratique des hermetiques* (Paris, 1626), p. 153.
72 Thomas Browne, *Religio medici*, bk. 1, par. 34.
73 Sennert, *Chymistry Made Easie*. See A. G. Debus, "Guintherius, Libavius and Sennert: The Chemical Compromise in Early Modern Medicine," in *Science, Medicine and Society in the Renaissance*, ed. A. G. Debus, 2 vols. (New York, 1977), I, 151–65.
74 These charges are repeated in Sennert, *Chymistry Made Easie*, pp. 19, 20, 21, 29, 30, 62, 110, 117, 124, 136 (confusion and inconsistency); pp. 37, 97, 99, 104, 115 (neologisms).
75 For the appeal for proof and demonstration, see, for example, ibid., pp. 26, 27, 32, 33, 51, 56, 82f., 98, 104, 115, 116, 126, 134, 135; for criticism of their invocation of an invisible realm, see pp. 31, 98.
76 For other contemporary objections to the Paracelsians' claim to be able to create a "homunculus" by chemical means alone, see Charles Schmitt, "John Case on Art and Nature," *Annals of Science*, 33 (1976), pp. 543–59, at pp. 556f.
77 See Sennert, *Chymistry Made Easie*, pp. 26, 27, 48, 82, 132ff., 138f.
78 Marginal note: "*Disp. 8*," i.e., Erastus, *Disputationes de medicina nova Paracelsi*.
79 All quotations are from J. B. van Helmont, *Oriatrike or Physick Refined*, trans. John Chandler (London, 1662), an English version of *Ortus Medicinae* (Amsterdam, 1648), reissued in 1664 as *Van Helmont's Works*. For a brief discussion of Van Helmont's critique of Paracelsus' use of analogy see now Walter Pagel, *Joan Baptista Van Helmont. Reformer of Science and Medicine* (Cambridge, 1982) pp. 46–9, 98, 206–7.
80 "I know that I do undergo Diseases, that I might shew a depraved and mortal nature" (ibid., p. 418).
81 *Transumptio* is the Latin name for *metalepsis*, a figure which "provides a transition from one *trope* to another," forming "a kind of intermediate step between the term transferred and the thing to which it is transferred"; Quintilian, *Institutes of Oratory*, VIII, vi, 37, trans. H. E. Butler, Loeb Library, 4 vols. (London, 1922), III, 323, where it appears as a form of metonymy. In other contexts it can be used to describe the improper use of synonyms: cf. Heinrich Lausberg, *Handbuch der Literarischen Rhetorik*, 2 vols. (Munich, 1960).
82 Johannes Kepler, *Epitome astronomiae copernicanae*, in *Gesammelte Werke*, ed. Max Caspar et. al. (Munich, 1937–), VII, 99f.; trans. Edward Rosen in "Kepler and the Lutheran Attitude Towards Copernicanism in the Context of the Struggle Between Science and Religion," *Vistas in Astronomy*, 18 (1975), pp. 317–37, at pp. 334f. *Gesammelte Werke* is cited hereafter as *GW*.

83 Johannes Kepler, *Harmonice mundi*, *GW*, VI, 366; cited by E. W. Gerdes, "Johannes Kepler as Theologian," *Vistas in Astronomy*, 18 (1975), pp. 339–67, at p. 343.

84 Johannes Kepler, *Ad Vitellionem paralipomena*, *GW*, II, 90; I am grateful to Dr. Judith V. Field for help in interpreting this passage. She points out the difference between Kepler's use of analogy in the mathematical works, where "this analogy between the conics has no ifs or buts or exceptions in it," compared with the magnetism–light analogy in the *Astronomia nova:* "Kepler has spotted real mathematical similarities and applied them to deduce mathematical facts about conics in general, whereas the physical analogy is an analogy in the weaker modern sense – an aid to visualization etc. For example, Kepler does not apply the inverse square law of light to his magnetic force, but he does ask where is the second focus of the parabola, and the answer is 'at infinity' – a new idea" (personal communication).

85 Kepler, *Ad Vitellionem*, *GW*, II, 92; I have benefited from Dr. Field's comments and from the recent French translation, *Paralipomènes a Vitellion*, trans. C. Chevalley, preface by R. Taton and P. Costabel (Paris, 1980). See also Gerd Buchdahl, "Methodological Aspects of Kepler's Theory of Refraction," *Studies in History and Philosophy of Science*, 3 (1972), pp. 265–98, at pp. 284–6, for comment on the use of analogy in this work.

86 For extended discussions of this book, see Alexandre Koyré, *The Astronomical Revolution: Copernicus–Kepler–Borelli*, trans. R. E. W. Maddison (London, 1973), pp. 159–279, and Gérard Simon, *Kepler astronome astrologue* (Paris, 1979), pp. 304–86.

87 Cited in Koyré, *Astronomical Revolution*, p. 163; from *GW*, III, 108. For Kepler's commitment to a science addressed to reality, see, for example, ibid., pp. 133, 176, 186, 198, 227, 274, 323, etc.

88 Ibid., p. 199.

89 Ibid., p. 204; from *GW*, III, 243.

90 Ibid., pp. 206, 207; from *GW*, III, 243, 245.

91 Ibid., p. 208; from *GW*, III, 245.

92 Ibid., p. 209, referring to *GW*, III, 247. I have replaced the concluding words of the Koyré-Maddison translation for *quasi* ("as if it were one") with "as it were."

93 Ibid., p. 210.

94 See Mary Hesse, *Models and Analogies in Science* (Notre Dame, Ind., 1966), and Koyré, *Astronomical Revolution*, pp. 224, 227, 229, 237, 241, 257, etc.

95 Koyré, *Astronomical Revolution*, p. 252; from *GW*, XV, 171f.: "Iam quilibet globus planetarum rursum statuendus est magneticus vel quasi (similitudinem enim volo, non pertinaciter rem ipsam)."

96 Ibid., p. 285; from *GW*, VII, 259.

97 See, for example, Koyré, *Astronomical Revolution*, pp. 108 n. 27, 139f., 153, 286, 440 n. 15, 451 n. 4; *Vistas in Astronomy*, 18 (1975), pp. 87, 288, 428, 434. On Kepler's insistence that reality can be understood only in quantitative and mensurable terms, see Koyré, *Astronomical Revolution*, pp. 347, 350, etc.; *Vistas in Astronomy*, 18 (1975), pp. 423, 522.

98 This is contained in Kepler, *GW*, VI, and subsequent page references in the text refer to this volume. See also the German trans., with useful introduction and notes, by Max Caspar: Kepler, *Welt-Harmonik* (Munich and Berlin, 1939).

99 For an elucidation of this distinction, see Chapter 8 of this volume.

100 Wolfgang Pauli, "The Influence of Archetypal Ideas on the Scientific Theories of Kepler," trans. Priscilla Silz, in *The Interpretation of Nature and the Psyche*, ed. C. G. Jung and W. Pauli (New York, 1955), pp. 147–240, at pp. 192–3, where Pauli also reproduces Fludd's illustrations. More of these are reprinted in Joscelyn Godwin, *Robert Fludd: Hermetic Philosopher and Surveyor of Two Worlds* (London, 1979), which, as its title suggests, is an enthusiastic account of Fludd's system, profusely illustrated.

101 D. P. Walker, "Kepler's Celestial Music," in Walker, *Studies in Musical Science in the Late Renaissance* (London and Leiden, 1978), pp. 53–5, citing *GW*, XVI, 154ff.

102 Robert Lenoble, *Mersenne ou la naissance du mécanisme* (Paris, 1943), pp. 96ff., 105ff.

103 Ibid., p. 80 and note.

4

Marin Mersenne: Renaissance naturalism and Renaissance magic

WILLIAM L. HINE

We now recognize that magic played a much greater role during the Renaissance than modern scholars at first were willing to recognize. Until recent decades, there was a tendency to think of magical thought as a kind of aberration in Western culture, one more appropriate to primitive societies than to the sophisticated European culture, and to feel that those magical ideas which did appear in the West should be referred to apologetically, if at all, as though admitting to a regrettable weakness. Even Lynn Thorndike's monumental *History of Magic and the Experimental Sciences*, although it gives evidence of a shift of attitude during the years it took him to write the work, for the most part speaks disparagingly about those who paid even lip service to magic. In the Introduction to the first volume Thorndike sounds apologetic about devoting so much effort to such fruitless ideas and justifies them in an antiquarian fashion.[1] By the time of his last volume, however, he is willing to endorse Keynes's description of Newton as the last of the magicians and the first of the moderns.[2] Indeed, in his discussion of Newton he seems almost to suggest that magic may have had some beneficial influences.

More recently, Renaissance magic has received a good deal of attention, spurred by such works as Frances Yates's *Giordano Bruno and the Hermetic Tradition* and D. P. Walker's *Spiritual and Demonic Magic*.[3] As a result of this and other work, we now recognize that magic was an important element in Renaissance thought and cannot be ignored. Despite the fact that there has been a reverse tendency to exaggerate its importance, we are now willing to try to understand its place in the thought of the period.

Early historians of science, disregarding the magical trend of thought, accepted the claims of Galileo, Bacon, and others that the rise of science was born of the clash of the modern scientific viewpoint

against the Aristotelian world view. In recent years, a third element
has been discovered in Renaissance thought, an element that has been
called Renaissance naturalism.

The twofold division of Aristotelian versus modern science made it
seem likely that Galileo should be ranked among the Platonists because
of his attack on Aristotle. An attempt to do so, however, revealed a
large body of Neoplatonic ideas that seemed far removed from modern
science.[4] Historians of science thus became aware of the Neoplatonic,
hermetic, cabalistic, or magical ideas of the Renaissance. The term
"Renaissance naturalism" was introduced into the literature to include
those ideas that were neither Aristotelian, in the scholastic sense, nor
yet "modern," in the sense of the mechanics of Galileo or the philosophy of Bacon.[5] Such a triune division allowed historians to gather up
the remaining ideas of the Renaissance and lump them together under
one rubric, despite the heterogeneous nature of such nonscholastic,
nonmodern attempts to understand the universe. Further research,
however, suggests that the term "Renaissance naturalism" needs clarification.

Discussions of this third category by recent historians have tended
to confuse at least two separate trends of thought. This chapter will
try to distinguish them, a task made difficult by the confusing nature
of the terminology that has been employed in the past as in the present.
Let us turn for guidance to the seventeenth-century French scientist,
Marin Mersenne, who makes just such a distinction.

Mersenne's first major work, *Quaestiones celeberrimae in Genesim*,
was written in three main sections.[6] The middle portion, which is in a
more traditional biblical commentary format, also contains discussions
of a variety of topics, including scientific ones. An example of the latter
is Question IX, discussing the possibility that the earth might move,
comparing and contrasting the arguments against its motion with Copernican arguments in favor of its motion.[7] Such a discussion represents,
in effect, the kind of division made by early historians of science who
saw history in Galilean terms as a conflict between Aristotelian and
"modern" science.

The first section of the work, however, is in large part a commentary
on, and a criticism of, two books by Julius Caesar Vanini, which are
primarily based on the neo-Aristotelian views of Pietro Pomponazzi.[8]
These ideas can quite properly be classified as Renaissance naturalism,
for neither Pomponazzi nor Vanini makes any appeal to powers pre-
sumed to be beyond the realm of nature. Their definition of nature does
not include angels and demons, whose existence cannot be demonstrated by natural reason and must be accepted by faith alone.

The last section of *Quaestiones celeberrimae in Genesim* is paginated separately and entitled *Observationes et emendationes ad Francisci Georgii Veneti "Problemata."* [9] It is a lengthy critique and correction of Georgio Veneti's *In scripturam sacram*, which had been placed on the Index for containing a list of theological errors, but which had nevertheless been reprinted in Paris the year before Mersenne's work appeared. [10] Being filled with opinions "Platonicorum, Rabinorum et Magorum" (which Mersenne attempts to refute), it was squarely within the tradition of Renaissance magic. [11] While some scholars today include magic in the same category as Renaissance naturalism, the two in reality belong in separate categories because magic envisages at least the possible use of supernatural powers and in particular attributes an important role to angels and demons.

Not only does Mersenne make a distinction between Renaissance naturalism and Renaissance magic in the organization of his work, he makes it explicit in the text by calling practitioners of the first atheists, their counterparts magicians (*magos*). The former, the naturalists or atheists, deny God's role in the world and "attribute everything to nature alone," while the magicians "worship demons" and attribute many activities to devils. [12] Mersenne claims that it is the magicians rather than the atheists who attribute miracles to spirits and use demonic arts to fabricate characters under certain constellations, in order to perform marvelous works, whether good or evil. [13] These statements are made in the section analyzing Vanini's works, and there are several reasons why Mersenne uses the term "atheist" in this context. First of all, the term is used by Vanini himself, who claimed that his works were attacks on atheism. However, Mersenne thought the term was applicable to Vanini, who had been convicted of atheism and burned at the stake in Toulouse. Mersenne felt that Vanini's execution was justified because he would acknowledge the existence neither of God nor of angels and demons: He "attributed all things to fate, and adored Nature as the bounteous mother and source of all being." [14] Further, Mersenne used the term because he thought such a line of reasoning tended to lead to atheism. Pomponazzi, with Vanini following in his wake, had given naturalistic explanations for many events reported in history that had been considered miraculous. In particular, they denied that the existence of angels could be demonstrated rationally, insisting instead that the reported appearance of angels could be explained by the use of various natural devices. [15] They also argued that certain miraculous events could be accounted for in terms of natural phenomena. Many of these events had religious connotations, and although Vanini was on safe ground in explaining away pagan miracles, he went further and argued that some marvels in the Christian tradition could be given a naturalistic explanation. [16] Mersenne could not accept this idea because he felt that miracles were a guarantee of the authenticity of Christianity.[17]

On the other hand, he accepted the warning issued by the church that magic too readily lent itself to commerce with devils, although magicians disclaimed any such intent in their works. Still, the *Picatrix* in the *Hermetica* had stated that idols in Greek temples, which were said to have foretold the future, were really talismans made by drawing down powers from the supercelestial or angelic realm by means of certain rituals and ceremonies. While this specific idea had been condemned by the church, because it was feared that devils might respond, pretending to be angels in order to entrap men, various magicians had discussed the possibility of drawing down such influences.[18]

Mersenne was, in fact, perturbed by both trends of thought. As a faithful son of the church, he believed in the angelic visitations and Christian miracles reported in the Scriptures and in church tradition. He felt he had to defend them both against those who would deny their existence by explaining them away and against those who tried to exercise control over spiritual powers. He saw them, therefore, as separate problems: rightly so, since their orientation and their philosophical background were quite different.

Pomponazzi, Vanini's source, was an Aristotelian, a fact that helps us distinguish his way of thinking from the Neoplatonic orientation of the magicians. This does not make Pomponazzi a scholastic, however, despite a modern description of him as "the last scholastic and the first man of Enlightenment."[19] Medieval scholars had "baptized" Aristotle by correcting those views of his that were not compatible with Christianity, such as his belief in the eternity of the world and the mortality of the soul. Vanini had studied in Padua and adopted many of Pomponazzi's ideas, making them the foundation of his two books, which were published in Paris in the period when Mersenne was working on his first major publication.[20] It was Vanini's misfortune, however, to carry these sophisticated Italian ideas into a provincial French town, which considered them far too radical. And it was the naturalistic tone of his books, along undoubtedly with Vanini's untimely end, that induced Mersenne to examine Vanini's work.

For both naturalists and the magicians the stars played a significant role in influencing the terrestrial world. For the former, however, the influence of the stars amounted to a form of determinism, providing a source and guarantee of regularity and order in the universe. In such a world the difficulty was to explain how man can possess and exercise free will. This problem, discussed by Pomponazzi in the *De fato* and in parts of the *De incantationibus*, was taken up by Vanini in the *Amphitheatrum*, which Mersenne deals with at length in the first part of *Quaestiones celeberrimae in Genesim*.[21] Mersenne follows Vanini's discussion in great detail, determined to turn every argument into a defense of the existence of God, his ostensible purpose in the first section of his work. There are some places, however, where Mersenne partially agrees with Vanini. He accepts the idea, for example, that good is its own reward, for he thinks that God rewards good and punishes evil in this world as well as in the next. But he is careful to state that the stars are the cause neither of evil nor of the punishment for it, since evil results from man's decisions.[22]

Vanini goes on to discuss the immortality of the soul, adopting Pomponazzi's opinion that this is to be accepted as an article of faith on the assurance of the church because it cannot be demonstrated rationally.[23] Mersenne retorts that natural reason *does* show the soul to be immortal, giving a list of arguments taken from Jacobus Carpentarius (Charpentier), *Platonis cum Aristotele in universa philosophia, comparatio*, and from Petrus Martinez, *In tres libros Aristotelis de anima commentarii*.[24] In case these arguments do not suffice to convince the reader, Mersenne also recommends further discussions by Leonard Lessius and Toletus.[25]

The problem of the immortality of the soul is followed by that of the sufficiency of natural causes, or whether everything can be attributed to influences from the stars. In response, Mersenne attempts to demonstrate that some things cannot be attributed to such a cause. Vanini had weakened his argument by proposing that although stars may not be instrumental in determining the future, they are at least portents of future events. Mersenne rejects this argument in a section entitled "stars are not signs of future events, especially those depending on our free will or on God alone."[26]

After his thorough examination of Vanini's *Amphitheatrum*, Mersenne takes up Book IV of Vanini's *De admirandis naturae deaeque*, a discussion of pagan and Christian miracles.[27] Mersenne is quite concerned about the matter, examining it at length. He accepts the basic point made by both Pomponazzi and Vanini that those "miracles" reported in antiquity that were not associated with the Judeo-Christian tradition were not true miracles and could be explained in a variety of ways. He insists, however, that most Christian miracles could not be so explained. For instance, Pomponazzi's suggestion, elaborated by Vanini, that the appearance of angels could be accounted for by mirrors reflecting an image at a distance leads to a discussion by Mersenne of the principles of optics, in the course of which he demonstrates that such mirrors cannot provide the kind of experience with angels reported in Scripture.[28] In discussing optics Mersenne draws on the latest information available to him, including the as yet unpublished work of a friend, Claude Mydorge.[29] Discussing cures of illnesses and resurrection from the dead, Mersenne agrees that such events reported in pagan literature may well have had naturalistic explanations, but affirms that those reported in Christian literature were true miracles.[30]

For Vanini both classes of events, whether the appearance of angels or dramatic cures, could be given naturalistic explanations, even if such explanations depended on a hidden or occult power, like that of the magnet. Pomponazzi had discussed some children who were cured of skin problems (a rash and a burn) by a man using words alone, without applying any medication to the skin itself.[31] In his explanation Vanini claimed this was a natural power that some men possessed, analogous to the power of the magnet. Although we cannot explain its cause, which is therefore hidden or occult, it is, nevertheless, not supernatural. Mersenne, who interjects a discussion of the magnet, based primarily on Gilbert's *De magnete*, in order to show the limitations of magnetism, does not agree that men can possess such powers. So concerned is he about this point that he devotes a whole chapter to his belief that "there is no virtue in man which can cure all illnesses and no idiosyncrasy of man to which we can attribute miraculous cures."[32]

When Mersenne turned his attention to Renaissance magic, however, as in the last section of *Quaestiones celeberrimae in Genesim*, devoted to the work of Georgio Veneti, his problem was just the opposite. Here he did not have to defend a belief in angels and demons, but rather to condemn too great a reliance on them. This was probably a last-minute addition, undertaken because Veneti's book had just been reprinted.[33] Although he had already discussed various topics relevant to magic here and there in his book, Mersenne added a detailed criticism of Veneti's magically oriented *In scripturam sacram*.

In contrast to the Aristotelianism of Renaissance naturalism, Renaissance magic rested on Neoplatonic thought, particularly as represented by the Hermetic corpus translated and used by Marsilio Ficino, to which Pico della Mirandola had added the cabala. Throughout the Middle Ages and early Renaissance, Hermes Trismegistus had enjoyed the reputation of an ancient Egyptian sage who had glimmerings of "Christian" thought long before the time of Christ. The historical accuracy of that assumption was questioned in the early seventeenth century by a Protestant scholar in England, Isaac Casaubon, who was engaged in a polemic against Roman Catholicism. As part of his attack he showed that the works ascribed to Hermes Trismegistus actually dated from the second century A.D. and that far from being the work of a prophet they revealed a very inadequate knowledge of Christianity.[34] We cannot tell whether Mersenne was aware of this when he wrote *Quaestiones celeberrimae in Genesim*, for although he refers to Casaubon at least twice in the section on Veneti, he does not mention Casaubon's criticism of Hermes. A few years later, however, he comments that Robert Fludd, a follower of the Neoplatonic tradition, would abandon his respect for the pseudo-Trismegistus if he were to read Casaubon:

> Since Fludd lists many authors, I refer only to those on whose authority he relies in his books. Among the first rank is the pseudo-Trismegistus, whose *Pimander* and other treatises he seems to think have equal authority and truth with Holy Scripture, and concerning whose value he would, I believe, change his mind, if he read the first *Exercitatione* of the *De rebus sacra*.[35]

Although some thinkers were drawn to the magical position because it seemed more compatible with religion, Mersenne considered the relationship a serious drawback. Indeed, he felt that magic and other related ideas were *too* closely associated with religion. Although the magi were considered by the ancient Persians to be wise in philosophy, because that philosophy was not joined to divine wisdom, that is, to the Judeo-Christian tradition, it was, in Mersenne's opinion, easily led astray. Consequently, magic was reproved in many places in Scripture and attacked by the church. Curiously enough, Mersenne's criticism of magic at this point is derived from Agrippa's *Vanity of Science*.[36]

Some kinds of magic were associated with tricks and illusions, and Mersenne dismissed these altogether. His attitude here was more extreme than that of some of his contemporaries, such as Athanasius Kircher, who was fascinated by "curiosities." Kircher built many magnetic devices, including some to convey messages secretly. He described them in loving detail in his book on magnetism, treating them as natural magic.[37] Not so Mersenne. On almost every occasion that Mersenne mentions magic he does so critically.

In contrast to the naturalist view, which emphasized natural law and ran the risk of determinism, magic was based on a certain conception of human freedom. Pico della Mirandola gave voice to that idea in his "Oration on the Dignity of Man," where he proposed that man stood apart from the great chain of being, could envision the heights as well as the depths of the universe, and thus to a certain extent was free to make of himself what he willed. In magic the question is not whether man's destiny is determined for him by his stars, but whether he can discover the stellar influences on his life and take steps to counteract them, if necessary, or direct them for his own benefit. The magical tradition contained explanations of ways in which this could be done. For example, music plays a role in ceremonies through which it draws down and employs heavenly influences. Marsilio Ficino recommended such practices, and Mersenne discussed Ficino's ideas.[38] To Ficino, music produced certain effects on an individual, depending on the constellations with which it is associated. Mersenne totally rejects this idea:

> That any influences whatever have been brought down from the stars by singing has been entirely repudiated, for this or that song does not provoke us to sadness or happiness because it is performed under this or that star, as is indicated by the fact that the same song has the same power when heard under various stars, as experience will confirm.[39]

Music does have certain powers; David cured Saul's illness by playing on a harp, and the Walls of Jericho fell down when the trumpets sounded, but in both cases the powers derived from vibrations set up in the air. These vibrations can cure illnesses by dissipating bad humors and exhilarating the mind. As for Jericho, since bombards when fired have been known to shatter windows by noise alone, and organ music has caused cathedral stones to vibrate, so likewise trumpets by hitting the right notes could shatter walls. At any rate, Mersenne concludes, Ficino was not acting like a good Catholic when he wrote magical nonsense in Book 3 of *De vita coelitus comparanda*.[40]

Other aspects of magical lore included the idea that various metals and stones had special sympathetic relationships with heavenly bodies. Mersenne deals with this claim in some detail in order to criticize it more effectively. He specifically rejects the theory that associated certain metals or stones with particular planets, since he insists that their association is substantiated neither by reason nor by experience:

> The leading alchemists list seven metals which they associate with the seven planets, for they assign lead to Saturn, tin to Jupiter . . . iron to Mars, copper to Venus . . . quicksilver to Mercury, and silver to the Moon . . . I think that what [they] . . . say about the sympathy of stones and metals with the planets is nonsense.
>
> I do not deny that metals have some power in medicines, but not because of the planets to which they are subject, for Saturn does not preside over lead, or Jupiter over tin, any more than does the Sun or Venus. Those metals are associated with their planets either by colour, weight, motion, or substance. Experience demonstrates that these associations are false, since iron is not the same colour as Mars, nor tin the same as Jupiter, nor quicksilver the same as Mercury. Lead is not the heaviest metal, although Saturn is the slowest and most remote planet. Indeed, quicksilver and gold are heavier than lead. Therefore I do not see why Saturn is associated with lead any more than with silver, for it cannot cause the pallid colour, since this colour is not caused by the planet itself, but by the various distances and diverse

media through which we discern the light of the stars and through which the various colours appear.[41]

Injecting a note of historical criticism into the argument, he maintains that the magical stories told about metals are entirely false. Nor does carving an image of any kind on them imbue them with any virtue.

> Whatever Aphrodesius, Porphyry, Artephius, Thebit, Venthorad, Apponensis, Albertus, Pomponazzi, Mizaldus, Ficino, Cardan and others say about characters and images impressing a virtue on metals, stones or other matter, is false, ridiculous and against all true philosophy.[42]

One can never attract heavenly influences by these means, and attempting to do so smacks of superstition. The metals and stones do not contain gods, souls, or heavenly spirits. He identifies these ideas variously as being associated with the magos, Platonists, Porphyry, Orpheus, Pythagoreans, and hermeticists. He cites Kepler instead, however, for the idea that heavenly bodies are made from the same material as earthly ones: "Kepler . . . thinks that the substance of the stars is the same as that of our world."[43]

The aspect of magic that most disturbed Mersenne in his analysis of Veneti's work derived from cabalism. Although he has already commented on it here and there in other parts of his work, Mersenne includes in this last section a copy of the *Sepher Jetzirah*, along with a selection from Postellus's commentary on it.[44] Resorting once more to historical criticism, he rejects its claim to great antiquity and Postellus's acceptance of it.

In his criticism of Vanini and in his defense of miracles and angels, Mersenne had given reasons for believing that they existed. Against the magicians, he now has to argue not that they exist, but that their powers are limited and that they should not be invoked lest men be drawn into evil pacts with devils. He is also critical of Agrippa and others for accepting revelations not recognized as a part of church tradition, as when they refer to a great many angels by name, since only three are named in Scripture.[45]

He also attacks other parts of the cabala, rejecting the idea of any power associated with the *Sephiroth* or the letters of the Hebrew alphabet and the words and names formed from them: "These things are false, since they rest on principles that are completely false."[46] Attempts to find secret meanings in the Scriptures by using letters as anagrams lead only, in his opinion, to an interest in mathematical permutations and combinations and do not reveal hidden messages in the Bible. "I am amazed at the ingenuity with which they derive a word from another equivalent one. They say anything they want to even when it is against Holy Scripture and divine law, and they argue for it."[47]

We thus find Mersenne entering into different kinds of discussions depending on whether it is naturalism or magic that he is confronting. There is a very real difference between the two in his mind, as there should be in ours. Naturalism is neo-Aristotelian, leans toward determinism, will have nothing to do with supernatural powers, and tries to explain away miracles by appealing to natural phenomena. Magic is Neoplatonic, emphasizes man's freedom, too readily attributes events to angels or demons, and mixes in too much religious language and terminology without a religious purpose. With naturalism, Mersenne's task was to explain the limitations of nature. With magic, he had to emphasize the limits of supernatural events and angelic powers.

Both traditions recognized hidden powers, like that possessed by the magnet in attraction or in its directional properties. Such powers must be classified as occult because one cannot explain them rationally, in terms of Aristotelian causes. Nevertheless, they do exist, as empirical evidence shows. Pomponazzi, Vanini, and their followers accepted them simply as occult but natural powers. For magicians such as Veneti, magnets represented magical powers and thus supported a Neoplatonic, hermetic cabalistic world view.

In conclusion, since both Renaissance naturalism and Renaissance magic used magnetic attraction as an illustration of their theoretical assumptions, it may well be that later scientists such as Newton, for example, saw in attraction a representation not of a hidden magical power, but of an occult, natural power. By distinguishing Renaissance naturalism from Renaissance magic; by redefining the term Renaissance naturalism to mean those explanations that were truly natural, even though they included occult causes; and by redefining Renaissance magic to mean a world view with a Neoplatonic, hermetic, and cabalistic orientation, we can begin to eliminate some of the confusion that has so far existed in discussions of this aspect of the relationship between science and the occult.

Acknowledgments

Research for this article was made possible by a Canada Council Research Grant S74-1956 and by an Atkinson College Minor Research Grant.

Notes

1 Lynn Thorndike, *A History of Magic and Experimental Science*, 8 vols. (New York, 1923–58), I, 2, 4–5.
2 Ibid., VIII, 589–603.
3 Frances A. Yates, *Giordano Bruno and the Hermetic Tradition* (Chicago, 1964); D. P. Walker, *Spiritual and Demonic Magic from Ficino to Campanella* (London, 1958).

4 See, for example, E. A. Burtt, *Metaphysical Foundations of Modern Science*, 2nd ed., rev. (New York, 1955), pp. 53–5, 58, 70, and Alexandre Koyré, "Galilée et Platon," in Koyré, *Etudes d'histoire de la pensée scientifique* (Paris, 1966), p. 168. Both writers believed that the scientific revolution grew out of a Platonic belief that the universe is basically mathematical in nature. In contrast, E. W. Strong, *Procedures and Metaphysics: A Study in the Philosophy of Mathematical-Physical Science in the 16th and 17th Centuries* (Berkeley, 1936), pp. 6, 10, 50–1, began his work with the assumption that the Neoplatonic tradition was the "metaphysical godfather" of modern science and was appalled to discover that the Neoplatonic-hermetic ideas were vastly different from what he expected to find.

5 See, for example, Robert Lenoble, *Mersenne ou la naissance du mécanisme* (Paris, 1943), Chap. III, pp. 83–167, and Richard S. Westfall, *The Construction of Modern Science* (New York, 1971), pp. 28–33.

6 Marin Mersenne, *Quaestiones celeberrimae in Genesim* (Paris, 1623). Cited hereafter as *QcG*.

7 Ibid., cols. 879–920.

8 There has been some debate about how closely Vanini's works follow those of Pomponazzi. Luigi Corvaglia, *Le opere de Giulio Cesare Vanini e le loro fonti*, 2 vols. (Milan, 1933), compares Vanini's writings with his sources in parallel columns. See Emile Namer, "Vanini n'est-il qu'un plagiaire?" *Revue philosophique de la France et de l'étranger*, 117 (March-April, 1934), pp. 291–5, or his discussion in *La Vie et l'oeuvre de J.C. Vanini* (Paris, 1980), pp. 234, 271–3.

9 Marin Mersenne, *Observationes et emendationes ad Francisci Georgii Veneti "Problemata,"* in *QcG*; cited hereafter as *Observationes*.

10 Francisco Georgio Veneti, *In scripturam sacram et philosophos tria millia problemata* (Paris, 1622).

11 Mersenne, *Observationes*, col. 39.

12 Mersenne, *QcG*, col. 570.

13 Ibid.

14 Ibid., col. 489.

15 Pietro Pomponazzi, *Les Causes des merveilles de la nature ou les enchantements*, trans. Henri Busson (Paris, 1930), pp. 185–202; Julius Caesar Vanini, *De admirandis naturae deaque mortalium arcanis*, in *Oeuvres philosophiques*, trans. X. Rousselot (Paris, 1842), pp. 229–36; Mersenne, *QcG*, cols. 469–538.

16 Vanini, *De admirandis*, *Oeuvres*, pp. 228, 262, 287; Mersenne, *QcG*, cols. 537, 483–4.

17 Mersenne, *QcG*, cols. 649–50.

18 Ibid., cols. 583–5. For other reactions to this idea, see D. P. Walker, "The Prisca Theologia in France," *Journal of the Warburg and Courtauld Institutes*, 17 (1954), pp. 234–40.

19 Ernst Cassirer, *The Individual and the Cosmos in Renaissance Philosophy* (New York, 1964), p. 81.

20 Julius Caesar Vanini, *Amphitheatrum aeternae providentiae divino-magicum, christiano-physicum, nec non astrologocatholicum, adversus veteres philosophos, atheos, epicureos, peripateticos, et stoicos* (Lyons, 1615); *De admirandis* (Paris, 1616). Both works were republished in French translation by Rousselot (Paris, 1842).

21 Because of the somewhat confusing organization of *Quaestiones*

celeberrimae in Genesim, it is not always apparent that Mersenne is examining Vanini's works. For that reason I previously made a careful textual comparison: see William L. Hine, "Mersenne and Vanini," *Renaissance Quarterly*, 29 (1976), pp. 52–69.

22 Vanini, *Amphitheatrum*, *Oeuvres*, p. 54; Mersenne, *QcG*, cols. 285–310.

23 Vanini, *Amphitheatrum*, *Oeuvres*, p. 100.

24 Jacob Carpentarius, *Platonis cum Aristotele in universa philosophia, comparatio* (Paris, 1573), pp. 5–13; Petrus Martinez, *In tres libros Aristotelis de anima commentarii* (Siguenza, 1575), pp. 482–3; Mersenne, *QcG*, cols. 363–76.

25 Leonard Lessius, *De providentia numinis et animi immortalitate libri duo adversus atheos et politicos* (Antwerp, 1613), pp. 233–321; Franciscus Toletus, *Commentaria una cum quaestionibus in tres libros Aristotelis de anima* (Lyons, 1580), pp. 519–33; Mersenne, *QcG*, cols. 376–77.

26 Vanini, *Amphitheatrum*, *Oeuvres*, pp. 27–8; Mersenne, *QcG*, cols. 379–90.

27 Mersenne examines book IV of the *De admirandis* in Objection XXV, *QcG*, cols. 457–652.

28 Ibid., cols. 475–8, 500–37.

29 Ibid., cols. 500, 510, 513, 516, 525, 531. Mydorge's book was not published until much later: Claude Mydorge, *Prodromi catoptricorum et dioptricorum* (Paris, 1641).

30 Mersenne, *QcG*, cols. 553–652.

31 Pomponazzi, *Causes des merveilles*, pp. 113, 131–5.

32 Ibid., pp. 131–42; Vanini, *De admirandis*, *Oeuvres*, p. 281; Mersenne, *QcG*, cols. 541–53.

33 Veneti, *In scripturam sacram*.

34 Isaac Casaubon, *De rebus sacra et ecclesiasticis exercitationes XVI* (London, 1614), pp. 70–87.

35 Mersenne, *Observationes*, cols. 40, 384. Mersenne refers explicitly to Casaubon's criticism of the traditional dating of Hermes Trismegistus in his introductory "Epistola" to Pierre Gassendi, *Theologi epistolica exercitatio* (Paris, 1630). See also *Correspondance*, ed. C. deWaard, R. Pintard, et al. (Paris, 1932), II, p. 445.

36 Mersenne, *QcG*, cols. 583–5, 589–90. Mersenne's discussion of Goetia and necromancy here is taken from Henri Cornelius Agrippa, *Paradoxe sur l'incertitude, vanité, et abus des sciences*, trans. Louis de Mayerne-Turquet (s.l. 1617), chap. 45, pp. 115–17.

37 Athanasius Kircher, *Magnes sive de arte magnetica* (Rome, 1641), bk. II, pt. iv, "Magia naturalis magnetica," pp. 324–94.

38 Mersenne, *QcG*, cols. 1704–5.

39 Ibid., cols. 1700–10; quotation from col. 1705.

40 Ibid., cols. 1148–64; quotation from cols. 1147–8.

41 Ibid., cols. 1147–8.

42 Ibid., cols. 1151–3; quotation from col. 1152; see also cols. 1163–4, 1162.

43 Ibid., col. 1162.

44 Mersenne, *Observationes*, cols. 255–60.

45 Ibid., cols. 31–2, 39; Mersenne, *QcG*, cols. 569–70.

46 Mersenne, *Observationes*, cols. 259–60; Mersenne, *QcG*, cols. 703–4, 1389.

47 For a discussion of Mersenne's ideas on permutations and combinations, see Ernest Coumet, "Des Permutations au XVIe siècle et au XVIIe siècle," *Permutations: Actes du Colloque, Paris, juillet 1972* (Paris, 1974), pp. 277–87.

5

Nature, art, and psyche: Jung, Pauli, and the Kepler–Fludd polemic

ROBERT S. WESTMAN

Thirty years ago, Wolfgang Pauli, the great Nobel quantum physicist and professor at the very university sponsoring this conference on occult and scientific mentalities in the Renaissance, published a famous essay entitled "The Influence of Archetypal Ideas on the Scientific Theories of Kepler." It appeared in a volume entitled *The Interpretation of Nature and the Psyche* in which Carl Gustav Jung, also a member of this university for many years, wrote a companion essay entitled "Synchronicity: An Acausal Connecting Principle."[1] Pauli's essay is often cited with great admiration by historians who write about Kepler or Fludd and by writers in the Jungian literature conscious of the respectability bestowed upon their work by the association of Jung with a renowned "hard scientist."[2] Pauli's historical study is, in all respects, a thoroughly professional historical analysis with scrupulous citation of texts, superior translations checked by the art historian Erwin Panofsky, and succinct interpretations. What is never mentioned by anyone in the history of science literature is the fact that the book appeared under the auspices of the C. G. Jung Institute in Zurich, and that the real subject of Pauli's article was not primarily Kepler as a historical figure but rather Kepler as an *illustration* of the problematic relationship between the observer and what is observed; or, in the language of Jung's analytic psychology, the relation between archetypal images and sense perception.

Until now, no one has asked publicly why Pauli wrote such an essay, why he encoded his analysis in Jungian terms, and what his relationship to Jung might have been. Nor, surprisingly, has anyone questioned the historical account for evidential accuracy or the terms of the analysis itself.

In this chapter I propose to look anew at the Kepler–Fludd polemic. I shall suggest that in order to make sense of the historical controversy, let alone Pauli's exegesis of it, we must attend closely to the rhetorical categories of the actors and their relationship to epistemology. It will emerge, quite surprisingly (as it did for me), that there is a profound unity in the questions with which Kepler and Fludd were concerned and that the center of a cluster of apparently diverse issues lay in the problem of establishing the connections between pictures, words, and things.

Historical origins of the polemic

In the autumn of 1617, Johannes Kepler, district mathematician in Linz and former imperial mathematician to Emperor Rudolph II in Prague, encountered a large, sumptuously illustrated work at the Frankfurt Book Fair. Its title, *A Metaphysical, Physical and Technical History of the Macro- and the Micro-Cosm,* could hardly have escaped Kepler's attention because he himself was then engaged in the final stages of a treatise of comparable scope and ambition, entitled *Harmonics of the World.*[3] Its author, Robert Fludd, describes himself as "Esquire" and "Doctor of Medicine in Oxford," emphasizing the priority of his noble birth: "Verily, for mine oune part, I had rather bee without any degree in Universitie than lose the honour was left me by my ancestors." As befitted a man of noble station, Fludd did not seek out professorial robes. Typical of many Englishmen of his class, he had completed a university degree and then set off on a "Continental Sojourn." On his travels, lasting about six years, he befriended and sometimes tutored a variety of men of high social rank: In France, he reports contacts with Charles de Lorraine (1571–1640), son of Henri I of Guise and his brother François, chevalier de Guise (1589–1614); in Italy, he met the humanist engineer Greuter (1560–1627), who evidently taught him some of the principles of simple machines. And finally, returning through Germany, he came into contact with some texts allegedly written by members of a secret religious fraternity calling themselves the Brothers of the Rosy Cross, or Rosicrucians.[4] These texts preached a message of reform of the disciplines of learning in preparation for the dawn of a new age prior to the end of the world. The leading printer of Rosicrucian treatises in Europe was Johann Theodore de Bry of Oppenheim. At about the same time that he published Fludd's works, he also produced several alchemical works by a German physician, Michael Maier (1566–1622), who had been attached to the court of Rudolph II in Prague. Maier could well have been the man who carried Fludd's manuscript to the De Bry press, and the work Kepler found at the Frankfurt Book Fair bore its colophon.[5]

Nothing very substantial is known today about the Rosicrucians as a social group, if indeed they existed at all. Intriguing claims about the connection between Fludd and Maier have been made by some, and the Rosicrucians have been elevated from historiographical obscurity into a "movement," a "scare," and even the predicate of a new age of "enlightenment." But, unfortunately, the evidence is extremely thin and we would do well to tread cautiously.[6] One way to proceed is to use Fludd himself as a source for the presentation of Rosicrucian beliefs. And since our interest here is in the dispute with Kepler we are not unjustified, perhaps, in following this narrower path.

Not long after the publication of Fludd's *Macrocosm*, dedicated to James I, he was accused by someone at the court of "mysticall learning" and collaboration with the Rosicrucians. In reply Fludd made a successful "Declaratio Brevis" before the king.[7] As a document of historical interest where a full outline of Rosicrucian doctrines might be expected, there is small return here from Fludd's notoriously uncontrolled verbosity. Nonetheless, when the "Declaratio" is raked with care, a few clues fall out. We learn that the R. C. Brothers are Calvinist, that they have access to profound secrets and truths in medicine and natural philosophy, and that they embrace all disciplines with the exception of jurisprudence. Apart from this rhetoric of truth, secrecy, and disciplinary competence, there is really only one phrase of propositional substance: "The true philosophy . . . will diligently investigate heaven and earth, and will sufficiently explore, examine and depict Man, who is unique, by means of pictures [*imaginibus depinget*]."[8] Now, though fragmentary, the phrase *imaginibus depinget* ("it will depict or portray by means of pictures") provides an important clue to the Kepler–Fludd polemic. Behind the strategies of personal polarization and the emotional language of contrasts and opposites that pervade the dispute with Kepler lay a real dispute about the meaning and reality of visual imaginings.

We pick up some hints of this theme in the polemical charges and countercharges that began when Kepler returned to Linz from the Frankfurt Book Fair and quickly penned a short comparison between certain sections of Volume I of Fludd's *Macrocosm* (the full work did not appear until 1621) and his own. This analysis appeared as an Appendix to his *Harmonics of the World* in 1619. Kepler immediately singled out for special criticism and comparison Fludd's philosophy of world harmony or *musica mundana*. This was but the opening salvo in a clash wherein Fludd wrote a lengthy rebuttal in 1621 entitled quite explicitly, *The Stage of Truth in which the Tragic Curtain of Error is Parted, the Smaller Stage Curtain of Ignorance is Raised and the Truth Itself is Brought Forth Publicly by its Minister, Or a Certain Analytic Demonstration*. The "analytic demonstration," which implies a treatise purporting to contain certitudes, consists of a sentence-by-sentence breakdown of Kepler's attack in the *Harmonics* with each Keplerian passage given in italics and then followed by lengthy and, not infrequently, redundant commentary.[9] To Fludd's tragic theater of truth and ignorance, Kepler replied with an *Apology* (1622) against each of Fludd's analytic propositions, and Fludd delivered a final blast in 1623 with his *Monochord of the World*.[10] In his two treatises Kepler accuses Fludd of indulging in "enigmatics [*aenigmata*]," of writing in an "occult and shadowy [*occultum et tenebrosa*] manner," of engaging in "dreams [*somnia*]," of creating "dense mysteries [*mysteriae profundissimae*]," of preferring the ancients to the moderns, and finally, of delighting in "pictures [*picturae*]" and "pure symbolism [*meri Symbolismi*]." Fludd, on the other hand, criticizes Kepler for being "excessively verbal [*multis verbis et longa oratione expressit*]"; for concerning himself only with "quantitative shadows [*umbras quantitativas*]" rather than with the "substance [*substantia*]," "exterior movements [*motus exteriores*]" rather than "internal and essential impulses [*actos internos et essentiales*]"; for seeing "effects rather than first causes [*ego causam principalem, ipse illius effectus animadvertit*]."[11]

The confrontational rhetoric of the combatants tempts us to bring out our own dichotomies: opposing mentalities and world pictures; scientific and nonscientific, modern and premodern, quantitative and qualitative modes of knowing. No doubt a case could be made for such an interpretation, but only at the expense of obscuring an important shared presupposition: Kepler and Fludd agree on the existence of an intelligible realm of nonsensible qualities and powers. For alongside Fludd's world of musical spirits, astral powers, and invisible spiritual illumination, we must place Kepler's world of invisible astrological and magnetic forces. Put otherwise, both are united in rejecting the medieval Aristotelian treatment of occult qualities either as altogether unintelligible or as mysterious effects with idiosyncratic causes.[12] However much each may have continued to invest in Aristotelian discourse and categories, both were trying *systematically* to construct accounts, albeit radically different, that would relate the invisible, insensible realm of being to perception. In this regard, the question arose: What kind of reality would be attached to visualizations of entities and relations that could not be seen directly? And how could one know for sure if one had the reality, even if one had the picture? The clash between Kepler and Fludd over the status of visual productions was also a dispute about those disciplines thought to possess the highest purchase on truth.

Fludd's visual epistemology: "ut pictura Genesis"

No one who has studied Fludd has failed to notice and frequently to reproduce the profusion of marvelous illustrations that populate all his treatises. Peter Ammann, taking Fludd's muscial philosophy as central to the whole system, speaks of "Fludd's predilection for graphic representation."[13] Wolfgang Pauli, whose essay we shall examine at some length later in this chapter, locates Fludd in "the alchemical tradition expressed in qualitative, symbolic pictures."[14] And Frances Yates pointed out that Fludd's pictures are frequently "an exact visual counterpart to the elaborate descriptions in the text" intended to serve both as symbols and mnemonic aids to the reader. She conjectured, reasonably, that Fludd himself prepared many of the original engravings for his books.[15] Most recently, a book of 124 of Fludd's engravings has been issued with an eye to both academic and popular consumption, by Joscelyn Godwin. On the back cover the publisher tells us that "Fludd had a genius for expressing his philosophy and cosmology in graphic form, and his works were copiously illustrated by some of the best engravers of his day."[16] It is not my purpose to dispute any of these contentions, all of which are true. But none go quite far enough. We must go on to look at the engravings not as *illustrations* but rather as *ways of knowing, demonstrating, and remembering*.

To look at Fludd in this way requires that we *invert* the author's strategy of presentation in the *Macrocosm*, and that we abandon secondary accounts that follow his own order. Consider first Fludd's schematic plan for the first volume of the *Macrocosm*:[17] (1) "Concerning the Metaphysics of the Macrocosm and the Origin of Its Creatures; (2) Concerning the Physics of the Macrocosm in Generative and Corruptive Advance." This is a standard Aristotelian division of subject matter, with the theoretical sciences of metaphysics and physics occupying pride of place. But now we turn to the second volume of the *Macrocosm*, which consists of a review of the disciplines man needs to know in order to make himself, imitatively (or, quite literally, by "aping"), like the cosmos outside himself. In a famous illustration, we have a large foldout of the macrocosm that Fludd entitles "The Mirror of the Whole of Nature and the Image of Art," and that is meant to "demonstrate in this Emblematic mirror" how these disciplines lead to knowledge of God, Demons, and the Creation. It is important to notice that Fludd labels these "The *More* Liberal Arts [*Artes Liberaliores*]" because they include subjects not included in the traditional trivium and quadrivium: timekeeping, geomancy, fortification, statics, perspectiva, and painting. It is the last art that translates for us, as it were, the conception of the entire work.

What, according to Fludd, is the pictorial art? Where does it fit into the classification scheme of the disciplines? Painting has both a theoretical and a practical component.[18] The theoretical part requires the good painter to know the general properties of lines, circles, plane and

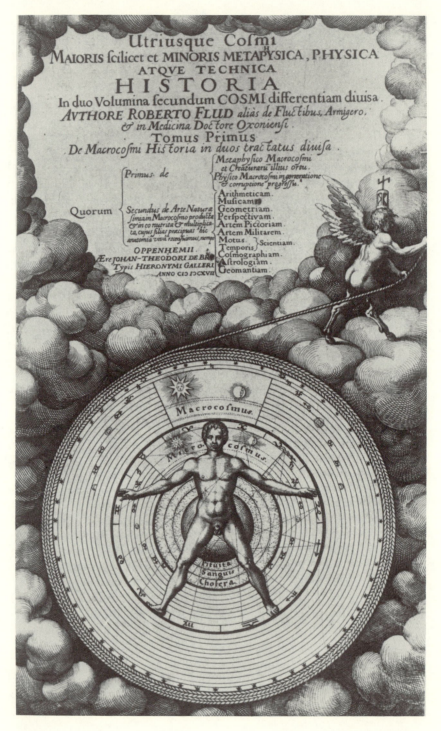

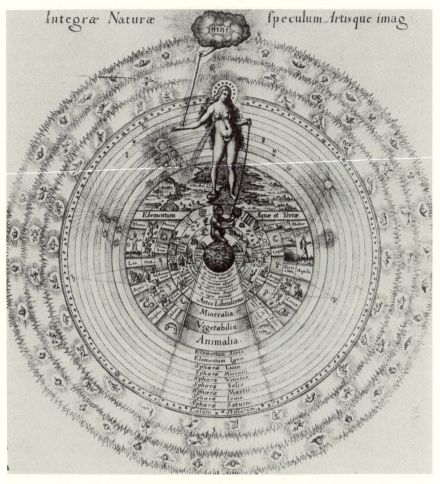

Figure 2. Art and nature mirrored. (Robert Fludd, *Utriusque cosmi . . . historia* [Oppenheim, 1617–21])

solid figures (geometry); the behavior of light (optics); and, especially, the projection of light rays into a plane (perspective). Above all, however, he must understand the meaning of *symmetria.*

By "symmetry," therefore, let us understand that most admirable proportional measure which we ought to love and contemplate not only in man himself but in all other natural

Figure 1. Microcosmic man, the macrocosm, and the scheme of the disciplines. (Robert Fludd, *Utriusque cosmi . . . historia* [Oppenheim, 1617–21])

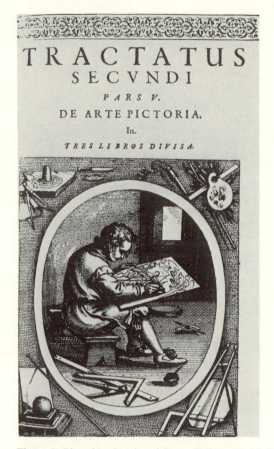

Figure 3. Picturing the pictorial art. (Robert Fludd, *Utriusque cosmi . . . historia* [Oppenheim, 1617–21])

things as well. For clearly, [symmetry] appears to be nothing but a certain most absolute harmonic instrument in all numbers such that everything is fitted together, divided everywhere into parts, by the most exact measure and, especially, among other noble beings, in the human body.[19]

The human form is the link between the theoretical and the practical aspects of visual representation, embodying, as it were, the elements of geometry and the principles of *symmetria*. The painter learns theory from books, but he learns it best by studying the human torso, and, especially, by the activity of pictorial *praxis* – drawing heads, eyes, limbs, and so forth. From the triangle he learns to draw the head in different poses; from there he can move on to circular objects (e.g., the sun, a goblet), living beings with irregular forms (e.g., a deer grazing

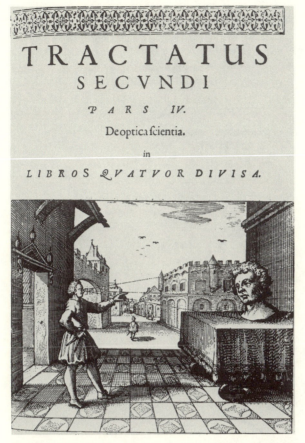

TRACTATUS
SECVNDI
PARS IV.
De optica scientia.
in
LIBROS QVATVOR DIVISA.

Figure 4. Picturing the science of optics. (Robert Fludd, *Utriusque cosmi
. . . historia* [Oppenheim, 1617–21])

in a forest), and distant scenes (e.g., a city). Study and depiction of
man's body is propaedeutic to further skill in visual portrayal, but even
more importantly, it is the first step *upward*, preparatory, that is, to
understanding the philosophy of nature and the creation. And this is
one of the primary meanings of the image that adorns the title page of
Fludd's *Macrocosm*: Man's shape, the "harmonic instrument," divides
the circular zodiac like a straight ruler according to the "symmetrical
number" four: elements, humors, limbs; meanwhile, winged time (personified) turns its visible agents, the sun and moon, in an eternal definition of macrocosm and microcosm. The message is conveyed in a
single emblem.

Anyone the least bit familiar with Renaissance iconography might
react at once that there is scarcely anything novel in Fludd's presen-

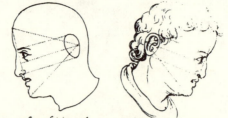

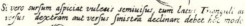

Si vero surfum afpiciat vulteus semiuifus, tum latus Triangli ai verfus dexteram aut verfus finistra declinare debet hoc modo

Vbi vero uultus directe a parte anteriori profpicit ibi latus Trian guli anterius debet orthogonaliter eleuari hoc modo:

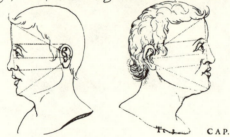

CAP.

Figure 5. Triangulating heads. (Robert Fludd, *Utriusque cosmi . . . historia* [Oppenheim, 1617–21])

tation. Is this not merely an expression of the well-known Renaissance love of emblems and hieroglyphs? Of those visual representations of virtues, vices, passions, and temperaments drawn from passages in the Bible, ancient mythology, medieval Christian allegory, and Egyptian pictorial writing?[20] There is no doubt that Fludd is indebted to this tradition of pictorial imagery, as he is to almost every other major tradition of Renaissance rhetoric and philosophy.[21] But his work was not merely an emblematic handbook for poets and artists; it was explicitly entitled a *historia*, a systematic account of the visible and invisible worlds from the time of the Creation (and even before!). And I suggest that the major iconographical source of this *historia emblematica* was Albrecht Dürer.

Figure 6. Capturing three dimensions in two through a matrix of threads.
(Robert Fludd, *Utriusque cosmi . . . historia* [Oppenheim, 1617–21])

There are several pieces of evidence that lock this claim firmly into place. First, Fludd, who is not overly generous in his citation of contemporary writers, refers explicitly to Dürer's *Four Books on Human Proportion*.[22] Second, several specific engravings are taken directly from Dürer. As an example, consider the woodcut showing the manner in which a scene of a distant city, viewed through a matrix of threads or *velo*, can be reduced to a two-dimensional surface. The technique was not original to Dürer, who had learned it during his studies in Italy, but it was he who made it known all over Europe.[23] Perhaps most remarkable of all is the direct use by Fludd of a Dürer figure to portray microcosmic–macrocosmic man. Here we see quite clearly the move

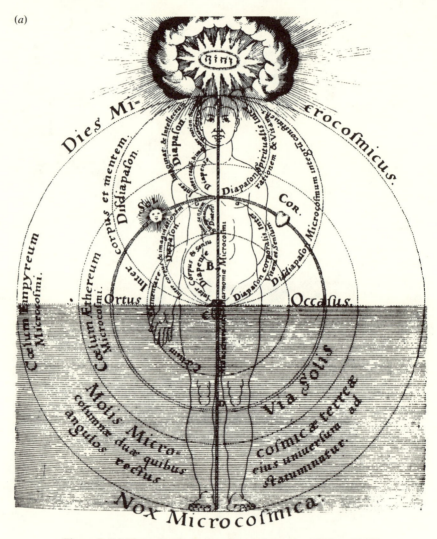

(a)

Figure 7. (a) Monochord of microcosmic harmony. (Robert Fludd, *Utriusque cosmi . . . historia* [Oppenheim, 1617–21]). (b) Elements of a well-proportioned human figure. (Albrecht Dürer, *Vier Bücher von Menschlicher Proportion* [1528])

from Dürer's practical text for the drawing of human figures. For Fludd, *praxis* is the prelude to *philosophia*. And in this image he shows us how the principles of his musical philosophy are to be found in human proportionality. Dürer's geometry dissolves into Fludd's figure, and Fludd then chooses three points as centers of harmony: the chin, the

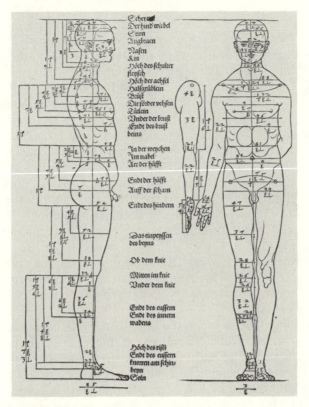

Figure 7. (*cont.*)

heart, and the genitals. The heart is the link between mind and body, the midpoint between the material and spiritual octaves. The picture lets us *see* what Fludd calls the "miraculous harmony" between the spiritual octave, linking imagination and intellect, and the material octave, linking body and imagination. God is directly in man as the axis of symmetry. In Fludd's diagram the axis is materialized into the "Monochord of Microcosmic Harmony," and makes possible a profusion of metaphors grounded in images of light, life, and love as centrally diffusive forces. Thus:

> We *see* how God is the player of *musica humana*, the player
> of the string of the monochord, the inner principle which,
> from the centre of the whole, creates the consonant effects
> of life in the microcosm. The string which by its vibration
> spreads the luminous effects of the Inspirer through macrocosm and microcosm as accents and sounds of love, as it
> were, is the luminous spirit which participates in the two extremes and which joins them together. This string equally

denotes the system of notation, or staff in man by which the
soul descends from the higher spheres and reascends towards them after death, when the ties of the body, the mea-
nest of all places, have been dissolved.[24] [my italics]

What Fludd is really doing here is extending the deeply Platonic
esthetic theory of Dürer, Alberti, and Vitruvius.[25] Among all the variety
of different forms that we observe, how are we ever to recognize what
is *eternally* beautiful? As Dürer remarks in an early fragment: "Our
judgment of what is beautiful is so uncertain that (for example) we
cannot state definitely why two persons are both beautiful although
they do not resemble each other in body, limb or measure. This demonstrates how blind we are in this respect."[26] But then, in the *Four
Books on Human Proportion* we find a succinct resolution: "For indeed, art is embedded in nature. He who can extract it [pull it out from
there] has taken hold of it. Once captured, it will save you from many
errors in your work. And a great deal of your work can be given certainty by geometry."[27] In *The Painter's Manual*, the predecessor of
his treatise on human proportionality, Dürer provided what Erwin Panofsky has aptly called

a revolving door between the temple of mathematics and the
market square. While it familiarized the coopers and cabinet
makers with Euclid and Ptolemy, it also familiarized the
professional mathematicians with what may be called
"workshop geometry." It is largely due to its influence that
constructions "with the opening of the compass unchanged"
became a kind of obsession with the Italian geometricians of
the later sixteenth century.[28]

In a sense, Fludd moved through Dürer's "revolving door," from the
marketplace through the temple of mathematics to metaphysics.

This brings us finally to the central image that appears in many of
Fludd's engravings, and that is reiterated over and over in his writings
as the basis of his entire philosophy: the interpenetrating pyramids.[29]
No commentator fails to describe this metaphysic because Fludd himself is eminently clear about its importance. The pyramids represent
an opposing dualism: light and darkness, *forma* and *materia*. All beings
constitute a proportionate mixture of *forma* and *materia*, and the point
lying exactly between the extremities of the pyramidal basis is the
equilibrium or *sphaera aequalitatis*. Stripped of this higher ontology,
what we have is nothing more than the *painter's light* – light striking
a dark room, the sun's rays playing on a building or a tree or an object
reflected in a mirror. A plate from Dürer's treatise on *symmetria*, used
by him to illustrate the divisions of the human form, shows two triangles, one inverted, the other resting on its base. If we compare this
woodcut with Fludd's illustrations of the metaphysical pyramids, it

does not stretch credulity too far to suggest that Fludd had transformed Dürer's instructional account of painting three-dimensional objects into a general account of the creation of *all* beings by the Creator.

There is an even more general way to construe the Fludd–Dürer link that amounts to a widening of our contextual field. The first point is that Fludd's major works are related to a prominent genre of Renaissance writing, namely the commentatorial tradition on the book of Genesis. Second, Fludd believed he had discovered a special and powerful mode of textual exegesis that would release the true wisdom of that book. This method amounted to a picturing or representing of the key terms of the text as three-dimensional images. Some discussion of these claims is now in order.

The period from the 1520s to the 1630s was something of a golden age for Genesis commentaries. In his excellent study of these works Arnold Williams found about forty such commentaries issuing from continental and English presses, and ranging in size from 300-page octavo to over 1,000-page folio.[30] Such was the diffusion of the "Genesis material" – an enormous accretion of exegesis, moral extension, scientific and legendary interpretation – that references to Moses can be found throughout many Renaissance literary genres, from historical, political, and agricultural works to treatises on poetry, philosophy, and astronomy. The Genesis commentators were an exceedingly learned group, the best scholars of their day. Among them were expert Hebraists, and they commanded an awesome knowledge of earlier exegetes: rabbinical, patristic, medieval, and contemporary. The most popular of all the commentaries was that of the Spanish Jesuit, Benedictus Pereira, whose *Commentariorum et disputationum in Genesin* appeared in four tomes between 1589 and 1598.[31] Aside from amassing information the major problem was an organizational one. Pereira's method was typical: Take up a verse, deal with linguistic problems, translation, and explanation; then divide the material into topics – disputations, questions, exercitations, or theses. It was in the disputations that different writers could reveal their special interests in Genesis, whether dogmatic, moral, devotional, or scientific. Pereira and Calvin are concerned, not surprisingly, with dogma, while Mersenne's interests were in physical meanings. According to Williams there was considerable consensus on hermeneutics. The commentators generally accepted the notion that Scripture could have multiple meanings and often cited a traditional distich:

> Litera gesta docet; quid credas, Allegoria: Moralis quid
> agas: quo tendas, Anagogica [The letter teaches the event:
> allegory what you should believe: Moral (or Tropological)
> what you should do: anagogic where you may go][32]

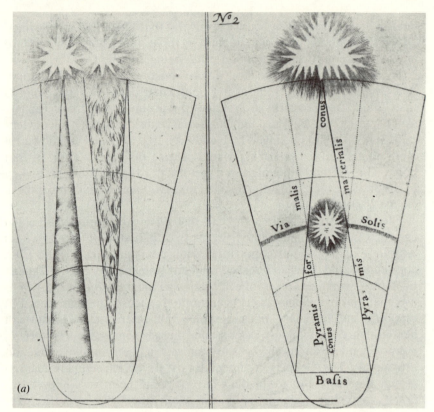

Figure 8. (*a*) Pyramids of dark and light principles (left) and interpenetration
of pyramids (right). (Robert Fludd, *Utriusque cosmi . . . historia*
[Oppenheim, 1617–21]). (*b*) Dürer's triangles for constructing well-
proportioned human figures. (Albrecht Dürer, *Vier Bücher von
Menschlicher Proportion* [1528])

The standard example is the word Jerusalem: "Literally, it means a
city; morally, it is virtue; allegorically, it is the church; and anagogi-
cally, it is heaven." But among the specific rules Pereira advocates for
the correct reading of Genesis, the tendency is to move away from
allegorical interpretation, to stress a historical reading (especially in
the early parts of Genesis), and to keep the miraculous "in bounds"
by seeking harmony with natural philosophy.[33]

Returning now to Fludd, we can see certain interesting differences
and resemblances to the dominant tradition of Genesis studies that we
have outlined here. A notable difference is that the major Genesis exegetes were university professors. Their works were organized according to the *disputatio/quaestio* method because they were often first
delivered as sermons or lectures. For Fludd, who was neither an ac-

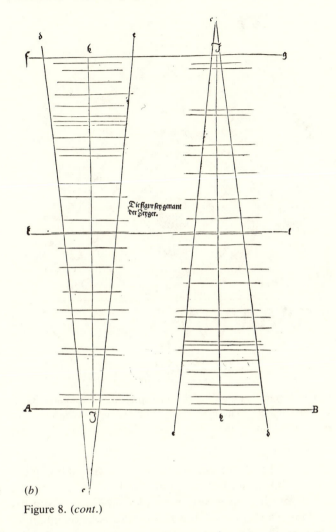

(b)

Figure 8. (*cont.*)

ademic nor a theologian, the goal of harmonizing Scripture and natural philosophy is not pedagogical in any formalized, academic sense. The balance of interest now tips away from the traditional academic disciplines into new areas such as art, alchemy, and geomancy. What Fludd proposed as a hermeneutic may be seen, in a sense, as a rejection of the high status language enjoyed in the universities. The picture now enjoys a special ontological status. At times the visual motif is an effort to *represent* the words of the text; at other times the *res picta* refers beyond itself; but always there must be a picture.[34] Thus we find references to "man" in the Bible, but no hint of "Vitruvian man"; we find light and dark in Genesis, but not interpenetrating pyramids. The

DEMONSTRATIO.

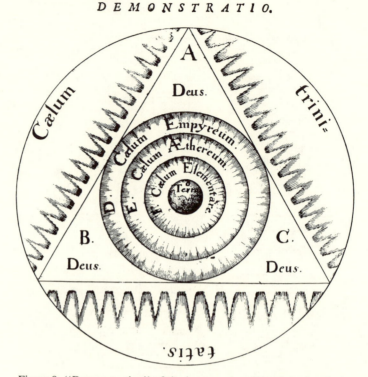

Figure 9. "Demonstration" of the incomprehensible trinitarian God creating and containing the quaternian world. (Robert Fludd, *Utriusque cosmi . . . historia* [Oppenheim, 1617–21])

Creation of the world, the creation of every individual being is like the act of painting. The principles of Light and Dark become the explanatory principles of all reality; the unity of God comes from his act of joining Light and Dark, just as the painter transcends language when he brings distant objects close or makes invisible entities visible. In different places, Fludd refers to the act of bringing the pyramids together as a *demonstratio*, sometimes a *descriptio*, at other times a *mysterium* or a *delineatio*.[35] All these terms, I suggest, have essentially the same sense: Picturing words produces in us a mysterious, expanded awareness that we did not previously possess. It is no wonder that Fludd would often speak of "the true philosophy" or his "philosophical key."

How does the "true philosophy" work out in practice? Consider the following Genesis problem: What was there before the Creation? According to Paracelsus, an uncreated, unformed, dimensionless, propertyless *materia prima*, aptly called the *mysterium magnum*.[36] This is,

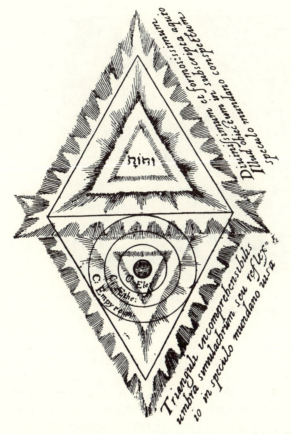

Figure 10. Mirroring of formal and material principles: the shadow of the incomprehensible trinity reflected in the world. (Robert Fludd, *Utriusque cosmi . . . historia* [Oppenheim, 1617–21])

of course, the *ex nihilo* problem. To call it a "great mystery" evades the very precise medieval discussions of whether God is coeternal with the place that he occupies and whether he can create form without matter.[37] Now God, unlike human artists, had no preexisting medium from which to educe form; hence, he must have had the power to create both form and matter from nothing. But what is "nothing"? Does it have a status equal to or higher than God? Fludd interprets "nothingness" as the darkest abyss extending without limit or form to infinity. In the *Mosaicall Philosophy* he tries to gloss the problem by referring to "a matter that was *in potentia ad actum*," supporting this interpretation with passages from Plato, Saint Augustine, and Hebrews 11:3. Consider Plato: "The nothing is like a vision in a dream, which when

Figure 11. The great darkness: *nihil ad infinitum*. (Robert Fludd, *Utriusque cosmi . . . historia* [Oppenheim, 1617–21])

a man awaketh proveth nothing save a mere imagination.''[38] Of course, if "nothing" is a dream vision, then we must wonder whether God's dream of the world had the same status. At any rate, Fludd pushes ahead without regard for such medieval subtleties, and offers a *picture* of the Great Darkness. In a sense this *res picta* contains within it the main presupposition of Fludd's epistemology: The occult, the mysterious, the textually obscure can be depicted in images and thereby grasped.

Fludd's entire series of woodcuts visualizing the Book of Genesis cannot be reproduced here, but it is easy to understand that once the Dark has been visualized it is possible, in principle, to move to the first act of creation, the FIAT LUX.[39] His engraving shows the appearance of light as a gradual emanation from a central dark core. The impregnation of the dark *materia prima* by the divine light leaves the dark central cloud passive but surrounded on the periphery by the "active fire of love." The two opposing principles are then mediated by a Spirit in between. And from this Trinity – Light, Dark, and Spirit – the four elements emerge, and the struggle between opposing (Aristotelian) qualities begins (hot/cold; wet/dry). This conflict of opposites achieves resolution when the four elements settle into concentric rings around a central light – the inverse of Light's first appearance. A succession of further pictures completes the cosmogenesis with por-

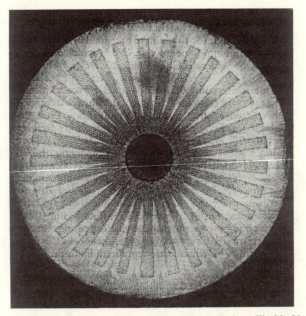

Figure 12. *In principio*: let there be light! (Robert Fludd, *Utriusque cosmi . . . historia* [Oppenheim, 1617–21])

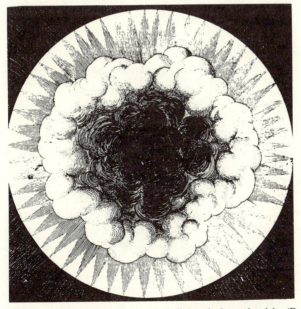

Figure 13. Emergence of the triad: light, dark, and spirit. (Robert Fludd, *Utriusque cosmi . . . historia* [Oppenheim, 1617–21])

Figure 14. Emergence and struggle of the four elements. (Robert Fludd, *Utriusque cosmi . . . historia* [Oppenheim, 1617–21])

trayals of the empyrean, ethereal, and elemental spheres, and the creation of visible celestial bodies. The details of this process are somewhat more involved than I have presented here, but the underlying principles are internally consistent throughout.

The Big Creation is the paradigm for all other creations. The principles that Fludd thinks he has found in Genesis can now be transferred as *explanans* to natural processes. Thus chemical operations[40] – and disquisitions on the human body[41] and the substance of wheat[42] (actually a discourse on the Eucharist) – all spell out the same terms of explanation: the interpenetrating pyramids, the triads emergent from the prior dualism (e.g., animal, vegetable, and mineral kingdoms), and the quaternaries following therefrom. This is obvious, by now, as a simple arithmetical sequence; namely, the sum of the first four integers. Fludd then has available to him nice possibilities for indulging in Py-

Figure 15. Resolution of strife: concentric settling of the four elements. (Robert Fludd, *Utriusque cosmi . . . historia* [Oppenheim, 1617–21])

thagorean speculations, and this is the basis for his extensive discourses on musical philosophy.[43] Since the Pythagorean tuning system consists of three perfect consonances – the octave = 2:1, the fifth = 3:2, and the fourth = 4:3 – *all* Fludd's musical pictures contain *only* the integers in the sum of the first four. In the Divine Monochord God's hand is shown reaching out to tune the Formal and Material Octaves. The monochord itself symbolizes God's unity, which makes possible the diversity and *symmetria* of Fludd's world harmony; it is also the central axis in Dürer's figures of human proportionality.

The imagination and the soul

If the Creation is picturable then it follows that whatever faculty has the capacity to make images of it must also have the ability

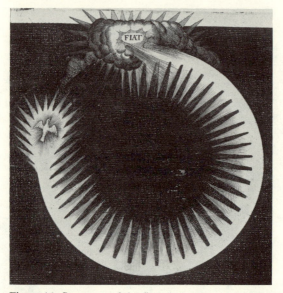

Figure 16. Summary of the first three days of the Creation, paradigm of all creations. (Robert Fludd, *Utriusque cosmi . . . historia* [Oppenheim, 1617–21])

to picture itself. And this, in fact, is what we find in a marvelous il-lustration of man's faculties. The Imaginative Soul links the Intellect and the Sensitive Soul by means of images. These images are "shad-ows" of the elements, that is, of the created world, and it is they that make possible the ground on which *Ratio* and *Intellectus* can be ex-ercised.[44] The soul itself is a unity although it possesses different fac-ulties. Fludd argues strenuously against Kepler that the soul does not come to know by "numbering," by dividing things into parts, but by searching out unities in the multiplicity of dark, occult experience, and it does so by creating pictures of unity.[45] As he puts it in his unpublished "Philosophicall Key": Knowledge of God is an ascent from "visible grosnes to visible subtilty by the degrees of nature," and "we must worke upon the invisible parts of man only with the eyes of contem-plation, for by it only must we learn to scale the blessed cone of unity by that central and most internal axeltree of the Pyramis."[46]

In concluding this section we may say that if Fludd had a strong interest in the created world of nature – perhaps much more so than preceding commentators on Genesis – his ultimate concern was still with Genesis itself. While nature itself sometimes threatens to become its own end, Fludd is clear that the purpose of studying the Creation is to retrieve the inner unity of the divine made possible in the original Creation.

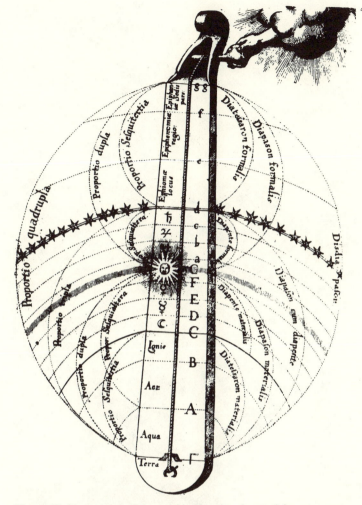

Figure 17. Monochord of the world: formal and material consonances brought into perfect Pythagorean harmony by the hand of God, "pulsator mundi." (Robert Fludd, *Utriusque cosmi . . . historia* [Oppenheim, 1617–21])

Kepler versus Fludd: from the mystery of the Creation to the cosmographic mystery

Although Kepler chides Fludd for indulging in "pictures forged from air," it would be a mistake to believe that Kepler assigns an insignificant role to pictorial representation. One of the most famous pictures in the history of science, after all, is Kepler's nest of poly-hedra, which purports to bear metaphysical, physical, and mathemat-

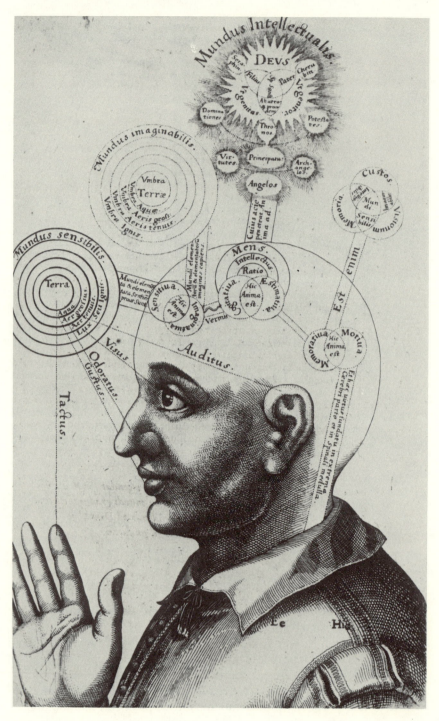

ical meanings.[47] To be precise, the polyhedra are images of God's archetypal Ideas, which he used to form the world in the best of all possible ways. Prior to these figures, both temporally and ontologically, came the image of God himself as a sphere (which Kepler makes no attempt to represent by pictures in the *Cosmographic Mystery*). But Kepler's and Fludd's Gods have *different* "archetypal scripts" for the world. Kepler's guiding image is a point that flows ceaselessly in all planes to form a sphere; Fludd's picture is a point that is the vertex of a pyramid of Light, and it is either mirrored symmetrically or opposed by an interpenetrating pyramid of Darkness.[48] Further, the two men were working from quite different kinds of texts: Fludd's moral and religious; Kepler's astronomical and physical. Not surprisingly, then, their hermeneutic preferences differ sharply. Fludd is always seeking moral, spiritual, and pietistic connotations in the geometry and arithmetic used to construct his pictures. Kepler's nose is always pointed in the opposite direction. The archetypal baggage he carries away from his deity unpacks into physical images and geometrical structures, whose purpose is to explain and to justify the Copernican world to which he is already committed. Kepler's pictures, as we shall see, must meet quite different standards in order to qualify as representations of reality.

One important ground for the Kepler–Fludd opposition lies in their differing use of epistemological authorities from the Platonic schools of late antiquity. Fludd is fond of Iamblichus and the Hermetic writings, but Kepler's greatest admiration was for Proclus. Kepler may well have possessed the latter's *Commentary on the First Book of Euclid's Elements* as early as the period during which he composed the *Cosmographic Mystery*.[49]

Proclus' work is a valuable source for the history and philosophy of Greek mathematics. As a Platonist, one of his central aims is to establish an epistemological foundation for "mathematicals," in opposition to Aristotle's notion that our mathematical ideas are both logically and temporally posterior to sensibles. In the middle of a long passage, cited in full by Kepler in his *Harmonics of the World*, Proclus expresses his main thesis in a beautiful image: "The soul was never a writing tablet bare of inscriptions; she is a tablet that has always been inscribed and is always writing itself and being written on by *Nous* . . . All mathematicals are thus present in the soul from the first."[50] Proclus argues for the truth of his position in the following way: Aristotle cannot

Figure 18. Man's faculties pictured: the world of images mediates between the world of the senses and the world of the intellect. (Robert Fludd, *Utriusque cosmi . . . historia* [Oppenheim, 1716–21])

be right when he claims that mathematicals are derived from sensibles, because nowhere in the sensible realm do we ever encounter an entity "without parts or without breadth or without depth" (i.e., a point). The unchanging and exact must come from some source that is of like nature to its subject. The soul is that source, "the generatrix of mathematical forms and ideas." But where does the soul get these ideas from? After all, the mathematical sciences may be superior to mere opinion in the stability of their ideas and their concern with immaterial objects, but they also make use of hypotheses that do not possess the certitude of the highest knowledge. The soul, then, is the *intermediary* between the perfect intelligible patterns of *Nous* and the imperfect, material sensibles. Certain patterns are innate in the soul and it, in turn, "projects" these patterns onto matter. In Kepler's usage, the soul "propagates" its patterns onto matter thereby in-forming it, embodying it in forms, yet also knowing what it embodies because it *matches* what is written on it by *Nous* with patterns of sensibles. This matching is what properly fills us with the conviction that the patterns in our soul are true of the sensible world.[51]

Now it is surely interesting to meet up with this sixth-century version of the correspondence view of truth, but there is a problem that Kepler is aware of: How do we know when we have more than an *arbitrary* correspondence between the soul's propagations and the configurations of sensible matter? And further, if the reality to which we are seeking a match is invisible, or if it occurred long ago – as long ago as the creation of the world – then how can we be sure that our visual imaginings are not dreams, poetry, or rhetoric? Kepler's answer is that the soul's picturing capacity is of a particular sort called *symbolisatio*.[52] Symbolization is an activity by which the soul matches intelligibles coeternal with God to sensibles. The image of the Trinitarian God is a point flowing forth in all dimensions to create the "symbolismus" of a sphere; and the sphere is, of course, the model of the world. This archetypal pattern is implanted in man's soul, which is like the diametral plane of a sphere. The soul is thus uniquely able to understand the created world by understanding the geometry of the sphere and the perfect solid figures inscribable within its bounds. It follows, then, that true symbolizations are geometrical shapes that possess archetypal properties of symmetry, equality, completeness, and so forth, and can be matched through measurement with relations between physicals. Such "fits" between intelligibles and structures that may be visible or invisible are true signs of divine intention and planning.

But what of the mind that merely pictures for its own sake? Kepler has a nice word for this: "play." In a marvelous passage from a letter of 1608, he writes:

> I too play with symbols and have planned a little work, Geo-
> metric Cabbalah which is about the Ideas of natural things in
> geometry; but I play in such a way that I do not forget that I
> am playing. For nothing is proved by symbols; things al-
> ready known are merely fitted [to them]; unless by sure rea-
> sons it can be demonstrated that they are not merely sym-
> bolic but are descriptions of the ways in which the two
> things are connected and of the causes of these connec-
> tions.[53]

Here Kepler is already trying to work out the distinction between
"playful picturing" – ideas subjectively pleasing to the senses and the
emotions, or beautiful images that appear before the mind – and "objective picturing," in which objects in the world are directly represented in the soul. The latter corresponds, in fact, to his definition of
visual perception: "Thus vision is brought about by a picture of the
thing seen being formed on the concave surface of the retina [*Visio
igitur fit per picturam rei visibilis ad album retinae et cauum parie-
tem*]."[54] As an astronomical epistemologist, Kepler wants to know, in
effect, what pictures have been formed on the surface of the "divine
retina." To the extent that our human eyes properly resolve the rays
coming in from the planets, we ourselves focus more clearly on the
divine picture. In short, the Dürer with whom Fludd identified in the
Four Books on Human Proportion allowed pictures to become vehicles
for inner images projected into the moral-spiritual space of the Genesis
text. Kepler, on the other hand, identified with Dürer as a geometer
and a geometrical optician whose interest is in radiations from the
visual field to the eye (for Kepler, the retina).[55]

This brief look at Kepler's epistemology provides a clue to his rhetoric in the Fludd polemic.[56] Fludd is accused of basing his harmonies
on "too many pictures," whereas Kepler's are based on "mathematical
demonstrations." Fludd's pictures are those familiar to what Kepler
calls "Chemists, Hermeticists and Paracelsians," while his own are
available only to "Mathematicians." Fludd appeals to the "Ancients";
Kepler to "Nature herself." Fludd's symbols are "Poetical and Ora-
torical"; Kepler's "Philosophical and Mathematical." If Kepler ex-
aggerates at all, it is *his own claim* to possess demonstrations, since,
as he admits in a less polemical context, the "philosophizing astronomer" has only conjectural knowledge of the divine ideas.[57] But apart
from this inflation of his own position Kepler's characterization of
Fludd is essentially on the mark. Kepler should be taken literally when
he says that Fludd's symbol pictures are poetical since they conform
to Aristotle's definition of metaphor in the *Poetics*: "the transposition
of a noun from its proper signification, either from the genus to the
species, or from the species to the genus; or from species to species,

or according to the analogous."[58] So in this sense, when Fludd speaks of spiritual pyramids or spiritual octaves he is transposing religious-philosophical nouns into predicates of mathematical and musical nouns, moving "emblematically" from the verbal to the visual. This is good for poets, Kepler allows, but not for philosophers. Aristotle explicitly tells us in his *Rhetoric* what he has already established in the *Posterior Analytics*, namely that necessary conclusions can be inferred only through syllogisms from universal premises. All other forms of reasoning can give us only probable knowledge and, at best, this can only be persuasive on such grounds.[59] By this standard, Fludd's claims relating pictures and words are *neither* probable *nor* persuasive; his unities are, in Kepler's words, "arbitrary." Thus, in the final passages of his Appendix, he writes:

> He [Fludd] picks out a few consonances and these he teases out from the interpenetration of his Pyramids which he privately carries around in his mind as a world drawn in pictures [*mundum pictum*], or he judges it [that world] to be represented by that [interpenetration of pyramids]. I have demonstrated that the whole corpus of tempered Harmonics is to be found completely in the extreme, proper motions of the planets according to measurements which are certain and demonstrated in Astronomy. To him, the subject of World Harmony is his picture of the world [*conceptus suus Mundi*]; to me it is the universe itself or the real planetary movements. From this brief discussion I think it is clear that, although a knowledge of the harmonious proportions is very necessary in order to understand the dense mysteries of the exceedingly profound philosophy that Robert teaches, nevertheless, the latter, who has even studied my whole work, will remain, for the time being, no less far removed from those perplexing mysteries than these [proportions] have receded [for him] from the accurate certainty of mathematical demonstrations.[60]

In Kepler's longer attack on Fludd, called the *Apology*, there are further articulations of this line of attack, which can now be read quite precisely: "I compared my diagrams to your pictures; I did not make my book as embellished as yours nor such as to appeal to the taste of the future reader; but I excused this defect on grounds of my profession, for I am a Mathematician." And: "He who is of a sort to seek out the mystical philosophy (which only treats things through riddles) wishes to feast the eyes on pictures; but he who seeks such things in my book shall not find them there."[61]

Turn now to the *disciplinary location* of the polemic. How are the five polyhedra to be classified in the matrix of the disciplines? The answer to this question again reflects interpretive differences. For Kepler the polyhedra are mathematical forms existing independently of matter, yet defining the distance *relations* between the sensible bodies of the planets. At the same time the polyhedra partake of divine intelligibility and purpose. We neither see the polyhedra nor hear the musical harmonics with the corporeal senses, but grasp them, instead, with the eye of the intellect. The physics of world harmonics then reduces to a discourse of purpose (metaphysics) and structure (mathematics). For Fludd, one speaks quite differently of polyhedral harmonies: Unlike Kepler, "in place of Geometrical words I use Physical words."[62] This physicalist language must be further explained: "I attribute my natural world harmony to forms which are actions hidden in matter [*ego vero formae actionibus occultis in materia mundana harmoniam meam naturalem attribuo*]."[63] The sense in which Fludd thinks that form "emerges" from matter is, of course, like the original account of Genesis; it is an act of creation, an image of *birth* in which the formal principle of light comes forth from the dark, shadowy womb of Chaos. This harmony of form and matter, Fludd insists, existed *before* the creation of the planets, and hence astronomy cannot take precedence in the hierarchy of disciplines. Fluddean, unlike Keplerian, physics reduces to a discourse of hermetic images (*erat umbra infinita in abysso*), Mosaic theology, and alchemy – images that are depicted by the divine painter.[64]

Such use of pictures to interpret Genesis – making, as it were, a cinematographic production of the Creation – touched a highly sensitive nerve in many of the great intellects of the seventeenth century, and may well explain the definitive reactions not only from Kepler but also from Mersenne and Gassendi. What the influence may have been on Descartes and Hobbes, whose natural philosophies pose interesting problems about the relationship between visibles and invisibles, has yet to be investigated.

The Jungian context of Pauli

The terms of Pauli's analysis will not be difficult to comprehend once we understand their Jungian origins. To begin with, no account of Jung's psychology would make sense without recognizing a central presupposition shared with Freud – the demarcation between consciousness and unconsciousness. "Everything of which I know," writes Jung,

> but of which I am not at the moment thinking; everything of which I was once conscious but have now forgotten; everything perceived by my senses, but not noted by my conscious mind; everything which, involuntarily and without paying attention to it, I feel, think, remember, want, and do;

> all the future things that are taking shape in me and will
> sometime come to consciousness: all this is the content of
> the unconscious."[65]

In one respect, Jung is here close to Freud: Painful feelings and
thoughts are intentionally removed from consciousness, and both the
intention itself and the memory of forgetting it are forgotten. Such
"contents" make up the unique, individual "personal unconscious"
of every individual.[66] But Jung departs from Freud in positing a substrate of universals in the unconscious. They are universal or "collective" in the sense that all human beings share them; they are not acquired by painful encounters with the world, but are rather inherited
characteristics of the species. The shared universals cannot be known
directly by an individual since they are not representable in themselves,
but are present rather like the axial system of a crystal which preforms
the crystalline structure in the liquid, a form without material existence.
Once materialized, these inner predispositions, urges, perceptual possibilities, or "archetypes," as Jung calls them, become available to
consciousness. They appear as recurring motifs in myths, fairytales,
dreams, and fantasies as images of powerful emotional force. Jung's
theory of archetypes as image formers clearly reminds us of Plato's
forms, Kepler's and Fludd's archetypes, and Bergson's *élan vital*. Yet
unlike the ancient and Renaissance formulations, Jung locates the archetypal ideas not in the divine *mens* or in an eternal realm but in the
human psyche. The study of the unconscious, then, becomes largely
a study of archetypal symbolism as manifested in the dreams and fantasies of individuals and in the literary, artistic, and scientific products
of human cultures – a kind of poetics and rhetoric of the psyche.

Among these archetypal symbols Jung places special emphasis on
the mandala or magic circle. This symbol is a representation of the
self, the totality of the personality that encompasses both the conscious
and the unconscious psyche. Other important symbols of the self are
the square and cross. Such quaternian structures represent the logic
of a complete judgment, much like Aristotle's four causes or the four
modes of biblical exegesis mentioned earlier (p. 191). Indeed, Jung
insists that the prevalence of four as a motif in Greek physics (four
elements), astronomy (four cardinal points), Eastern mysticism (four
ways of spiritual development), and Christianity (the three evangelists
plus Saint Luke) cannot be accidental. Four is the conscious structure
of full psychic functioning:

> In order to orient ourselves, we must have a function which
> ascertains that something is there (sensation); a second func-
> tion which established *what* it is (thinking); a third function
> which states whether it suits us or not, whether we wish to
> accept it or not (feeling), and a fourth function which indi-

cates where it came from and where it is going (intuition).
When this has been done, there is nothing more to say . . .
The ideal of completeness is the circle or sphere, but its nat-
ural minimum division is a quaternity.[67]

If the quaternity is particularly significant because it represents the
goal of life, the full expression of the four aspects of the self's indi-
viduality, it follows, then, that where one or more parts of the self are
denied expression in conscious life these parts form a lower, incompatible, repressed, "occulted" part of the personality. The "shadow,"
as Jung terms it, consists not only of tendencies that are antisocial and
morally reprehensible, but also of positive qualities: "normal instincts,
appropriate reactions, realistic insights, creative impulses, etc."[68] Individuation, which Jung considered "the central concept of my psychology,"[69] is a process whereby the different selves are brought more
and more together into a coherent center, the shadow is steadily de-
creased by illumination, the person comes to be aware of the wholeness
of his or her self.

Jung's quadripartite theory of psyche functions was complemented
by a psychology of consciousness, of "attitude types."[70] All the judgments we make about people and things are relative to the cluster of
attitudes, the habitual reactions, the styles of behavior that dispose us
to concentrate either outward onto external objects (extroversion) or
inward onto subjective psychic objects (introversion). Each of the four
psychic functions will be directed or experienced in an inward or an
outward way, depending upon the personality type. And the general
cast of the personality will be determined by which psychic functions
predominate: for example, an extroverted feeling type, an introverted
thinking type, and so on.

At first glance, many of Jung's writings would appear to be of enormous interest and relevance to historians of science and ideas. With
monumental erudition his books integrate perspectives from mythology, folklore, history of religion, alchemy, and ethnology. But Jung's
work has failed to capture the sympathy of either internalist or externalist historians because his aim is an "introverted" one (somewhat
like Fludd's ubiquitous discoveries of interpenetrating pyramids),
namely, to study religious, alchemical, and early scientific texts for
evidence of archetypal patterning. In much of Jung's writing culture
is a *repository of the universal*, and its purpose is thus to deepen the
hermeneutics of the self. No wonder that his discourse is an amalgam
of early philosophical and religious languages: A Gnostic discourse of
equal and opposite principles, Light and Dark; a neo-Pythagorean discourse of circles, squares, triangles, and special numbers; a Christian-
Platonic terminology of God images and archetypes; an Eastern mystical terminology of mandalas and mana; an alchemical language of spir-

itual weddings and mysterious unions of opposites. This rhetoric of the psyche allows Jung to "amplify" meanings by passing back and forth in search of parallels between the inner images of dreams and the outer images of alchemy and theology, in quest of "bridges" and "doors" to the unconscious psyche.

Pauli pursues just such a strategy of amplification in his analysis of Kepler and the polemic with Fludd. As he tells us in the opening paragraph of his essay:

> Although the subject of this study is an historical one, its purpose is not merely to enumerate facts concerning scientific history or even primarily to present an appraisal of a great scientist, but rather to illustrate particular views on the origin and development of concepts and theories of natural science in the light of one historical example.[71]

Exactly what "particular views" does Pauli wish to illustrate? "Many physicists," he writes, "have recently emphasized anew the fact that *intuition* and the *direction of attention* play a considerable role in the development of the concepts and ideas, generally far transcending experience" (my italics). These intuitions are described variously by Pauli as "inner images preexistent in the human psyche," "primordial images," "instincts of imagination," a "preconscious, archaic level of cognition," "images with strong emotional content, not thought out but beheld, as it were, while being painted," "ordering operators," "image formers," "symbolical images," and, of course, "archetypes."[72]

Pauli can now move effortlessly from a psychogenic account in the context of discovery to an ontogenic recapitulation in the history of science. But it is important to see that his use of history is quite the opposite of Jung's, proceeding from general premises about the unconscious sources of human cognition to particular propositions about phases in scientific change and explanations for the holding of particular beliefs. Hence, the supposed phase of "modern, quantitative-mathematical descriptions of nature" is preceded by a "magical-symbolical description of nature," and in Kepler both stages are clearly visible as a "remarkable intermediary stage."[73] The archetypal symbolism that lies behind and orders all Kepler's physical theories is the famous image of the trinitarian Christian godhead: a sphere in which the image of God the Father is in the center, the Son in the outer surface, and the Holy Ghost in the ever-equal relation between circumference and central point. This is Kepler's *mandala*, which, as Pauli is quick to point out, lacks any hint whatsoever of the quaternity. The elements of Kepler's mandala are three – point, radius, and surface; motion is directed away from the center in a straight line.[74] The heliocentric universe thus becomes the "bearer of the mandala-picture, the earth being related to the sun as is the ego to the more embracing self."[75] And then Kepler's undeniably religious fervor in the heliocentric system is a *direct consequence* of the background archetypal image and "by no means the other way around, as a rationalistic view might cause one erroneously to assume."[76] Finally, in Pauli's view, it was precisely this extroverted trinitarian mandala symbolism that Kepler brought to full consciousness as a *way of thinking* and "produced that natural science which we today call classical."[77]

In Robert Fludd Pauli found an intellectual "counter world" to Kepler. Intuitive and feeling functions predominate, and qualitative hieroglyphic pictures "preserve the unity of the inner experience of the observer."[78] It is as though Fludd's pictures, which *appear* to be about nature, are really pictures of psychic states; they are visualizations of intuitions and feelings projected onto the world, but lacking any sufficient criterion of correspondence to an external reality. Their internal coherence as images is the only warrant offered of correspondence to nature. And the archetypal symbolism predominant in Fludd is "fourness." Pauli singles out for special treatment in an Appendix a passage from Fludd's 1621 reply to Kepler, sections of which I quote here:

> Here the dignity of the quaternary number will be discussed and I shall defend it with might and main as far as my weak intellect allows, spurred on by the insolence of the author [Kepler] . . . Sacred theology extolled the paramount superiority of this number above all others . . . this quadratic number is likened to God the Father in whom the mystery of the whole sacred Trinity is embraced . . . Indeed all nature can be comprehended in terms of four concepts: substance, quality, quantity and motion. In fine, a quadruple order constantly pervades the entire nature, namely seminal force, natural growth, maturing form and the compost. By this we can clearly demonstrate that this number 4 should rather be chosen to distinguish and divide the humid (primal) matter than the number 3 or the number 5.[79]

For Pauli, Fludd's wholeness of contemplation in his inner life meant that he paid a price in being unaware of the quantitative side of nature and its laws. In Kepler, on the other hand, the thinking function predominates and thus his images of both inner and outer reality are expressed in terms of the *measurability* of nature and psyche. The immeasurable side of experience, the imponderables of the emotions, are less conscious. Thus the stage is set for a "collision" of two worlds or an opposition between two types of minds – the one considering quantitative relations between parts to be crucial; the other experiencing the qualitative indivisibility of the whole as central.

The resolution of the deep opposition between trinitarian/quantitative and quaternian/qualitative thinking is the true theme of the two essays that make up *The Interpretation of Nature and the Psyche*. As a historical-psychological-epistemological thesis, Pauli contends that the trinitarian symbolism of seventeenth-century classical physical science – with its categories of space, time, causality, and its emphasis on the *measurable* side of experience – created a world of psychic incompleteness. In quantum physics, however, that absolute deification of measurement was dethroned:

> There is a basic difference between the observers, or instruments of observation, which must be taken into consideration by modern microphysics, and the detached observer of classical physics . . . whose influence can always be eliminated by determinable corrections. In microphysics, however, the natural laws are of such a kind that every bit of knowledge gained from a measurement must be paid for by the loss of other, complementary items of knowledge.[80]

Physical nature, previously considered to be an "objective order," was now relativized with respect to the means of observation. The nineteenth-century psychophysical program of reducing psychic phenomena to quantitative measurements was subject to the same limitation.[81] By contrast, Pauli writes, Jung's psychology made possible the knowledge of "an unconscious psyche of considerable objective reality" through methods of introspection and amplification.[82] In his companion essay entitled "Synchronicity: An Acausal Connecting Principle," Jung proposed a new quaternian schema – a kind of quantum interpretation of the psyche – "thanks to the friendly interest which Professor Pauli evinced in my work."[83] Now the archetypes are interpreted as psychic probabilities, an "acausal orderedness" in which psychic and physical events coincide in a meaningful way without causal connection, as when, for example, a dream, vision, or premonition corresponds to some external reality.[84] In the end the goal of the Jung–Pauli formulation is self-evident: Physical and psychic ordering-schemas are *complementary* aspects of the same reality, rather than an opposition, as exemplified in the historical clash between Kepler's and Fludd's pictures.[85]

The Jung–Pauli relationship: a final note

The Jung–Pauli complementarity thesis at once raises many historical questions regarding the existence of a *personal* "complementarity" in their relationship. When and where did they work together? How and why were each other's views of psychology, physics, and history affected? Was Pauli ever in analysis with Jung? Full answers are beyond the scope of the present chapter and must remain tantalizing issues for future research. But I am able here to shed a bit of light on this fascinating relationship thanks to a few clues that are available.

First, we know that both men were deeply concerned with problems of *visualization*. "For almost twenty years, I have occupied myself with the psychology of the pictorial representation of psychic processes," wrote Jung in a 1932 commentary on the work of Picasso.[86] And further on:

> Those pictorial elements which do not correspond to any "outside" must originate from "inside." As this "inside" is invisible and cannot be imagined, even though it can affect consciousness in the most pronounced manner, I induce those of my patients who suffer mainly from the effect of this "inside" to set them down in pictorial form as best they can. The aim of this method of expression is to make the unconscious contents accessible and so bring them closer to the patient's understanding. . . . In contrast to objective or "conscious" representations, all pictorial representations of processes and effects in the psychic background are *symbolic* . . . The possibility of understanding comes only from a comparative study of many such pictures.[87]

Here we see quite the opposite of what we have seen in the Renaissance. Man, the subject, no longer imitates the external object; now inner, invisible "objects" assume priority in pictorial representation. Physics, stood on its head, becomes psychology; art becomes therapy.

Wolfgang Pauli, too, was profoundly involved with visualization in his professional work during the period of the genesis of the quantum theory (although a comprehensive analysis of his physical work is not here proposed). Arthur I. Miller has characterized the period 1913–27 in physics with the perceptive phrase "visualization lost and regained." He cites a number of poignant phrases from major participants engaged in creating a consistent account of atomic phenomena. Thus Niels Bohr in 1929 spoke of the "conscious resignation of our usual demands for visualization"; and Pauli, looking back on this period from 1955, referred to "a brief period of spiritual and human confusion caused by a provisional restriction to *Anschaulichkeit*," that is, visualization through pictures or mechanical models.[88] But, following the characterizations of Heisenberg, Miller believes that the times were even more extreme than Bohr's and Pauli's phrases would lead one to believe. It was a situation of "despair and helplessness because of their loss of visualization and of their distrust in customary intuition. It was a period when such time-honored concepts as space, time, causality, substance and the continuity of motion were separated painfully from their classical basis."[89] The physical world lost touch with common sense, and the planetary electron became an unvisualizable entity.

The central theme of this chapter – pictures, texts, and things – now connects directly with a second concern of Jung and Pauli: opposites and complementarity. For Jung the problem of opposites surfaces explicitly during the period of his break with Freud in the theory of psychological types. "This work [*Psychological Types*]," wrote Jung, "sprang originally from my need to define the ways in which my outlook differed from Freud's and Adler's. In attempting to answer this question, I came across the problem of types; for it is one's psychological type which from the outset determines and limits a person's judgment."[90] In Jung's framework, Freud's perspective was extroverted because it placed the focus on external objects and events; Adler's was introverted because it emphasized the "will to power."[91] Among the many historical influences on the formation of his typology is another figure critical to our story, William James. James, wrote Jung, was "the first to draw attention to the extraordinary importance of temperament in colouring philosophical thought. The whole purpose of his pragmatic approach is to reconcile the philosophical antagonisms resulting from temperamental differences."[92] As James himself put it:

> Of whatever temperament a professional philosopher is, he tries, when philosophizing, to sink the fact of his temperament . . . Yet his temperament really gives him a stronger bias than any of his more strictly objective premises. It loads the evidence for him one way or the other, making for a more sentimental or a more hard-hearted view of the universe, just as this factor or that principle would. He trusts his temperament. Wanting a universe that suits it, he believes in any representation of the universe that suits it. He feels men of opposite temper to be out of key with the world's character, and in his heart considers them incompetent and "not in it," in the philosophic business, even though they may far excel him in dialectical ability.
>
> Yet in the forum he can make no claim, on the bare ground of his temperament, to superior discernment or authority. There arises thus a certain insincerity in our philosophic discussions; the potentest of all our premises is never mentioned.[93]

James proceeded, then, to construct a table of opposites as follows:

Tenderminded	*Toughminded*
Rationalistic	Empiricist
(going by "principles")	
Intellectualistic	Sensationalistic
Idealistic	Materialistic

Optimistic	Pessimistic
Religious	Irreligious
Free-willist	Fatalistic
Monistic	Pluralistic
Dogmatical	Sceptical

Jung does not criticize the strategy of typologizing, but his criticisms illuminate his own attitude toward the problem of opposites. The main deficiency is that James's categories are restricted to the level of the thinking function: They are about the quality of "mindedness." This results in an excessive one-sidedness that fails to allow for a sufficient degree of balance in the schema. Thus, one might have dogmatic or religious empiricists or tender-minded types who are fatalistic or sceptical. Just as James's categories are constructed exclusively from "thinking" qualities, so his solution to their opposition is too intellectualistic. Pragmatism, which interpreted "truth" in terms of its practical efficacy and usefulness, could serve, at best, only as "a transitional attitude preparing the way for the creative act by removing prejudices."[94] In 1921 Jung proposed a biological image: "The solution of the conflict of opposites can come . . . only from a positive act of creation which assimilates the opposites as necessary elements of co-ordination, in the same way as a co-ordinated muscular movement depends on the innervation of opposing muscle groups."[95] By the end of his life, in a book that Jung considered to be the crowning insight of his entire work (*Mysterium conjunctionis,* 1962), he moved from biology to alchemy, assembling a massive quantity of information from ancient texts in which were displayed symbols of separation, opposition, and syntheses of opposites: Rex and Regina, Adam and Eve, Sun and Moon, and, then, *unio mentalis, nirdvandva,* the *unus mundus,* and *conjunctio.*[96] Alchemy became Jung's royal road to the archetypal unconscious.

If James was the principal subtext for Jung's account of opposites, his importance for Niels Bohr's thought was similarly remarkable. In a brilliant essay exploring the roots of complementarity in Bohr's work, Gerald Holton calls attention to some critical passages in William James's *The Principles of Psychology* (1890)[97] which provide striking analogies to Bohr's own formulations. James writes: "Consciousness does not appear to itself chopped up in bits; it flows. Let us call it the stream of thought, of consciousness, or of subjective life." And later: "Like a bird's life, [thought] seems to be made of an alternation of flights and perchings. The rhythm of language expresses this, where every thought is expressed in a sentence and every sentence closed by a period . . . Let us call the resting places the 'substantive parts,' and the places of flight the 'transitive parts,' of the stream of thought." And again, in a remarkable image: "The attempt at introspective analysis . . . is . . . like . . . trying to turn up the light quickly enough to see how the darkness looks." It is as though, with Heraclitus, one cannot step into the same river twice; the flowing of thoughts and the introspective bracketing or analyzing of thoughts are, as Holton puts it, like "two mutually exclusive experimental stations."[98] Shortly after these passages, James explicitly formulates a concept of complementarity:

> It must be admitted, therefore, that in *certain persons*, at least, *the total possible consciousness may be split into parts which coexist but mutually ignore each other*, and share the objects of knowledge between them. More remarkable still, they are *complementary*. Give an object to one of the consciousnesses, and by that fact you remove it from the other or others. Barring a certain common fund of information like the command of language, etc., what the upper self knows the under self is ignorant of, and *vice versa*.[99]

There is excellent evidence, assembled by Holton from various sources – including an interview with Bohr by Thomas Kuhn and Aage Petersen the day before he died! – that Bohr considered James to be one of his favorite authors.[100]

Now Jung certainly knew James's *Principles of Psychology* as early as 1902, when he wrote his doctoral dissertation ("On the Psychology and Pathology of So-Called Occult Phenomena"), and he met James at the Clark University conference in 1909. But nowhere does he make reference to James's important notion of complementarity.[101] We may speculate that he was not yet "ready" to see it, that he saw only "opposites" in James at a time when he was experiencing the break with Freud. It was only much later, in conversations with Pauli – apparently after the war – that Jung began to consider the resolution of opposites in the language of complementarity.[102] And it seems fair to assume that Pauli's first encounter with complementarity came through Bohr's formulation. For Pauli, then, the central move was from *physis* to *psyche* and *historia*, a *rediscovery of the psychological meaning of complementarity*.[103]

How exactly Pauli first encountered Jung is not clear at present. We know that Pauli held the professorship in theoretical Physics at the ETH from 1928 until his death in 1958.[104] Jung was professor at the same institution from 1935 until 1943, when he became professor at Basel.[105] But there is some likelihood that the relationship may have begun as early as 1932, when Jung began to attend sessions of the famous Eranos discussion group, which had been formed in the 1920s by Frau Olga Froebe-Kapteyn in the grounds of her residence at the northern end of Lake Maggiore near Ascona, Switzerland.[106]

Sometime between 1932 and 1935, Jung delivered two lectures to the Eranos group entitled "Dream Symbols of the Process of Individuation" and "The Idea of Redemption in Alchemy." These were published in the *Eranos-Jahrbuch* for 1935 and 1936, respectively.[107] Subsequently the two essays were expanded, an Introduction and Epilogue added, and the entire work was issued in 1944 under the title *Psychologie und Alchemie* (Volume XII of the *Collected Works*). The centerpiece of this study was Jung's analysis of the symbol of the mandala. This account was based upon the study of a series of some 400 dreams, all from the reports of one dreamer. The dreamer was not in analysis with Jung. As he tells us: "It goes without saying that while the dreamer was under the observation of my pupil he knew nothing of these interpretations and was therefore quite unprejudiced by anybody else's opinion."[108] Elsewhere Jung reiterated his own noninvolvement in the direct collecting of the dream materials: "I even refrained from observing this particular case myself and entrusted the task to a beginner who was not handicapped by my knowledge – anything rather than disturb the process. The results which I now lay before you are the unadulterated, conscientious and exact self-observations of a man of unerring intellect, who had nothing suggested to him from outside and who would in any case not have been open to suggestion."[109] From one of Jung's biographers, Frieda Fordham, we are able to learn a bit more about the dreamer and his analyst:

> A young intellectual . . . had come to Jung with a severe
> neurosis. An interesting point is that this young man was
> only seen by Jung for a short interview, after which he re-
> corded his dreams and visual experiences for five months
> with a pupil – a woman doctor who was then a beginner –
> and then continued his observations alone for another three
> months.[110]

Finally, we have it from Jung's own account that only 59 of the 399 dreams were actually used in the construction of his account of the "psychic fact" of the mandala, "because the dreams touch to some extent on the intimacies of personal life and must therefore remain unpublished. So I had to confine myself to the impersonal material."[111]

Two months after the end of this conference, as I worked on the revisions of this chapter, a colleague from my department at UCLA, Professor Peter Loewenberg, informed me that he planned a visit to the Jung Institute in connection with his own research. I asked him if he would mind inquiring there about "anything concerning the Jung–Pauli relationship." Loewenberg generously obliged and relayed to me the following extraordinary information. On 14 September 1982, he learned from Jung's successor at the ETH, Professor C. A. Meier, that Pauli had been in psychoanalysis with a woman analyst named Rosenbaum, a pupil of Jung. Three days later, he interviewed Jung's former secretary, editor, and biographer, Aniela Jaffé, and learned that the mandala dreams were those of Wolfgang Pauli![112]

Even without any other explicit personal information, the historian is now in an unusually privileged position to make sense of certain creative features of the Jung–Pauli relationship. The putatively archetypal material from Pauli's dreams constituted the main evidence on which Jung built the case for his public, scientific validation of the mandala as a symbol of the self. That the dreams came from a man of demonstrated brilliance in scientific research would have encouraged Jung in his conviction that what he had in Pauli's dreams was *independent confirmation* of what he had found elsewhere in other clinical material. Second, the material allowed Jung to move beyond the "problem of neurosis in puberty," which had constituted the substance of his interest in *Symbols of Transformation* (1911–12), and to approach what he called "the broader problem of individuation."[113]

If we now look at the dreams and their interpretations for evidence of themes that later appear in Pauli's essay, we shall not be disappointed. Jung's text is liberally interspersed with alchemical and cosmological *pictures* that he chose as independent, comparative material to show visual counterparts to the dreamer's reports. The picture that appears above the section entitled "The Initial Dreams" is none other than Fludd's famous – and nowadays often reproduced – representation of the cosmos. It shows man, the ape of nature, seated on the earth, in his left hand a golden chain linking him to the female figure of the *anima mundi*, and her right hand, in turn, linked to the hand of God. In Dream 40, Jung offers the following interpretation: "The idea of the *anima mundi* coincides with that of the collective unconscious whose centre is the self. The symbol of the sea is another synonym for the unconscious."[114] In his essay Pauli argues that for Kepler, the *anima mundi* is "no more than a kind of relic" in contrast to the "magical symbolical attitude" of Fludd, who is a feeling-intuitive type.[115] Pauli must have believed that Fludd's pictures represented symbols of the collective unconscious and the self, and that by studying Fludd he was gaining access to the *Fluddean part of himself*. Similarly, and by contrast, he portrays Kepler's conception of the soul "almost as a mathematically describable system of resonators."[116] One cannot help being reminded here of the powerful opposition between Heisenberg's nonvisual, mathematical equations and Schrödinger's "intuitive pictures [*anschaulichen Bilder*]" or Bohr's "visual ideas [*Vorstellungen vor Augen*]"[117] Put in Jungian terms: The crisis of visualization in quantum mechanics apparently resonated with the conflict in Pauli between the "feminine," intuitive, emotional, picturing part of himself and the "masculine," measuring, quantifying, critical part. Those who knew

Pauli observed that he had a hypercritical streak in him. Markus Fierz describes him as one who

> radiated a very strong personal force. One was immediately impressed by his sharp and critical judgment. In discussions he was in no way willing, and perhaps completely unable, to accept unclear formulations. He seemed hard to convince, or he reacted in a sharply negative manner . . . [and there was] his often caustic way of jumping at his discussion partner, which put many into disarray.

At Pauli's death Viktor Weisskopf characterized him as "the conscience of theoretical physics." Yet when Pauli walked into the laboratory all kinds of "misfortunes" would occur, so that some joked about the "Pauli effect."[118]

In Dream 59 of Jung's *Psychology and Alchemy*, the World Clock Dream, the dreamer reported "the most sublime harmony." Here is the dream:

> There is a vertical and a horizontal circle, having a common centre. This is the world clock. It is supported by the black bird.
>
> The vertical circle is a blue disc with a white border divided into $4 \times 8 = 32$ partitions. A pointer rotates upon it.
>
> The horizontal circle consists of four colours. On it stand four little men with pendulums, and round about it is laid the ring that was once dark and is now golden (formerly carried by the children).
>
> The "clock" has three rhythms or pulses:

1	The small pulse:	the pointer on the blue vertical disc advances by 1/32.
2	The middle pulse:	one complete revolution of the pointer. At the same time the horizontal circle advances by 1/32.
3	The great pulse:	32 middle pulses are equal to one revolution of the golden ring.[119]

Jung's analysis of the impersonal, archetypal content in this dream is extraordinarily complicated and his "parallels" often difficult to follow. But this text, we now know, would have had very special personal meaning for Pauli. In this respect the following piece of Jungian interpretation is quite significant for our problem in this chapter because it brings together in a single Jungian alchemical image all the themes we have found in Pauli's historical work: "We shall hardly be mistaken if we assume that our mandala aspires to the most complete union of

opposites that is possible, including that of the masculine trinity and
the feminine quaternity on the analogy of the alchemical hermaphrodite."[120]

Acknowledgments

For their supportive comments and suggestions, I wish to thank Rachel N.
Klein, J. E. McGuire, Irenka Taurek, and Brian Vickers.

Permission to reproduce illustrations was kindly granted by the following
institutions: for Figures 1–6, 7a, 8a, and 9–18, UCLA Biomedical Library;
for Figures 7b and 8b, The Huntington Library, San Marino, California.

Notes

1 The entire volume was originally published as *Naturerklärung und Psyche*,
Studien aus dem C. G. Jung-Institut, IV (Zurich, 1952), and appeared in
revised form in 1955 in the English translations of R. F. C. Hull (Jung) and
Priscilla Silz (Pauli). Pauli informs the reader that he consulted Jung and
(his present successor) C. A. Meier on the psychology of scientific
discovery (p. 149).
2 Among historians, see especially Gerald Holton, "Johannes Kepler's
Universe: Its Physics and Metaphysics," in *Thematic Origins of Modern
Science* (Cambridge, Mass., 1973; first pub. *American Journal of Physics*,
24 [1956]; pp. 340–51), p. 82; Allen G. Debus, *The Chemical Philosophy:
Paracelsian Science and Medicine in the Sixteenth and Seventeenth
Centuries*, 2 vols. (New York, 1977), I, 256–60; Frances A. Yates,
Giordano Bruno and the Hermetic Tradition (London, 1964), pp. 442ff. For
Jungian writers, see Aniela Jaffé, *From the Life and Work of C. G. Jung*,
trans. R. F. C. Hull (New York, 1971; first pub. 1968), p. 43; Vincent
Brome, *Jung: Man and Myth* (London, 1978), pp. 289, 291; Barbara
Hannah, *Jung: His Life and Work* (New York, 1976), p. 305, quoting
another Jung associate, Marie-Louise von Franz (*Number and Time*
[Evanston, 1974], pp. 6ff.): "The chance to combine his paper on
synchronicity with Wolfgang Pauli's work on Kepler was . . . exceedingly
welcome as he hoped it would make scientists take this new idea more
seriously."
3 Robert Fludd, *Utriusque cosmi maioris scilicet et minoris metaphysica,
physica atque technica historia* (cited hereafter as *UCH*). The publication
sequence of Fludd's work was not intended to ease the lives of librarians.
We shall use abbreviation conventions following the useful and accessible
work of Joscelyn Godwin (*Robert Fludd: Hermetic Philosopher and
Surveyor of Two Worlds* [London, 1979], p. 93):

UCH I,a: *Tomus primus de macrocosmi historia* (Oppenheim, 1617)
UCH I,b: *Tractatus secundus de naturae simia seu technica macrocosmi historia in partes undecim divisa* (Oppenheim, 1618)
UCH II,a,1:   *Tomus secundus de supernaturali, naturali, praeternaturali
et contranaturali microcosmi historia in tractatus tres dis-
tributa* (Oppenheim, 1619)
UCH II,a,2:   *Tomi secundi tractatus primi sectio secunda, de technica microcosmi historia in portiones VII divisa* (?Oppenheim,
?1620)

UCH II,b: *Tomi secundi tractatus secundus; de praeternaturali utri-usque mundi historia in sectiones tres divisa* (Frankfurt, 1621)

UCH I,a and *UCH* I,b = vol. I; *UCH* II,a,1 and 2 and *UCH* II,b = vol. II; tractatus II, sec. 2, and tractatus III never appeared.

Johannes Kepler, *Harmonice mundi* (Linz, 1619), in Kepler, *Gesammelte Werke*, vol. VI, ed. Max Caspar (Munich, 1940). Cited hereafter as *GW*.

4 On Fludd's travels and the reference to his social status, see J. B. Craven, *Doctor Robert Fludd (Robertus de Fluctibus); The English Rosicrucian: Life and Writing* (Kirkwall, 1902; repr. New York, n.d.), pp. 24–5; C. H. Josten, "Robert Fludd's Theory of Geomancy and His Experiences at Avignon in the Winter of 1601 to 1602," *Journal of the Warburg and Courtauld Institutes*, 27 (1964); pp. 327–55. On the Rosicrucian problem, see Frances Yates, *The Rosicrucian Enlightenment* (London, 1972); Debus, I, 206ff.; Serge Hutin, *Robert Fludd (1574–1637): alchimiste et philosophe rosicrucien* (Paris, 1971), pp. 44ff.

5 The Fludd–Maier connection is by no means certain. My statement is a conjecture based upon Fludd's and Maier's common connection with the De Bry press and Maier's presence in London sometime before 1618. J. B. Craven, Maier's biographer, hardly inspires confidence when he writes: "But the most distinguished friend in England whom Maier had was the famous Doctor Robert Fludd. How they became acquainted we do not know, but it appears that when in England Maier 'lived on friendly terms' with Fludd. It is said that it was at Maier's instigation Fludd wrote, or at least published, in 1617 his most excellent 'Tractatus Theologo-Philosophicus,' dedicated to the Brethren of the Rosy Cross. We are told that Maier, having become a member of this mysterious order, admitted Fludd to its privileges when in England. The whole matter is, however, buried in obscurity, if not in contradiction" (*Count Michael Maier* [Kirkwall, 1902], pp. 6–7). This murkiness in the historical account is hardly improved by Craven's failure to cite supporting references. Frances Yates has some interesting remarks about Maier's alchemy, but her attempts to connect Fludd and Maier through Count Frederick V, elector of the Palatinate, are equally speculative (*Rosicrucian Enlightenment*, chap. VI).

6 Frances Yates has argued that the period between the Renaissance and the so-called scientific revolution can be viewed as a "Rosicrucian Enlightenment" (chap. VI). For a trenchant and telling attack on her proposal and on the historical methods employed to substantiate it, see Brian Vickers, "Frances Yates and the Writing of History," *Journal of Modern History*, 51 (1979), pp. 287–316.

7 The entire work is translated with a fine accompanying analysis by William H. Huffman and Robert A. Seelinger, Jr., "Robert Fludd's 'Declaratio Brevis' to James I," *Ambix*, 25 (1978), pp. 69–92.

8 Ibid., p. 82.

9 Robert Fludd, *Veritatis proscenium; in quo aulaeum erroris tragicum dimovetur, siparium ignorantiae scenicum complicatur, ipsaque veritas a suo ministro in publicum producitur, seu demonstratio quaedam analytica* (Frankfurt, 1621); the passages in Kepler are from *GW*, VI, 373–7, 556.

10 Johannes Kepler, *Pro suo opere harmonices mundi apologia adversus demonstrationem analyticam CL. V. D. Roberti de Fluctibus medici oxoniensis: in qua ille se dicit respondere ad appendicem dicti operis* (Frankfurt, 1622), in *GW*, VI, 383–457; Robert Fludd, *Monochordum mundi* (Frankfurt, 1623).

11 Kepler, *GW*, VI, 374, 396–9, 431, 446; Fludd, *Veritatis proscenium*, pp. 5, 12, 13, 36.
12 See Keith Hutchison, "What Happened to Occult Qualities in the Scientific Revolution?" *Isis*, 73 (1982), pp. 233–53.
13 Peter Ammann, "The Musical Theory and Philosophy of Robert Fludd," *Journal of the Warburg and Courtauld Institutes*, 30 (1967), p. 205.
14 Pauli, p. 205.
15 Frances Yates, *Theatre of the World* (Chicago, 1969), p. 74. Yates's speculations about Fludd in this work are much more accurate, I think, than in her other works.
16 Godwin, *Robert Fludd*.
17 The disciplines are presented in a Ramist-style classification scheme, and pictorially in the sections of a zodiacal wheel (*UCH* I,a, pp. 3–6). Owen Hannaway has made an interesting and coherent attempt to differentiate Paracelsian alchemy in the work of the Rudolphine Prague physician Oswald Croll (ca. 1560–1609) from the *Alchymia* of Andreas Libavius (ca. 1550–1616), teacher at the Coburg Gymnasium. Hannaway suggests that Libavius's critical move in the creation of the discipline of chemistry was his didactical rearranging of alchemical terms such that the key reference words could be tied down in a unique system of classification, thereby ridding them of "additional" symbolic meanings: "Where before there was echo, affinity, sympathy," he writes, "now there is definition, division and distinction" (*The Chemists and the Word: The Didactic Origins of Chemistry* [Baltimore and London, 1975], pp. 142–9, esp. pp. 148–9). One difficulty with this thesis is that we find in Fludd, surely a "Crollian," the same impulse to Ramist methodizing that we find in the allegedly revolutionary schoolmaster Libavius. The critical differences at issue here may reside less in the relationship between Ramist classification trees and text than in the controlling function of the *pictures* as both representations and interpretations of the text.
18 Fludd, *UCH* I,b, pp. 317–41 ("De arte pictoria").
19 Ibid., p. 320.
20 Cf. the influential emblematic handbook of Cesare Ripa (?1560–?1623) entitled *Iconologia overo descrittione dell'imagini universali cavate dall'antichita et da altri luoghi . . . opera non meno utile, che necessaria a poeti, pittori, & scultori, per rappresentare le virtu, vitij, affetti, & passioni humane* (Rome, 1593); an excellent modern English-language edition of the eighteenth-century German edition by Johann Georg Hertel is available in paperback (*Baroque and Rococo Pictorial Imagery: The 1758–60 Hertel Edition of Ripa's "Iconologia,"* introduction, translations, and 200 commentaries by Edward A. Maser [New York, 1971]). As Maser indicates, one of the great questions of the Renaissance was the *ut pictura poesis* problem: "whether poetry, 'written imagery,' or painting, 'depicted poetry,' came first" (p. viii); see also Peter M. Daly, *Literature in the Light of the Emblem: Structural Parallels between the Emblem and Literature in the Sixteenth and Seventeenth Centuries* (Toronto, 1979).
21 Fludd was the supreme synchretist. He borrowed openly (and sometimes not so openly) from a multitude of sources. Ammann has identified Marsilio Ficino, Guido d'Arezzo, Agrippa of Nettesheim, and Francesco Giorgi as sources of Fludd's musical philosophy (pp. 219–23); Debus has found in him deep commitments to Paracelsus and his followers (I, 226ff.); Yates has stressed Fludd's indebtedness to John Dee, Vitruvius, and the corpus

hermeticum (*Theatre of the World, Giordano Bruno*). No doubt further scholarship will divulge still other influences at play.
22 Fludd, *UCH* I,b, p. 320.
23 See Erwin Panofsky, *Albrecht Dürer*, 2 vols. (Princeton, 1943), I, 247–53; cf. David C. Lindberg's discussion, which demonstrates the reliance of quattrocento linear perspectivists on the rudiments of medieval visual theory (*Theories of Vision from Al-Kindi to Kepler* [Chicago and London, 1976], pp. 150–4).
24 Fludd, *UCH* II,a,1, pp. 274–5; quoted in Ammann, p. 209.
25 On Dürer see, besides Panofsky, Walter Strauss's useful commentary in his translation of Dürer's *Underweysung der Messung* (1525): *The Painter's Manual: A Manual of Measurement of Lines, Areas, and Solids by Means of Compass and Ruler Assembled by Albrecht Dürer for the Use of All Lovers of Art with Appropriate Illustrations Arranged* (New York, 1977), pp. 10–12.
26 Hans Rupprich. *Dürers Schriftlicher Nachlass*, 3 vols. (Berlin, 1965–9), II, 129: Sloane MS. 5230, fol. 32; quoted in Strauss, p. 12.
27 Rupprich, III, 295; quoted in Strauss, p. 12.
28 Panofsky, I, 257.
29 Fludd, *UCH* I,a, pp. 20–1; II,a,2, pp. 179–91 ("De speculativa pyramidum metaphysicae et physicae scientia"); see also Godwin, pp. 42–53.
30 Arnold Williams, *The Common Expositor: An Account of the Commentaries on Genesis, 1527–1633* (Chapel Hill, N.C., 1948), p. 6. I should like to thank Brian Vickers for alerting me to this book.
31 Ibid., pp. 8ff.
32 Ibid., pp. 20f.
33 Ibid., pp. 22–3.
34 Fludd may well have been aware of hexameral literature which, like his own work, sought to picture the creation. Among works that he could have known are: Charles de Bouelles, *Libellus de nichilo*, in *Liber de intellectu* (Paris, 1510), fol. 63; Hartmann Schedel, *Liber chronicarum* (Nuremberg, 1493), fols. 2^v–5^v; Miles Coverdale, trans., *The Bible* (Antwerp, 1535), fol. 1; Gregor Reisch, *Margarita philosophica* (Freiburg, 1503), fol. 16v. For a convenient illustration of select aspects of these works, see S. K. Heninger, Jr., *The Cosmographical Glass: Renaissance Diagrams of the Universe* (San Marino, 1977), pp. 14–30.
35 The extent to which the meaning of the *res picta* expands beyond a direct mirroring of the text varies from one picture to another. This is especially true of the engravings captioned "experimentum," which appear to have explicitly physical connotations. In these illustrations it is the text that expands the meaning domain of the picture; in other illustrations (e.g., the music temple), the picture has mnemonic, pedagogical, and moral-spiritual functions. The interested reader could construct an inventory of the rhetorical subscripts used by Fludd for his pictures. Consider, as an example, the following: "demonstrationem physicae nostrae pyramidis tam materialis, quam formalis hoc modo describimus" (*UCH* I,a, p. 166); "demonstrationes proprietatum sanctae trinitatis per icones et exempla factae" (ibid., p. 26); "experimentum est tale," showing fire raising water from a lower to a higher flask through a siphon (ibid., p. 33); a plate showing how to draw human faces is called "delineatio" (*UCH* I,b, p. 334), but the practice of "delineatio" transfers easily into demonstration, as in the portrayal of macro-microcosmic man ("praecedentium demonstrationem

in duplici pyramide, in formali scilicet et materiali delineavimus," [ibid., p. 242]).

36 Fludd, *UCH* I,a, p. 23.

37 See Nicholas H. Steneck, *Science and Creation in the Middle Ages: Henry of Langenstein (d. 1397) on Genesis* (Notre Dame, Ind., 1976), pp. 27ff.

38 Robert Fludd, *Mosaicall Philosophy: Grounded upon the Essential Truth, or Eternal Sapience* (London, 1659), p. 45.

39 This series occupies much of the first treatise of Fludd's *Macrocosm* (*UCH* I,a, pp. 26ff.); cf. Godwin, pp. 24–8.

40 See, for example, Fludd's adaptation of his weather glass for the examination of urine: καθολικον *Medicorum* κατοπρον: *in quo, quasi speculo politissimo morbi praesentes more demonstrativo clarissime indicantur, et futuri ratione prognostica aperte cernuntur, atque prospicuntur* (Frankfurt, 1631), p. 271; Godwin, p. 64.

41 See Fludd's *Anatomiae amphitheatrum effigie triplici, more et conditione varia designatum* (Frankfurt, 1623), title page; Godwin, p. 74.

42 See Robert Fludd, "A Philosophicall Key or Ocular Demonstration, Opening and Decyfering a Great Deale of the Hidden Mysteryes of Nature, Partly by an Experimental Conclusion, as Also by an Intellectual Speculation," transcribed with an introduction by Allen G. Debus, *Robert Fludd and His Philosophicall Key* (New York, 1979), pp. 63–156. The meaning of "ocular demonstration" should now be clear in the light of our interpretation.

43 Ammann has convincingly shown the coherence of Fludd's musical philosophy (pp. 198–227), but although music was obviously an extremely important discipline to Fludd, it could be argued that alchemy and medicine were equally important (cf. Debus, *Chemical Philosophy*, I, 226ff.). Our contention is that *all the Fluddean disciplines* were intended to lead back to the same metaphysical presuppositions, so that no matter where one started one got back to the unity of God through the interpenetrating pyramids. This gnosis was possible only through Fludd's symbolic picture language.

44 In some respects Fludd's account of the structure of the psychic apparatus stays so close to traditional medieval versions that he fails to develop adequately the epistemological grounding for visualizing the invisible. Thus "tenuous air" is a sensible transmitted through hearing; it is the basis of Fludd's physics of music. One can *picture* these musical streams as Fludd does in his musical temple, but he does not explain to us in his section on the human souls how we can have an "imaginable shadow," i.e., a picture of the cause of an auditory effect: "In anteriori porro interioris montis seu capitis Microcosmi parte residens anima dicitur *imaginativa*, vel ipsa phantasia et imaginatio; quia rerum corporalium et sensibilium, non quidem veras imagines, sed similitudines et quasi umbras intuetur. Unde mundi et rerum mundanarum ideas et icones speculatur, quatenus est imaginatio et res in abstracto, non autem realia, vel res in concreto, atque prout sunt, apprehendit" (*UCH* II,a,1, p. 218).

45 This is Fludd's conclusion in his final argument against Kepler (*Replicatio in apologiam ad analysin XII* [Frankfurt, 1622], pp. 20f.). The entire section is translated with facing Latin in Pauli, pp. 213–25.

46 Fludd, "Philosophicall Key," p. 142.

47 Kepler, *GW*, VI, p. 396: "Nam et picturis ex aere abundat excusus liber, et tute ipse Analysi III modo recensuisti, quibus utaris picturis in vicem sermonis, Templo, Columnis, Hieroglyphicis, Speculo, Turri, Triangulorum

figuris." On the polyhedral hypothesis, see Eric J. Aiton's introduction to
Kepler's *Mysterium cosmographicum*, trans. A. M. Duncan as *The Secret
of the Universe* (New York, 1981), pp. 17–31; also Robert S. Westman,
"Kepler's Theory of Hypothesis and the 'Realist Dilemma,'" *Studies in
History and Philosophy of Science*, 3 (1972), pp. 247–64.

48 Kepler, *Mysterium* (Duncan trans.), pp. 92–3; cf. Kepler, *Apologia,* in *GW*,
VI, 441: "Tuam quidem distributionem dierum creationis inter personas
Trinitatis Sacrosanctae transmitto Theologis: ego sat habeo si in ipsa figura
Mundi inque praecipuis eius membris, quandam exhibeam similitudinem
sacrosanctae Trinitatis."

49 Kepler points to the contrasting preference for Proclus in the *Apologia,*
GW, VI, pp. 395, 435, 451.

50 Proclus, *A Commentary on the First Book of Euclid's Elements,* trans. with
introduction and notes by Glenn R. Morrow (Princeton, 1970), p. 14.

51 Kepler quotes a long passage from Proclus in the *Harmonice mundi*, from
which he builds his own position (Proclus, pp. 10–15; Kepler, *GW*, VI,
218–21).

52 Kepler, *GW*, VI, 224; see D. P. Walker's splendid discussion of Kepler's
epistemology in *Studies in Musical Science in the Late Renaissance*
(London and Leiden, 1978), pp. 44–57.

53 Kepler, *GW*, XVI, 158: "Ludo quippe et ego Symbolis, et opusculum
institui, Cabalam Geometricam, quae est de Ideis rerum Naturalium in
Geometria: sed ita ludo, ut me ludere non obliviscar. Nihil enim probatur
Symbolis, nihil abstrusi eruitur in Naturali philosophia, per Symbolas
geometricas, tantum ante nota accommodantur: nisi certis rationibus
evincatur, non tantum esse Symbolica sed esse descriptos connexionis rei
utriusque modos et causas." If for the word "symbols" we substitute the
word "pictures" the passage coheres with the kinds of objections Kepler
makes against Fludd a decade later.

54 Kepler, *GW*, II, 153. Lindberg (p. 280) has observed that Kepler uses the
terms *pictura, idolum, imago* and *species* interchangeably. This domain of
usage would allow Kepler to move between physical and metaphysical
contexts.

55 Stephen M. Straker has suggested that Dürer's *Underweysung der Messung*
was the critical context for Kepler's theory of radiation through small
apertures, by analogy with Dürer's method of replacing lines with threads
passing from a luminous source to the surface on which the image was
formed ("Kepler's Optics: A Study in the Development of Seventeenth-
Century Natural Philosophy," unpublished doctoral dissertation, Indiana
University, 1970, pp. 267–71, 370–93). David Lindberg shows that Kepler's
solution was geometrically equivalent to the solution advanced by
Francesco Maurolico some eighty years earlier (pp. 187, 277). My
contention here is that what Kepler took from Dürer underlines in another
way his epistemological and disciplinary differences with Fludd: not how to
use light in order to paint pictures, but how to use pictorial *praxis* to
understand the geometry of radiation! Cf. Kepler's use of Dürer in the
Harmonice mundi for the construction of the heptagon (*GW*, VI, 55). A
very recent work, which I have seen only as this chapter goes to press,
argues fascinatingly that Kepler's theory of vision established for Northern
Renaissance art a new way of picturing the world: "The issue is not 'record
of fact' versus the 'look' of things, it is not different ways of perceiving the
world, but two different modes of picturing the world: on the one hand the

picture considered as an object in the world, a framed window to which we bring our eyes, on the other hand the picture taking the place of the eye with the frame and our location thus left undefined" (Svetlana Alpers, *The Art of Describing: Dutch Art in the Seventeenth Century* [Chicago, 1983], p. 45).

56 Kepler, *GW*, VI, 374.
57 Ibid., III, 19 (*Astronomia nova*).
58 Aristotle, *Poetics* 1475b 7–9, in *The Poetics of Aristotle*, trans. T. Buckley (London, 1869), chap. 21, p. 452.
59 *Aristotle's Treatise on Rhetoric*, trans. T. Buckley (London, 1869), I, ii, pp. 11–24.
60 Kepler, *GW*, VI, 376–7.
61 Ibid., VI, 396–7.
62 Fludd, *Veritatis proscenium*, p. 39: "Nostra ergo historia mundi Physica haec omnia introducit, licet mutatis vocabulis gratia subjecti voces physicas pro mathematicis accipiens. Nonne comprehenditur hoc totum in verbis Mercurij Trismegisti."
63 Ibid.
64 Ibid.
65 C. G. Jung, *The Structure and Dynamics of the Psyche;* in *Collected Works,* 20 vols. (Princeton, 1967–   ), VIII, 185; hereafter cited as *CW*. Of the vast literature on Jung I have found three sources to be uniquely valuable: Jung, *Memoirs, Dreams, Reflections* (New York, 1965; first pub. 1961); Aniela Jaffé, *From the Life and Work of C. G. Jung* (New York, 1971); Peter Homans, *Jung in Context: Modernity and the Making of a Psychology* (Chicago, 1979).
66 See esp. Henri Ellenberger, *The Discovery of the Unconscious* (New York, 1970), pp. 705–7; Homans, pp. 43–114.
67 C. G. Jung, *Psychology and Religion: West and East,* in *CW*, XI ("A Psychological Approach to the Dogma of the Trinity"), 167.
68 C. G. Jung, in *Aion,* in *CW*, IX, ii, 266.
69 Jung, *Memoirs*, p. 209.
70 Jung worked out the theory of types in the period 1913–17 as a consequence of the split with Freud. In *Psychological Types* (*CW*, VI), which appeared in 1921, he classifies and surveys what we might call "typologizing strategies" in such areas as early theological thought, Schiller's poetry, esthetics, William James's psychology, Indian and Chinese philosophy.
71 Pauli, p. 151.
72 Ibid., p. 153.
73 Ibid. p. 154. This looks something like Ernst Haeckel's "biogenetic law," to which appeal was frequently made by late nineteenth- and early twentieth-century psychologists. Frank Sulloway has argued for its prominence as a justificatory foundation in Freud (*Freud, Biologist of the Mind* [New York, 1979], pp. 150ff). Cf. Ernst Haeckel, *Generelle Morphologie der Organismen,* 2 vols. (Berlin, 1866), II, 300. One can also find a whiff of this view, perhaps influenced by Pauli's essay, in Frances Yates's proposal for a new historiography of the scientific revolution (see R. S. Westman, "Magical Reform and Astronomical Reform: The Yates Thesis Reconsidered," in R. S. Westman and J. E. McGuire, *Hermeticism and the Scientific Revolution,* [Los Angeles, 1977], p. 10).
74 Pauli, pp. 174–5.

75 Ibid. pp. 175–6n.
76 Ibid., p. 171. Gerald Holton, commenting on this passage, presents Pauli's interpretation while removing its explicitly Jungian presuppositions. He then presents his own influential characterization of Kepler: "To make the point succinctly, we may say that in its final version *Kepler's physics of the heavens is heliocentric in its kinematics, but theocentric in its dynamics,* where harmonies based in part on the properties of the Deity serve to *supplement* [my italics] physical laws based on the concept of specific quantitative forces" (*Thematic Origins*, p. 82). Elsewhere Holton explains the meaning of "supplement" with the more general, systematic notion of "thematic hypothesis" (pp. 47–68). These hypotheses are neither verifiable nor falsifiable; they are accepted "as a bridge over the gap of ignorance" while the grounds of belief in them are suspended (p. 53). Themata are not to be justified but rather catalogued, as the folklorist classifies cultural practices and traditions. Although here Holton explicitly dissociates his proposal for a kind of folklore level of the categories of the scientific imagination from Jung, Kant, and others, one may regard his position, not unreasonably, as an original development of the Pauli–Jung theory of culture and psyche.
77 Pauli, p. 175.
78 Ibid., p. 207.
79 The full text in English translation with facing Latin is taken from Fludd's *Veritatis proscenium* and given in Pauli's appendix II (pp. 226–36).
80 Ibid., p. 211; cf. Wolfgang Pauli, "Die philosophische Bedeutung der Idee der Komplementarität," *Experientia*, 6 (1950), pp. 2, 72.
81 Pauli, "Influence of Archetypal Ideas," p. 209.
82 Ibid., p. 210.
83 Jung, "Synchronicity," p. 136.
84 The new quaternity produces the following pair of opposites: indestructible energy/space-time continuum; constant connection through effect (causality)/inconstant connection through contingency, equivalence, or meaning (synchronicity) (ibid., pp. 136–7; cf. Jung, *Memoirs*, p. 400).
85 Pauli and Jung, *Interpretation*, pp. 140, 210.
86 C. G. Jung, *The Spirit in Man, Art and Literature*, in CW XV, 135.
87 Ibid., p. 136.
88 Arthur I. Miller, "Visualization Lost and Regained: The Genesis of the Quantum Theory in the Period 1913–1927," in *On Aesthetics in Science*, ed. Judith Wechsler (Cambridge, Mass. and London, 1978), p. 73. Miller contrasts this with the word *Anschauung*, which means "intuition through the pictures constructed from previous visualizations of physical processes in the world of perceptions."
89 Ibid., p. 74. Cf. J. C. Maxwell, who, as M. Norton Wise has demonstrated, received powerful guidance in his work on the development of the electromagnetic field theory from the visual image of mutually embracing curves ("The Mutual Embrace of Electricity and Magnetism," *Science*, 203 [30 March 1979], pp. 1310–18).
90 Jung, *CW*, VI, v.
91 See Frieda Fordham, *An Introduction to Jung's Psychology* (Bungay, Suffolk, 1982; first pub. 1953), pp. 30–1.
92 Jung, *CW*, VI, 319.
93 William James, *Pragmatism: A New Name for Some Old Ways of Thinking* (London and New York, 1911), pp. 7f.; quoted in Jung, *CW*, VI, 300.

94 Jung, *CW*, VI, 319–21.

95 Ibid., VI, 321.

96 Ibid., XIV, 499.

97 William James, *The Principles of Psychology* (New York, 1950), I, 203–6; Gerald Holton, "The Roots of Complementarity," in Holton, *Thematic Origins*, pp. 139–40.

98 Holton, "Roots of Complementarity," p. 140.

99 James, *Principles of Psychology*, p. 206; quoted in Holton, "Roots of Complementarity," p. 142.

100 Holton, "Roots of Complementarity," pp. 137–8.

101 See Jung's account of the Clark conference in a letter of 23 July 1949 to Virginia Payne (*C. G. Jung Briefe*, herausgegeben von Aniela Jaffé, 2 vols. [Olten and Freiburg, 1972], II, 157–60).

102 Jung to Prof. Pascual Jordan, 1 April 1948, ibid., p. 118: "Wir diskutieren hier zusammen mit Pauli die unerwarteten Beziehungen zwischen Psychologie und Physik. Die Psychologie erscheint im physikalischen Gebiet, wie zu erwarten, auf dem Feld der Theorie-Bildung. Die im Vordergrund stehende Frage ist eine psychologische Kritik des Raum-Zeit Begriffes. Zu dieser Frage habe ich gerade dieser Tage eine merkwürdige Entdeckung gemacht, die ich aber zuerst noch mit Pauli von der physikalischen Seite her überprüfen möchte." There is much still to be learned on this phase of the Jung–Pauli relationship, on which I hope to contribute in the future.

103 Just in the period when Pauli was making the problem of visualization in physics a problem for psychology and history, Karl Popper was dissociating philosophy of science from picturing: "Bohr's theory was based on a very narrow view of what *understanding* could achieve. Bohr, it appeared, thought of understanding in terms of pictures and models – in terms of a kind of visualization. This was too narrow, I felt; and in time I developed an entirely different view. According to this view what matters is the understanding not of pictures but of the logical force of a theory: its explanatory power, its relation to the relevant problems and to other theories" (*Unended Quest: An Intellectual Autobiography* [La Salle, Ill., 1974], p. 93).

104 Markus Fierz, "Wolfgang Pauli," *Dictionary of Scientific Biography*, p. 423.

105 Michael Fordham, "Carl Gustav Jung," *Dictionary of Scientific Biography*, p. 192.

106 See Brome, p. 214, and the magnificent April 1955 issue of the Swiss monthly magazine *Du*, devoted entirely to the Eranos group. The list of people associated with the Eranos group reads like a Who's Who of interdisciplinary cultural studies in the twentieth century: Erich Neumann, Mircea Eliade, Karl Kerenyi, Gershom Scholem, Andreas Speiser, and Adolf Portmann, to name but a few. I am grateful to Irenka Taurek for making available to me her copy of this issue of *Du*.

107 See Jung, *CW*, XII, vii.

108 Ibid., XII, 46.

109 Ibid., XII, 102.

110 Frieda Fordham, p. 67; cf. Jung, *CW*, XII, 42: "In order to avoid all personal influence I asked one of my pupils, a woman doctor, who was then a beginner, to undertake the observation of the process. This went on for five months. The dreamer then continued his observations alone for three

months. Except for a short interview at the very beginning, before the commencement of the observation, I did not see the dreamer at all during the first eight months. Thus it happened that 355 of the dreams were dreamed away from any personal contact with myself. Only the last forty-five occurred under my observation. No interpretations worth mentioning were then attempted because the dreamer, owing to his excellent scientific training and ability, did not require any assistance. Hence conditions were really ideal for unprejudiced observation and recording.''

111 Jung, *CW*, XII, 215.
112 Personal communication. I have not sought to discover further information about Pauli's analyst.
113 Jung, *CW*, V, 215.
114 Ibid., XII, 188.
115 Pauli, ''Influence of Archetypal Ideas,'' pp. 157, 206.
116 Ibid., p. 207.
117 Miller, pp. 88, 93.
118 All the above information is based on Fierz, pp. 424–5.
119 Jung, *CW*, XII, 203–4.
120 Ibid., p. 205.

6

The interpretation of natural signs: Cardano's De subtilitate versus Scaliger's Exercitationes

IAN MACLEAN

It is sometimes the case that historians of science neglect the vigorous humanistic tradition of science – Aristotelian physics and Galenic medicine – which is represented in the Renaissance by a bibliography many times greater than that of the experimental literature to which they direct their attention.[1] Such neglect can disguise to modern readers the nature of the conceptual problems encountered to some degree by all Renaissance thinkers and can suppress differences perceived by them, even if not apparent to us today. Sixteenth-century scientific debates share a vocabulary, a mode of expression, and a conception of argumentation and genre: They are divided by issues in virtue of which a generation of thinkers formulated their individual conceptions of the world and its workings. This chapter is devoted to the study of one such debate which was widely known and often quoted: that which opposed Girolamo Cardano (1501–76) to Julius Caesar Scaliger (1484–1558). Of the two, Cardano has attracted more attention[2] because his writings (and especially the *De subtilitate*) lie on the fringes of occult and experimental literature; Scaliger's answer to the *De subtilitate* belongs squarely to the humanistic tradition of science. Cardano explicitly rejects Aristotelianism as a synthetic explanation of the universe and thus is seen as forward looking; Scaliger represents, in the traditional view, that dead bough of the tree of knowledge usually labeled scholasticism, which is characterized by empty verbiage, obscurantism, and incongruity with the real and the natural. As a continuator of the philological tradition of science (i.e., the belief that the correct exegesis of authoritative ancient natural philosophers can yield reliable information about the world and the practical problems arising from man's presence in it), Scaliger is dismissed in most accounts of Renaissance scientific inquiry. I shall argue that both Cardano and Scaliger have a place in any history of sixteenth-century mentalities; that they both

231

are involved in an institutional history which should not be divorced from the reception of their works by contemporaries and later generations of readers; that both encounter similar problems of expression and description; that similar epistemological and interpretive impasses may be uncovered in both; and that, finally, if the prediction of coming developments in thought is accepted as a criterion of success as a thinker, Scaliger rather than Cardano should be commemorated.

Girolamo Cardano's *De subtilitate* appeared first in partial form in Nuremberg in 1550, before the completed edition in twenty-one books was published in Paris in the following year. Revised editions appeared in 1554 and 1560. At the time of its publication, Cardano's reputation as a practicing doctor, commentator on medical texts, writer in the astrological tradition, and producer of horoscopes stood at its zenith;[3] the many reimpressions of his work in the decade following its first appearance testify to a receptive public, as does its translation into French by a professional translator (Richard Le Blanc)[4] in 1556. According to Cardano's biographer, Henry Morley, the *De subtilitate* offers the reader "a comprehensive and philosophical survey of nature, and an account of the subtle truths which underlie the wonderful variety of things which fill the universe"; Cardano sets out to "describe the circle of the sciences and (expressing each by those of its facts which were most difficult of comprehension) to apply his wit, or his acquired knowledge as a philosopher, to the elucidation of them."[5] Although he conceived of the *De subtilitate* as an encyclopedic and comprehensive work, Cardano nonetheless published in 1557 a work entitled *De rerum varietate* to "complete" the *De subtilitate*.[6] Both works in turn refer to a prior unpublished work by Cardano, the *Arcana aeternitatis*, in which matters omitted from the *De subtilitate* and its supplement are included; and a further work, the *De fato*, is referred to by Cardano in his account of his own writings (the *De libris propriis*) as a fourth member of a coherent quartet.[7]

It would be misleading, however, to suggest that the *De subtilitate* (even taken in conjunction with other texts) offers a systematic account of the universe. Although the division into twenty-one books follows an approximately traditional disposition of material, the text itself deals with such random topics as how to beget male children, the recipe for an elixir concocted by Cardano's father to ensure long life and to prevent graying hair, why a siphon works, why the stars sparkle, why a rose has thorns, why bastards are more robust than legitimate children, and why philosophers are melancholic.[8] The books are interspersed with diagrams and illustrations, which often come at unexpected places and seem in some cases to be digressions from a loosely knit argument that purports to deal with the universe systematically, but does so (if at all) with the help of backtracking, free association, and meander-ings.[9]

In 1557 Julius Caesar Scaliger, the established neo-Aristotelian scholar (best known today for his exposition of the *Poetics*, but editor of and commentator on parts of the peripatetic corpus, including the *De plantis*, some books of the *Historia animalium*, and Theophrastus's *De causis plantarum)*[10], produced a refutation of the *De subtilitate* entitled *Exotericae exercitationes de subtilitate*. The title refers to the well-known division of Aristotle's work into two categories: "popular" (exoteric) and "less accessible."[11] In describing his book as exoteric, Scaliger is claiming to write accessible practice pieces (*exercitationes*), not rigorous scholarship.[12] There are, in fact, 365 exercises (some sub-divided): They follow the text of the *De subtilitate* sequentially, ex-tracting quotations[13] from it that are refuted with a mixture of philol-ogical learning, references to common sense, *reductiones ad absurdum*, scorn, and flippancy. Scaliger does not want his reader to be in any doubt about the tone of his refutation; he therefore includes as marginalia indications of how the text is to be read: "castigat," "ludit," "contradicit," "urbane," "pulcherrime," "acutissime," and, predictably, "subtile," "subtilius," "subtilissime." Scaliger also does not want his reader to miss any of his jokes, so he duly enters "jocus" in the margin where appropriate. An example of one of these may indicate why Scaliger did not have confidence in his reader's ability to identify them: When Cardano says that men and animals are at their most beautiful when naked, Scaliger retorts that if birds are in question, they are better naked and trussed on the table than feathered in flight. [In the same exercise, however, Scaliger makes a passing allusion to a more promising (probably obscene) joke about a hairy boy, but with-holds it.][14] For all this flippancy, the *Exercitationes* contain many se-rious objections to Cardano's description of the world, principally based on Aristotelian method and physics. These serious objections were in turn subjected to a critique by Cardano in the 1560 edition of the *De subtilitate*, entitled *Actio prima in calumniatorem librorum de subtilitate*.[15]

The publishing history of these works throws an interesting light on their status and impact. The *De subtilitate* was published at least six times in the decade following its first appearance, as well as being translated into French; thereafter, it appeared only sporadically. Sca-liger's *Exercitationes*, on the other hand, were not reissued for nearly twenty years after their first publication; they were then taken up by a university publisher and reprinted many times. Their popularity in German academic circles is attested by the use made of the exercises for providing topics for dissertations in which Scaliger's oversimpli-fication and polemical misrepresentation of Aristotle are exposed.[16] It may be inferred from this that Cardano's work was thought, by French printers at least, to have a potentially wide readership who would be receptive to a vulgarizing work of popular science; Scaliger's, on the other hand, was recognized to be a useful textbook for university institutions, which still maintained a solid core of peripatetic teaching on their syllabuses.

The fortunes of Cardano's and Scaliger's books corroborate this inference. Cardano's work has its emulators – polymaths like himself, aiming at a general reader[17] – but it does not seem to play a part in the more serious academic debates of the late Renaissance.[18] Scaliger, by contrast, has a well-defined place in the intellectual pantheon of seventeenth-century Continental universities, whereas Cardano is either forgotten or seen as marginal to intellectual debate. The institution of the university guarantees survival to the one, repeated republication, and quotation by other authors, even where the seriousness of Scaliger's text is in doubt; the other's works reappear only sporadically until Gabriel Naudé revives interest in him by publishing his autobiography in 1643.[19] In the two cases publication seems to fulfil very different roles; yet it is also clear that both authors view publication (even of "popular" works) as a means of establishing the authority of their ideas. In the context of intellectual debate, it is possible therefore to see all publication as a claim to seriousness as well as a simple means of communication or a way to make money. But not all publication is universally accepted to be serious in intent; in spheres where the dominant institutions of academic life withhold respectability from authors by refusing to recognize their place in the republic of letters, a sort of parallel publication seems to take place that apes the conditions of the established mode of debate. Examples of this may be found in the occult tradition in the Renaissance, in which introductions, liminary letters and verses, reference to other texts, indices, and scholarly apparatus are found in the same way as they are found in conventional academic works.[20]

The *De subtilitate* allies itself to some degree with the genre of occult writing; indeed, its very title suggests strongly an occult subject and approach. In the case of this work and of the *De rerum varietate*, Cardano claims that the source of his writing – its structure, tone, and contents – was revealed to him in a dream, embodying, as it happens, elements of traditional disposition and numerology.[21] But any book purporting to lay bare the workings of cosmos, man, sciences, arts, and celestial orders must ground itself in an authority greater than a dream and in a genre that is familiar, in some respects at least, to a contemporary reader. As writers in the occult tradition face similar problems in establishing the authority of their texts, it is not surprising that Cardano should refer at appropriate moments to a well-known example of this tradition, the *De occulta philosophia* of Cornelius Agrippa.[22] Agrippa's work contains many features characteristic of the genre. He begins by producing a genealogy of occult writing that represents a sort of stemma of authority, leading back to ancient Egypt;[23] this parallels (or mimics) humanists' references to the classical canon as an authoritative source. The occult, however, claims to represent the hidden interior of philosophy, whose exterior is "communis opinio" and the empty verbiage of scholasticism.[24] This radical division between hidden and accessible knowledge is reinforced by a moral division in the potential readers of the different kinds of philosophy: Accessible knowledge is for the common herd; hidden knowledge for the wise and virtuous. In a letter published at the beginning of the *De occulta philosophia*, Johannes Trithemius advises Agrippa to follow this principle, and the author of the *Liber secretorum alchemiae* derives it from God, who, he claims, has hidden the secrets of alchemy from all his people save those who are virtuous and who confess his goodness and omnipotence.[25]

This division in readership between initiates and uninitiates has, of course, strong religious overtones in the age of the Reformation. Access to the word of Scripture was a major point of contention; moderate reforming Catholics such as Erasmus call for Holy Writ to be generally accessible; the post-Tridentine church is explicitly opposed to this view and objects strongly to the unimpeded diffusion of the Bible. Whereas Erasmus's Greek New Testament of 1516 calls in its Preface for its readership to include women, agricultural laborers, and weavers, the Louvain Bible of 1550 objects to vernacular versions of Scripture and to its diffusion among the common people. The connection between the theological and the occult is made by Paul Skalich de Lika in his *Occulta occultorum occulta* of 1555:

> Although I have made clear, manifest and unambiguous the
> knowledge which my predecessors have handed down
> wrapped up in enigmas and fables, or expressed in confused
> or crude language, yet have I, as it were, locked it up with
> the most secure key, lest the arcane and secret doctrine of
> the wise should fall into the hands of fools, and should allow
> unlettered or biassed men, or even women, or butchers, or
> artisans, or farm workers to enter into disputes about the
> highest mysteries of faith, and thereby profane everything
> (as, alas! now happens daily).[26]

In this passage may be perceived a strange but characteristic rhetorical gesture in occult texts: On the one hand, they claim to open up a hidden universe to the uninitiated; on the other, they protect this hidden universe from the eyes of the vulgar by a number of expressive and argumentative ploys. These include the technique of introducing chaff

as well as wheat, nonsense as well as excellent doctrine, in their treatises: "I must confess," writes Agrippa, "that my book gives accounts
of many dazzling but useless tricks besides my magic."[27] Where this
technique is not found, one may instead encounter the use of language
that makes no sense unless the reader has already some knowledge of
the occult: Skalich de Lika, for example, sets out his lore in "canons"
that all have the same conditional formulation: "Whoever understands
already . . . will be able to understand further."[28] Yet other writers
employ the technique of referring to a body of literature not accessible
in published form, known only to initiates, and held back for fear of its
desecration. Copernicus himself, in the Preface to the *De revolutionibus orbium caelestium*, wonders publicly whether "it were better to
follow the example of the Pythagoreans and others who were wont to
impart their philosophic mysteries only to intimates and friends, and
then not in writing but by word of mouth."[29] As biblical authority can
also be found to justify such withholding of information,[30] it is not
surprising that those claiming to deal in arcane material should be
tempted to withhold part of it, just as the writers of nostrums often
left out crucial information in the recipes they published. Thus, with
the use of such techniques, writers in the occult tradition often claim
to have elucidated better than ever before the mysteries of the universe,
while at the same time to have protected such sacred knowledge from
misappropriation. This double gesture guarantees the survival of the
genre by its forever-deferred promise of explanation and clarity and
its inbuilt need to be itself subjected to the exegesis that it has inflicted
on the stemma of texts that precedes it.

Several problems arise from this gesture. Where are the true limits
to be placed on human cognition? What are the specific aims of any
exegesis? How can exegesis of hidden things be verified and author-
ized? Cardano's answers to these problems indicate the limited extent
to which he accepts the conditions of writing of the occult tradition.
It is true that he refers to the need not to cast pearls before swine, and
cites as an authority at crucial moments an inaccessible text (his own
unpublished *Arcana aeternitatis*);[31] but both he and his translator, Richard Le Blanc, are clear that his is a philosophical task in the line of
Aristotle, Pliny, and Albertus Magnus, and that the purpose of undertaking it is not partial mystification but general (if qualified) demystification.[32] Furthermore, Cardano believes that limits should be placed
on human knowledge, unlike the occult tradition, which supports the
view that all things are knowable to the wise.[33] In the *Arcana aeternitatis* he refers to three possible sources of knowledge of hidden
things: intuition or innate ideas, senses and reason "which mislead us
to a great degree," and ecstasy.[34] The source of the *De subtilitate* is
apparently the third category (since Cardano says he dreamed it all),

but its means of expression lies in the second, fragile category of the sensible and the intelligible. The difficulty of expression, more than the inherent difficulty of his subject, is of concern to Cardano; he declares that there is no point in writing about what is known already, but that if one chooses to enter new territory one encounters problems of elucidation, perception, and reasoning.[35] There are subjects which it is not legitimate to explore (such as the fabrication of poisons, certain forms of divination, and the nature of the Godhead);[36] but, in general terms, it is man's birthright to inquire into the mysteries of nature, and God has instilled in him the desire and the ability to do this.[37]

Cardano, by declaring that he wishes to expound clearly that which is hidden or obscure, distinguishes himself from the occult tradition in the *De subtilitate*, even if some features of this genre of writing can be said to survive in his text. He allies himself explicitly with Galen in terms both of disposition of material and of method.[38] As well as bearing the marks of Galen's advice in his *Isagoge*, Cardano's text also presents what look like Ramist dichotomies *avant la lettre*.[39] This feature is produced by Cardano's pretension to universal science – a pretension that both Scaliger and, later, the neo-Aristotelian Rudolf Göckel sharply attack.[40] For them, the *De subtilitate* is little more than a declamation, a hotchpotch of disparate and uncoordinated facts, explanations, and erroneous beliefs. Yet they also subscribe to a total system – approximately that of Aristotle; and together with Cardano, they can be distinguished by this belief from the less ambitious local investigations of anatomists and experimental scientists, which are identified as the beginnings of modern scientific method by most historians.[41] Cardano's and Scaliger's explanations, on the other hand, are based on universal principles and carry with them their own version of metaphysics.

Like Galen, Cardano claims that experience is in the end the only convincing and trustworthy authority: As he sets out to tell his reader that which the reader does not yet know, he can command consent and belief only if experience upholds his arguments. Where the experience is repeatable and measurable, Cardano sounds like an experimental scientist; but he includes in the category of experience some aspects of his reading (notably anecdotes),[42] unlike Fallopius, Vesalius, Harvey, or Malpighi. For them, observation is what we now consider it to be; for Cardano and for near-contemporaries such as Johannes Schenk von Grafenberg, it includes hearsay and what others have observed.[43] Yet Cardano claims not to speak as a philologist in the *De subtilitate:* "A reader may be surprised that I express a different opinion (to this) in my *Contradicentia medica*; in that book, I set out to follow the opinion of the ancients, in this one, the truth."[44] In rhetorical terms, Cardano is here playing his last card in authorizing and verifying his own discourse; he is giving his own voice in the text an indisputable truth value. Cardano's approach in the *Contradicentia medica* – the discussion of disagreements among ancient medical texts – is essentially philogical and identical to that of Scaliger; his practice in the *De subtilitate*, to modern eyes at least, looks remarkably similar, because the same mixture of argument from experience and argument from authority is present. Both Cardano and Scaliger lay claim to methodical exposition and neutral scientific discourse; each accuses the other of confusing fact and authority; both seem, to modern readers, to be guilty of the offense of which they accuse each other.

These are some of the problems Cardano faces because of his choice of genre and expression; before I pass to the epistemological and interpretive limits of his undertaking, it would be wise to provide a clearer idea of his concept of subtlety, and Scaliger's critique of it. Cardano offers a formal definition at the beginning of his book: "Subtilitas est ratio quaedam, qua sensibilia a sensibus, intelligibilia ab intellectu, difficilè compraehenduntur [Subtle things are those which are sensible to the senses, or intelligible to the intellect, but with difficulty comprehended]."[45] Subtlety is sited in substances, accidents, and representations.[46] The various sorts of subtlety in substances are listed (thinness, smallness of quantity, fluidity, divisibility, or any combination of these qualities); only the first belongs to the domain of traditional physics.[47] Subtlety in accidents is, of course, more varied; that in representations is described principally in Book xv ("De inutilibus subtilitatibus") and includes such things as acrostics, poems hidden in poems, and mathematical conundrums. All subtlety lies at the very edge of perceptibility and intelligibility; thus a series of related concepts is attracted to it – difficulty, rarity, thinness, implausibility, and unexpectedness: Subtlety is frequently "praeter communem opinionem."[48] But is is not an occult recuperative device such as "spiritus," which is able to explain anything and may be endowed with any attributes whatsoever; Cardano insists that it is identified by a true method and supported by evidence drawn from experience.[49]

Scaliger's attack on this is to some degree predictable: He claims that subtlety is not a coherent category applying to substances, accidents, and representations; that it does not fit into an Aristotelian category as accident or quality; and, most damningly, that it is not sited in nature at all but in the mind perceiving nature. Indeed, Scaliger is able to identify moments at which Cardano himself, in spite of his claims, locates subtlety in the mind and not in the object it perceives, and so is able to show that Cardano's practice of subtlety is different from his doctrine.[50] Here we have a neo-Aristotelian using peripatetic arguments to support a radical epistemology that implies that knowledge of the real is impossible. A century before Scaliger, Lorenzo Valla, the Italian jurist and opponent of scholasticism, argued in a similar way that reality is no more or less than a construct of human linguistic categories, as Ernst Cassirer and Donald Kelley have shown.[51] Scaliger's dichotomy of nature and perception places firm limits on man's ability to know the external world, and, furthermore, threatens the opposition of intelligible and sensible in a radical way.[52] What is already explicit in the *Exercitationes* is underscored in a yet more rigorous way by the logician Göckel in a series of striking corollaries.[53] Scaliger's dichotomy is not simply a feature of neo-Aristotelianism, to be dismissed (as Cardano himself dismisses it) as the hollow triumph of words over things or a late flourish of nominalism.[54] It reappears in other, more significant, contexts; notably in English empirical philosophy of the late seventeenth century, as evidenced by Boyle's strictures on the limitations of taxonomy as a science or by Locke's distinction between nominal and real essences.[55] If this were not enough, subtlety is even threatened by its own inherent logic: If subtlety is that which is perceived with difficulty, it follows that it is yet more difficult to perceive its causes; and its causes, being difficult to perceive, generate the possibility of an infinite regress both of the perceived object and the perceiving agent. As Michel de Montaigne says, possibly with Cardano in mind, "en subdivisant les subtilitez, on apprend aux hommes d'accroistre les doubtes."[56] Scaliger chooses a different version of the same critique: To perceive subtlety, there must be a subtle faculty of the mind whose own subtlety, being more arcane and less easily discovered, requires a yet finer faculty to perceive it; thereby, a spiral of ever-increasing difficulty threatens to come into being.[57]

Scaliger's criticism raises a central problem about Cardano's *subtilitas*: How can he be sure that he is dealing in this and not simply in obscurity or ambiguity? Cardano acknowledges that this is a problem at the very beginning of the *De subtilitate* and agrees that he has to convince his reader that he knows what he is talking about by providing him with compelling evidence in the form of trustworthy authority, a fundamental discipline that informs his study of nature, a set of clearly defined first principles, and a clear science of taxonomy and etiology. These features of the *De subtilitate* will be examined briefly in turn.

I have already suggested that there is a possible institutional definition of authority; in the case of Renaissance universities, this is often neo-Aristotelianism, and it is expressed through the publication of text and commentary. Cardano rejects Aristotle, but accepts the criterion of publication or composition; as in the occult tradition, his references to other writings by himself create an alternative guarantee of veracity, as was noted above.[58] He also refers to God as ultimate authority, describing God in both the *De subtilitate* and the *De rerum varietate* as the true author of his works.[59] Elsewhere, however, his position is manifestly less secure. In Book xix he answers the question, Do demons exist? with a positive affirmation, yet within a few lines admits that he has never encountered one.[60] The reader is therefore led to believe that the testimony of his father (together with other anecdotes) is crucial; yet he has said already that anecdotal historians of nature such as Pliny and Albertus Magnus "obviously lie and are in error."[61] Not all anecdotes are true; but it is possible, apparently, to affirm when they are true and when not. Clearly there is a grave problem here at the level of textual authority.[62]

This problem is not necessarily solved at the level of ratiocination or experience. Cardano appeals at times to "clear arguments,"[63] but it is equally obvious that he and Scaliger do not agree on what constitutes the ground rules of argumentation. Even "common sense" can produce opposite conclusions from the same evidence, demonstrating thereby its lack of community. Like Aristotle and Galen before him, Cardano emphatically declares that no textual authority can oppose conclusions drawn from experience;[64] in this, he is joined by the most conservative of philological scientists, such as Jean Riolan the Elder.[65] Yet Cardano deals in explanations that relate to general causes, to first principles – in short, to universals: His "proofs" at the level of experience are particulars (very often in the form of anecdotes or examples). Furthermore, his explanations purport to relate to total coverage of the knowable universe.[66] All this is very far from the cautious approach to the knowledge of particulars and its relation to universals found in Aristotle's *Metaphysics* and used by Scaliger to deflate Cardano's pretensions as an encyclopedic writer.[67]

Cardano claims Euclidian geometry to be the founding discipline of the study of nature; in this he has been hailed as forward looking, and one writer has even placed him, by virtue of his forays into mathematics, among the forerunners of probability theory.[68] But Scaliger, and later Göckel, contest the claims of geometry and uphold metaphysics in a traditional peripatetic way.[69] In doing this, they argue that Cardano's text is not even faithful to geometry because it contains a patent mixture of the true and the false, the probable and the possible, the serious and the flippant, the exact and the approximate, the logical and the rhetorical. The Renaissance genre characterized by such a hotchpotch is the declamation, of which the most famous example is Agrippa's *De incertitudine et vanitate omnium scientiarum et artium* of 1531.[70] Cardano vigorously rejects this devaluation of his work in the *Actio prima*; but he also uses the polemical ploy of accusing other texts of being no more than declamations. This is clear from his description of alchemy:

> The chemical art contains much that is remarkable, much that is absurd, yet more that is doubtful, but some things also which are beautiful, beneficial for health and efficacious; other things which are of no moment at all, or which are very speculative; lastly, and in greatest abundance, things of great detriment and danger.[71]

On the question of first principles and opposites in nature, Cardano is characteristically iconoclastic; he refers to five principles (matter, form, spirit, place, and movement), three elements (earth, fire, and air), and two qualities (hot and moist); in this he is explicitly anti-Aristotelian and antioccult.[72] In describing cold and dry as the privation (steresis) of qualities, Cardano explicitly contradicts Aristotle, who declares that cold is not the privative opposite of hot[73] and enmeshes himself thereby in a complex argument concerning privation – a principle in Aristotelian physics – which he appears to misrepresent.[74] He does not appear to subscribe to the fixed oppositions characteristic of the occult, which may be remotely derived from the Pythagorean parallel list quoted by Aristotle (male/female, odd/even, right/left, at rest/moving, etc.);[75] nor does he subscribe to the fourfold peripatetic set. At one point in the *De subtilitate* he seems to suggest that contraries with middle terms constitute the only possible category of (real) opposite.[76] Yet elsewhere in the same text, contraries with excluded middle terms are used (e.g., potency/act, reason/experience, universal/particular, intelligible/sensible, substance/accident),[77] and I have already mentioned his use of the privative opposition. It seems as though Cardano, for all his claims to novelty, is still reliant on the substratum of Aristotelian metaphysics.

Scaliger, of course, is not slow to point this out;[78] but he gains greater relish from pouring scorn on Cardano's endeavors in the domain of taxonomy. It is true that Cardano disaligns the humors and the elements, which might suggest that he rejects the numerological basis of occult science; but this rejection is chiefly a result of his reduction of the number of elements to three, which causes him to abandon the traditional schema.[79] He begins both the *De subtilitate* and the *De rerum varietate* with a series of binary oppositions that would permit a logician to draw hypotyposes to account for all his work; all existing things, he avers in the *De subtilitate*, are either substances or accidents; if substances, they are either corporeal or incorporeal; if incorporeal, they are either independent or depend on something else; if they depend on something else, they are either causes or not causes. Such dichotomies in the manner of Ramus have an authoritative and persuasive ring to them.[80] But they are not characteristic of his taxonomy as a whole. He is far more interested in creating new botanical and zoological classes, based on novel criteria; these all seem to a modern eye to be essentialist in spirit, and not to be precursors of the work of classical taxonomists such as Cesalpino or Ray.[81] Cardano also dabbles in arbitrary occult-sounding alignments of metals, tastes, colors, and planets, which are reproduced in tabular form in the text. He does not offer any scientific explanation for the alignments he makes; he merely points out that in his system, the number of the metals, tastes, colors, and planets is the same, and that he therefore felt the urge to correlate them.[82] His other bursts of numerology – usually based on odd numbers, although four is also a favored quantity – are equally unmotivated. Men, for Cardano, fall sometimes into three classes (the divine, the human, and the bestial), sometimes into four (the honest, the prudent, the effeminate, and the bestial). There are, for him, four excellent things in nature: man, elephant, diamond, and gold. There are five sorts of stones, seven sorts of human calamity, nine sorts of animals, eleven antithetical pairs of human passions. One could give a much longer list of such divisions.[83] Cardano is very close to the occult tradition in this feature of his writing; he shows the same tendency to create apparently arbitrary subsets and the same use of numerology as a heuristic device that permits the inquirer to postulate correspondences between classes of similar number. Scaliger, an orthodox peripatetic botanist, takes much delight in exposing the weakness of Cardano's taxonomy by producing plants and animals that escape his categories or exist simultaneously in several; he also attacks Cardano's numerology, calling to his aid Aristotle himself, who denied, according to the author of the *Exercitationes*, that similarity in number is significant in the establishment of genera and species.[84]

A final word should be said about etiology. Cardano makes no reference to the scholastic science of causes, although his work is principally engaged in uncovering to the wondering eye of his reader the hidden causes of things. He rarely dwells on the philosophical problems of causation[85] and seems not to distinguish with any rigor such concepts as *virtus, vis, proprietas, causa*, and *ratio*.[86] He mocks neo-Aristotelians for finding empty names to fill the gaps that should be occupied by proper explanation,[87] but may be accused himself in turn of similar tactics. This is particularly evident in his use of the notions of sympathy and antipathy, which can "explain" the effect of one thing on another only by simple affirmation of their presence in the relationship in question.[88] Most prominent, however, are Cardano's mechanistic explanations; many machines are described and illustrated, and their workings are accounted for by Cardano's physical theories. At some points in the text the universe itself is conceived of as a grandiose divine machine obeying mathematical laws.[89] Cardano locates purpose firmly in nature, making it a sort of generalized final cause in scholastic terms,[90] and does not trouble to distinguish among material, efficient, and formal causes. Although spirit is one of his five natural principles, he does not use it as it is used by some occult writers (i.e., as a term that describes an otherwise inexplicable or unnameable force or property of an object). Instead, he seems to prefer to apply the argument from function, which is a feature of the writing of Galenists in the Renaissance[91] and may well arise from his own interest in and veneration of Galen.

These epistemological issues seem to separate Cardano radically from his neo-Aristotelian critics; yet it is clear that they are able to sustain some sort of dialogue, a fact brought about by their shared conceptions of argument and interpretation. Cardano roundly denounces Aristotle (as do academics and sceptics) for obscurity, ambiguity, word spinning, and remoteness from nature; yet we have seen that he also falls prey to a number of similar accusations. Like Scaliger's, Cardano's authority is at crucial points in his argument located in texts; he is as unable as the neo-Aristotelian to cross the divide separating words from things. His specific instances of explanation, which are still admired by Naudé in the following century, are nonetheless parasitic on general categories that, by virtue of their very nature, can never be proved by experience and that lead him at times to produce absurd propositions. For example, he says that gold must taste better than silver because of its preeminence in the hierarchy of metals, yet admits that although silver has a taste, gold has none whatsoever.[92] His appeals to logic can be, and are, contested on the methodological level by others who do not subscribe to his view of it.[93] He offers no method of identifying what is trustworthy or not trustworthy in the accounts of other natural philosophers, and no authority for separating other sciences such as alchemy into true, doubtful, and erroneous elements, beyond the authority of his own voice in the text, which he baldly states is truthful.[94] By such use of rhetoric, his own book takes on a problematic status somewhere between scientific writing and declamatory literature. His account of causality is constantly threatened by vapidity or tautology; by the use of such terms as "virtue," "property," "power," the hidden cause is made identical to its manifestation. Where the hidden cause and its manifestation are displaced, he has recourse to notions of sympathy and antipathy, by which anything can be explained.[95] His attempts to show the deep numerological patterns of the universe are either arbitrary (by his own confession) or unprovable; they can only be affirmed. Scaliger can be forgiven for pouring scorn on such enterprises, among which is a project to offer an exhaustive account of the proportions between parts of the human body, from which apparently, as Scaliger points out, the less noble parts have been excised.[96]

Even Cardano's status as a self-authorized speaker of truth is impugned in the *Arcana aeternitatis* in which there is a chapter proclaiming "that there is some falsehood in all truth, and some truth in all falsehood."[97] Through such writing we have now passed into the vertiginous Renaissance world of global paradox and *coincidentia oppositorum* and have come close again to the occult tradition with its tenebrous metaphysics and complex conceptions of truth. The portmanteau term *subtilitas* directs an enterprise that sets out to link particular explanations of natural phenomena with the general laws elaborated to account for them; Cardano, its author, presides over theatrical conjuring tricks performed on intellectual riddles. This image is one that he himself evokes in Book xviii of the *De subtilitate* ("De mirabilibus, et modo repraesentandi res varias praeter fidem")[98] and in the *De rerum varietate*; the magician is like the interpreter, himself *hors jeu*, directing the action and offering at times to let the audience in on the secret. The similarity with the position of Scaliger is striking (he, like Cardano, orchestrates his authorities and bends them to his own will); the divergence from experimental science and its resolutive or inductive method, with its concessions to the authority of the evidence and its heuristic use of analogy,[99] is clear.

Cardano's *De subtilitate* reflects the Renaissance desire for a new (or revised) encyclopedia that would allow man to become, in Descartes's words, the master and possessor of nature. Its publication and consumption lead it to be classed with other works of similar scope, which apparently oppose in important ways the continuing Aristotelian synthesis, but which share with this tradition central problems of epistemology and interpretation. In one way, and to a limited extent, even experimental science at this time falls prey to these problems, insofar as the observations it records are structured already by the expectations of the observer and his concept of his role; but experimental scientists do not always look for new universal explanations and are careful to limit the field of their inquiry and to use mathematical bases for their demonstrations. In the *De subtilitate*, Cardano mixes his evidence and makes extensive claims for his new physics and metaphysics – claims that are countered in a similar spirit by Scaliger. But Scaliger, in dividing nature from man's perception of it, and in locating reality in the human mind, comes closer than Cardano to predicting the preoccupations of subsequent philosophers. This is not, however, the reason for the continuing reappearance of his textbook in the seventeenth century. That phenomenon can be most plausibly explained by the demands made by conservative university syllabuses on his publishers and by the convenience of Scaliger's text as a hunting ground for peripatetic dissertations and exercises.

Whatever their differences of opinion, Cardano and Scaliger belong to one "universe of discourse" whose contours can be perceived through the formulation of their polemic. The substratum of that polemic is made up inexorably of a metalanguage grounded in neo-Aristotelianism; it proceeds by an interpretive method that does not, in the final analysis, distinguish between object and word, world and text; it embodies a concept of argumentation that allows the certain and the probable to be mixed and the true and the false to interpenetrate. Nature and its workings are to be explained; but that explanation is parasitic on prestructured perception and is not in the end to be measured against the evidence so much as against the language in which such perception is expressed. It is in the context of such a mentality that the rejection of Aristotelianism as emblematic of a philological approach to nature may best be judged. When Galileo pillories Aristotle through Simplicio in the *Dialogo sopra i due massimi sistemi del mondo*, he is not so much concerned with the failure of the peripatetic system to account successfully for natural phenomena as with its claims to be universal. The fact that Aristotelianism contains as a central tenet the notion that all human taxonomy is a construct and that the real is inaccessible to man except through the operations of his mind, is polemically suppressed by the Italian physicist. Nonetheless, his rejection of metaphysics in the *Lettere intorno alle macchie solari* could well serve as a final judgment on the work of both Cardano and Scaliger:

> Either we strive, by our speculations, to attain the true and intrinsic essence of natural substances, or we are satisfied with the knowledge of some of their properties [*accidenti*]. I hold the search for essences to be equally impossible and futilely exhausting in the case of elementary substances which are to hand, as for celestial substances which are very distant . . . I do not understand the true essence of earth or fire any better than I understand that of the moon or the sun: such knowledge awaits us when we have come to the state of heavenly bliss, and only then.[100]

Notes

1 On the continuing vigor of the peripatetic tradition, see Charles B. Schmitt, "Towards a Reassessment of Renaissance Aristotelianism," *History of Science*, (1973), pp. 159–93.

2 See, for example, Henry Morley, *The Life of Jerome Cardan of Milan, Physician*, 2 vols. (London, 1854); Michel Foucault, *Les Mots et les choses* (Paris, 1966), pp. 39, 43. The most recent work on Cardano is by Alfonso Ingegno, *Saggio sulla filosofia di Cardano* (Florence, 1980).

3 See Morley, II, 56–70.

4 Le Blanc is also the translator of Plato's *Io*, Hesiod, Virgil, Ovid, Chrysostom, and Filippo Beroaldo the Elder.

5 Morley, II, 58.

6 Girolamo Cardano, *De libris propriis*, in *Opera omnia*, ed. Charles Spon, 10 vols. (Lyons, 1663), I, 71, 74. This edition is hereafter cited as *OO*. Cardano's *De rerum varietate* (hereafter *DRV*) appeared in 1557, 1558, 1580, and 1581.

7 Cardano, *De libris propriis, OO*, I, 71, 79, 109. Only Cardano's account of the chapter headings of the *De fato* survives (*OO*, I, 99–100); the *Arcana aeternitatis* was published from a manuscript by Spon (*OO* X, 1–46).

8 Cardano, *De subtilitate* (hereafter *S*), *OO*, III, 556, 390, 363, 412, 577, 557, 558.

9 These are sometimes acknowledged as such by Cardano: e.g., *S*, ii, *OO* III, 400: "Transtulit nimis nos longè a proposito orationis continuitas."

10 These works appeared in 1556, 1584, and 1566, respectively (the latter two posthumously). On Scaliger, the most recent work is Vernon Hall, *Life of Julius Caesar Scaliger, 1484–1558* (Philadelphia, 1950); see also M. Billanovich, "Benedetto Bordone e Giulio Cesare Scaligero," *Italia mediovale e umanistica*, 11 (1968), pp. 187–256.

11 Cicero, *De finibus*, v. 12: "Duo genera librorum sunt Aristotelis: unum populariter scriptum quod ἐξωτερικόν appellabant, alterum limatius."

12 In his refutation of Cardano, Scaliger in fact refers to his "nobiles exercitationes" as a work of serious scholarship (*Exotericarum exercitationum liber XV de subtilitate* [Frankfurt, 1582], x, p. 58; li, p. 196; lxi, p. 219: hereafter *E*). These appear to have existed, but never to have been published, according to Johannes Crato à Crafftheim's liminary letter to Joseph Justus Scaliger in the 1576 edition (ã 3r: "utinam vero, iterum utinam Nobiles illius atque Familiares Exercitationes publice extarent"). Paganinus Gaudentius is clear that the *Exotericae exercitationes* are not altogether serious: "Exotericas [Scaliger] appelavit, non acroamaticas, indicâsse non semper se ex animi sententia locutum; sed indulsisse sibi ipsi et inseruire voluisse ὑποθέσει'" (*De nonnullis quae non peripatetice dixisse videtur Iul. Caesar Scaliger in opere de Subtilitate*, in *De Pythagoraea animarum transmigratione, Aristoteles veterum contemptu et alia* [Pisa, 1641], p. 201). Cf. Cornelius Agrippa's definition of the Renaissance genre called the declamation, quoted below, note 70.

13 Scaliger sometimes misquotes, apparently for polemical purposes: e.g., the substitution of "apprehenduntur" for "compraehenduntur" in Cardano's formal definition of *subtilitas* (*E*, i.l, p. 1).

14 Ibid., cclv, p. 790: "Hic erat historia ponenda, de piloso puero. Sed supra satis."

15 *OO*, III, 673–713. Both the National Union Catalogue of American Libraries and the Bibliothèque de l'Arsenal at Paris record editions of the *De subtilitate* containing the *Actio prima* printed in Basel in 1553; these editions are, in fact, according to their colophon, Basel, 1560. The wrong attribution of date is due to the medallion portrait of Cardano on the title page, which is independently dated 1553.

16 On Scaliger's publisher, see R. J. W. Evans, *The Wechel Presses: Humanism and Calvinism in Central Europe, 1572–1627*, The Past and Present Society, supp. 2 (Oxford, 1975); for an exemplary set of dissertations, see those produced under the presidency of Johannes Sperling

in Wittenberg between 1645 and 1647, which are listed in the catalogues of the British Library and the Bibliothèque Nationale in Paris.

17 One example of such a writer is Pierre de la Primaudaye, *L'Academie françoyse* (1580).

18 Robert Lenoble's *Mersenne ou la naissance du mécanisme,* 2nd ed. (Paris, 1971), pp. 121–33, 503–5, indicates that Cardano was better known in the seventeenth century for his astrological works than for the *De subtilitate*.

19 Naudé's edition of the *De vita propria* may well have a connection with the republication of Le Blanc's translation of the *De subtilitate* at Rouen in 1642. Naudé, who played a part in Spon's edition of 1663 (*OO*, I, ẽ4ʳ), may have been moved to celebrate Cardano because of his remarkable anticipation of the advances in science witnessed by Naudé (*OO*, I, 13ʳ).

20 Cornelius Agrippa's *De occulta philosophia* (Paris, 1567) offers an excellent example of this, with its preface and exchange of letters with Johannes Trithemius, the abbot of the monastery of Saint James "in suburbio Herbipolis (= Würzburg)."

21 There are 21 books in *S*, and 100 chapters in *DRV*. The hierarchical arrangement of material is similar in both books: first principles and elements; the heavens; light; life forms in ascending order of excellence (metals, stones, plants, animals, man); man's arts and sciences; spirits; demons; angels; God.

22 For example, *S*, xv, *OO* III, 518; xvii, *OO*, III, 627–8; xviii, *OO*, III, 646, where, however, Agrippa is described as "impius" and "homo vanissimus."

23 Agrippa, *De occulta philosophia,* α4ʳ, p. 2; cf. also Paulus Scaliger (Skalich de Lika), *Occulta occultorum occulta* (n.p., 1556), passim.

24 See J. F.'s preface to his translation of Agrippa's *Three Books of Occult Philosophy* (London, 1657), α7ᵛ: "There is the outside and the inside of philosophy: but the former without the latter is but an empty flourish."

25 A commonplace: see Agrippa, *De occulta philosophia,* α6ᵛ, p. 499; Calid filium Iazichi, *Liber secretorum alchemiae,* in *De alchemia,* ed. Chrysogonus Polydorus (Nuremberg, 1541), p. 338.

26 Skalich de Lika, p. 5.

27 Agrippa, *De occulta philosophia,* α2ᵛ: "Fateor praeterea magiam ipsam multa supervacua et ad ostentationem curiosa docere prodigia." See also Trithemius's letter, α6ᵛ.

28 For example: "Qui scit, quomodo non occultum, quod est non occultum non occulti, et occultum, quod est occultum occulti, sibi invicem non contradicunt: sciet, quomodo Academici, Stoici, Peripatetici et Epicurei: potissimum autem ex his Plato et Aristoteles concordent" (Skalich de Lika, p. 6).

29 Quoted by R. Mandrou, *From Humanism to Science,* trans. B. Pearce (Hassocks, 1978), p. 38.

30 The *locus biblicus* of this gesture is found in 2 Esdras (4 Ezra) 14:26: "Perfectis quaedam palam facies, quaedam sapientibus absconce trades." Skalich de Lika uses this as his epigraph. See also Calid filium Iazichi, p. 338.

31 Or rather sacred things before dogs: "Quia stupidis et vulgo haec non sunt aperienda, et (ut dici solet) sanctum dare canibus, frequentioribus uterer exemplis" (*DRV*, c, *OO*, III, 349). For references to the *Arcana aeternitatis,* see *S*, i, *OO*, III, 358; xii, *OO*, III, 562; *DRV*, c, *OO*, III, 349.

32 *S*, i, *OO*, III, 358; xviii, *OO*, III, 650; *DRV*, c, *OO*, III, 348–9; Le Blanc,

"Epistre," in *Les Livres de la subtilité* (Paris, 1556), *ii^{r-v}. On demystification as the end of philosophy, see Aristotle, *Metaphysics,* A.1. There is a degree of mystification that remains – "quaedam grata obscuritas," (*S*, xviii, *OO*, III, 650); this is justified by the need of the book to appeal to the public (cf. also *S*, xiv, *OO*, III, 583–4).

33 In this, the occult differs from the apocalyptic tradition, which explicitly allows for areas of incomprehension that will be illuminated in the future by the unfolding of the divine plan: see Michael E. Stone, "The Metamorphoses of Ezra: Jewish Apocalypse and Medieval Vision," *Journal of Theological Studies*, 33, (1982), pp. 1–18. I am grateful to Professor George Caird for having indicated this article to me and for having located for me the verse in 2 Esdras quoted in note 30.

34 *OO*, X, 3: "Firma omnis cognitio nostra triplex . . . aut à principiis animae ab initio inditis, aut à sensibus atque ratione, quae nos longius abducit, aut afflatu." Cf. Skalich de Lika, p. 9.

35 Cardano specifies four difficulties: "rerum obscuritas, incertorum dubitatio, causarum inventio, recta earum explicatio" (*S*, i, *OO*, II, 357).

36 *S*, ii, *OO*, III, 398; *DRV*, lxviii, *OO*, III, 268; *S*, xxi, *OO*, III, 671.

37 *S*, ix, *OO*, III, 545: "Palam est igitur naturam in cunctis solicitam mirum in modum fuisse, nec obiter, sed ex sententia omnia praevidisse, hominesque quibus hoc beneficium Deus largitus est, ut causam rerum primam inveniant, participes esse illius primae naturae: neque alterius esse generis naturam, quae haec constituit, ab illorum mente, qui causam eorum, car ita facta sint, plenè assequi potuerunt." This transition from nature to God is found in other Renaissance texts: see Ian Maclean, "Montaigne and Philosophical Speculation," in *Montaigne*, ed. I. D. McFarlane and Ian Maclean (Oxford, 1982), pp. 110–12.

38 *S*, xvi, *OO*, III, 608; *DRV*, c, *OO*, III, 348.

39 Notably *S*, i, *OO*, III, 357–8; *DRV*, i, *OO*, III, 1.

40 *E*, preface, ã7; Rudolf Göckel, *Analyses in exercitationes aliquot* (Marburg, 1599), i. 3, pp. 9–11: "An dialectica disputet de omni ente?"

41 On this distinction, see p. 245.

42 See especially *S*, xix, *OO*, III, 655–61.

43 Schenk von Grafenberg's books, *Observationes medicae*, published between 1584 and 1597, are frequently quoted as sources of scientific information, although much of his material is, by his own admission, hearsay. Cf. Foucault's comments on the equivalence of "lire" and "voir" in Renaissance thought (*Les Mots et les choses*, pp. 56–8). It may be inaccurate to assert that Harvey and others *never* accepted observations made by others, just as it may be inaccurate to assert that they never indulged in "thought experiments," but always recorded experiments they had actually carried out. But the distinction made here would seem to be defensible in general terms.

44 *S*, ii, *OO*, III, 390: "Admirebitur forsan aliquis, quòd in Contradicentium libris aliter senserim. Sed ubi opiniones antiquorum sequi propositum fuit, his verò docere veritatem."

45 *S*, i, *OO*, III, 357.

46 Ibid.

47 Ibid., ii, *OO*, III, 383; see also Scaliger, *E*, i.1, p. 1; Aristotle, *De generatione et corruptione*, ii. 2 (329b32–330a5).

48 *S*, i, xxi, *OO*, III, 357, 671; Emilio Parisano, *De microcosmica subtilitate pars altera* (Venice, 1635), p. 27.

49 *S*, i, *OO*, III, 357: "Cum nulla sit authoritas adversus experimenta scribentibus."

50 *E*, i. 2, pp. 4–5; cccxxii, pp. 1025–6; ccxl, p. 1068.

51 Ibid., i.1, pp. 1–4; Ernst Cassirer, *Das Erkenntnisproblem in der Philosophie und Wissenschaft der neueren Zeit* (Berlin, 1922–3), I, 120–52, esp. pp. 122–4; D. R. Kelley, *The Foundations of Historical Scholarship* (New York and London, 1970), pp. 29–32. On reality as a construct of the human mind (or as a function of Platonic Ideas), see Robert Westman on Proclus (Chapter 5 of this volume) and Brian Vickers on Van Helmont (Chapter 3).

52 For example, *E*, cccxxi, p. 1025: "Si subtilitas sit in difficultate cognitiones essentiarum, et caussarum: subtiliores eae erunt scientiae, quae longius à sensu distant." The problem of the articulation of senses and intellect (or nature and convention, or real and nominal) is too complex to be discussed adequately in the context of this chapter, but see, in general, Tzvetan Todorov, *Théories du symbole* (Paris, 1977).

53 Göckel, pp. 1–7.

54 Se *Actio in calumniatorem*, *OO*, III, 679 (a reply to *E*, i.1): "Hac in parte nescio quid magis demirer, an stuporem, an livorem, an ineptiam deducit me ad subtilitatis interpretationem ex Cicerone, cum ego non de verbo librum faciam, sed de significato, quòd ego ex primo hoc nomine tanquam proximiore huic multo quam sua intelligibilitas, quo homo Latinissimus, ac Ciceronianus pro intelligentia utitur . . . propterea declaravi quid intelligi vellem . . . etenim parum videtur hic assuetus lectioni Galeni, qui toties clamitat non debere nos de verbis litigare, modò de re constet." Cardano makes reference here to Scaliger's *De causis linguae Latinae* and his attack on Erasmus, and to Galen, *Methodus medendi*, xiv.

55 See Robert Boyle, *The Origin of Forms and Qualities* (1666), in *Works* (London, 1744), II, 466; John Locke, *An Essay Concerning Human Understanding* (London, 1690), iii.6 ("Of the Names of Substances"); and Richard I. Aaron, *John Locke*, 2nd ed. (Oxford, 1965), pp. 121ff.

56 Michel de Montaigne, *Essais*, iii.13, in *Oeuvres*, ed. A. Thibaudet and M. Rat (Paris, 1967), p. 1043. See also Ian Maclean, "Montaigne and Cardano," *French Studies*, 37 (1983), pp. 143–156, on the connection between these writers.

57 *E*, i.1, p. 2, and the many taunts addressed to Cardano on his lack of subtlety (e.g., civ. 1, p. 382: "De subtilitate loquentem non subtiliter loqui dedecet"; cclv, p. 790: "Licet spectare te in subtilitate scena non subtiliter agentem"). Cardano had almost invited this critique by admitting that writing about subtlety was more difficult than subtlety itself (*S*, i, *OO*, III, 357).

58 See note 31. There is usually a logical aporia in these references; Cardano refers his reader to the inaccessible *Arcana aeternitatis* on those matters that are "supra humanam mentem," yet apparently known to Cardano and used by him as principles or axioms (e.g., *S*, i, *OO*, III, 358).

59 Ibid., xxi, *OO* III, 672; *DRV*, c, *OO*, III, 349.

60 *S*, xix, *OO*, III, 656: "Daemonas ipsos esse et vagari . . . ego qui numquam daemonas vidi." See also *DRV*, xciii, *OO*, III, 317–36.

61 *S*, i, *OO*, III, 357: "palam mentiantur"; cf. the reference to "ambiguae et fabulosae authoritates" (*S*, xv, *OO*, III, 588).

62 See Foucault, pp. 50–8.

63 *S*, i, *OO*, III, 359: "ab evidentibus rationibus demonstrari."

64 Ibid., III, 357, quoted in note 49.
65 "Cum ergo stultum sit ratione pugnare contra sensum et experientiam, pro antiquitatis reverentia" (Jean Riolan, *Ad librum Fernelii de procreatione hominis commentarius* [Paris, 1578] fol. 17ᵛ, quoted by H. B. Adelmann, *Marcello Malpighi and the Evolution of Embryology* [Ithaca, N.Y., 1966], II, 753).
66 *S*, i, *OO*, III, 357-9; *DRV*, i, *OO*, III, 1.
67 *E*, passim, and "Excusatio," p. 1130: "In hac humana caligine rerum omnium ignaros esse nos." It should be pointed out, however, that at one point (*S*, xii, *OO*, III, 562) Cardano concedes the peripatetic point about the infinity of particulars.
68 Ian Hacking, *The Emergence of Probability* (Cambridge, 1975), pp. 54-6.
69 *DRV*, xcix, *OO*, III, 346; *S*, xvi, *OO*, III, 598; *E*, cccxxi, pp. 1025-6; Göckel, pp. 9-13.
70 Cf. Cornelius Agrippa's definition of a declamation: "Proinde declamatio non judicat, non dogmatizat, sed quae declamationis conditiones sunt, alia joco, alia serio, alia falsè, alia severè dicit: aliquando mea, aliquando aliorum sententia loquitur, quaedam vera, quaedam falsa, quaedam dubia pronunciat . . . multa invalida argumenta adducit" (*Apologia adversus calumnias, propter declamationem de vanitate scientiarum . . . intentatas,* xlii, in *Opera* [Lyons, n.d.], II, 326-7). Cf. also the remarks made earlier in this chapter about the rhetorical strategies found in occult writing.
71 *S*, xvii, *OO*, III, 615.
72 *E*, xvi-xviii, pp. 76-91; Agrippa, *De occulta philosophia,* iii, pp. 4-6.
73 *S*, ii, *OO*, III, 382; *Categoriae,* v, 4a30-1; *E*, xviii, pp. 86-91.
74 *S*, ii, *OO*, III, 381; cf. the disagreement between Cardano and Scaliger on the nature of the vacuum (*S*, i, *OO*, III, 359; *E*, v.2, p. 15). See also Göckel, p. 8.
75 See Ian Maclean, *The Renaissance Notion of Woman* (Cambridge, 1980), pp. 1-4; Skalich de Lika, p. 77 (the list is attributed here to Iamblichus and Proclus).
76 *S*, ii, *OO*, III, 372: "Natura enim semper extrema mediis iungit." Cardano's terms here are, of course, realist, unlike those of Scaliger, as he locates opposites in nature and not in the conventional categories of words.
77 *S*, i, passim, which also contains a version of privation or steresis (*OO*, III, 359: "materia prima qualitatem quandam retineat, quam indefinitam vocamus").
78 *E*, i-v, pp. 1-16; ccxliv.2, p. 1074 (on sympathy/antipathy).
79 *S*, ii, *OO*, III, 373; *E*, cccvii, pp. 917-97 (vs. *S*, xiv, *OO*, III, 582) attacks Cardano for other revolutionary gestures in his text, such as the readjustment of faculty psychology.
80 *S*, i, *OO*, III, 357-8; *DRV*, i, *OO*, III, 1.
81 See David Hull, "The Effect of Essentialism on Taxonomy," *British Journal for the Philosophy of Science,* 15 (1965), pp. 314-26, and Phillip R. Sloan, "John Locke, John Ray and the Problem of Natural Systems," *Journal of the History of Biology,* 5 (1972), pp 1-53.
82 *S*, xiii, *OO*, III, 571: "Aristoteles [coloris genera] in septem dividit, eisque totidem coaptat sapores, ut iucundissimi iucundissimis, tristissimi tristissimis, medii mediis respondeant. Nos postquam ad septem redigisse conspeximus cum nullum numeris tribuat honorem, rati è numero erraticarum sumpsisse, erraticis colores, et sapores dicavimus" (cf. *E*, ccxcviii. 10, pp. 378-80). Also *S*, vi, *OO*, III, 452 (the alignment of metals

and planets): "Haec [metalla] septem esse iuxta planetarum numerum quis existimabit?" See also *DRV*, lxxxix, *OO*, III, 309.

83 *S*, xi, *OO*, III, 557; *DRV*, xlvi, *OO*, III, 177; *S*, xii, *OO* III, 561; *S*, vii, *OO*, III, 459; *Arcana aeternitatis*, ix, *OO*, X, 14–16; *S*, ix, *OO*, III, 520; *S*, xiv, *OO*, III, 585–6.

84 *S*, ii, ix, x, *OO*, III, 400, 507, 522, opposed by *E*, cxxxix, pp. 463–6; clxxxii.i, pp. 597–9; cxcvii, pp. 638–40; also *E*, clxxxii, p. 598, on numbers. There is a certain irony in an Aristotelian attacking others for proliferating meaningless classes; see Hull, "Effect of Essentialism."

85 But *S*, xxi, *OO*, III, 671 distinguishes "causa," "principium," and "occasio" in relation to God.

86 Le Blanc ("Epistre," *ii) seems to hold that these terms are synonymous: "Cardanus . . . decrit les causes occultes, raisons, vertus et propriétés de diverses matières non vulgaires."

87 *S*, ii, *OO*, III, 383.

88 *S*, xviii, *OO*, III, 638: "Sympathiam voco consensum rerum absque manifesta ratione: velut antipathiam dissidium." See also *E*, cccxliv.2, p. 1074, and Göckel, who apparently concedes that there are sympathetic events (*Analyses*, p. 4: "Huc refero vim occultam: unde multa admiratione dignissima existunt"). The examples he gives are a corpse bleeding as its murderer goes near it and a wound being healed by anointing the weapon that caused it.

89 See *S*, iii, xxxi, passim.

90 *S*, x, *OO*, III, 545, quoted in note 37.

91 See Maclean, *Renaissance Notion of Woman*, pp. 33, 45.

92 *S*, vi, *OO*, III, 454; *E*, civ. 5, p. 385 (trans. Morley, II, 178).

93 For example, *E*, xvi.2, p. 81: "Quae sequuntur, ostendunt, quod et in tuis antilogiis observavimus, te dialecticas leges, ut levissimè loquar, neglixisse."

94 See above, note 43. Cardano accuses Aristotelians of this (*S*, ii, *OO*, III, 386: "Aristotelici, quo audacter in his, in quibus coargui non possunt, litigant"). In accusing Cardano of the same usurpation of authority, I am exposing my own text to the possibility of a similar accusation, and to the perils of infinite regress.

95 *S*, xviii, *OO*, III, 638; cf. the "indefinite quality" of prime matter (*S*, i, *OO*, III, 359).

96 *S*, xi, *OO*, III, 555–6; *E*, cclxvii, pp. 811–12.

97 "In omni falso, veri, in omni vero, falsi aliquid contineri"; the title is given in *De libris propriis*, *OO*, I, 68.

98 *S*, xviii, *OO*, III, 635ff.; the aporia inherent in "praeter fidem" (How can you believe what is beyond belief?) causes Le Blanc as translator to attenuate this title to "Des Inventions merveilleuses, et de la manière de representer choses diverses, *presque* incredibles" (my italics).

99 Interestingly, Cardano seems to identify analogy with tautology: "Ergo quonam pacto motus calefacit, inquirunt Aristotelici, ac multa nugantur: tandem verò ad id redeunt, ut calor sit effectus motus: estque hoc ac si dicerent, Nescimus. Idem enim per idem ostendere, certum nugacis atque imperiti argumentum est" (*S*, ii, *OO*, III, 381). The last sentence is translated by Le Blanc as: "Car demonstrer une chose par chose semblable, c'est un vrai argument d'un blasonneur ignare" (*Les Livres de la subtilité*, fol. 32ᵛ).

100 Galileo Galilei, *Opere*, Edizione Nazionale (Florence, 1964–6), II, 462; quoted by J. P. Larthomas, "A Propos de la Méthode résolutive chez Galilée," *Université de Nice: cahiers du séminaire d'épistémologie et d'histoire des sciences,* 9–10 (1980), p. 33.

7

Kepler's attitude toward astrology and mysticism

EDWARD ROSEN

Among the friends of Johannes Kepler (1571–1630), the name of David Fabricius (1564–1617) stands out for two reasons. First, as an observational astronomer he initiated the study of variable stars. Second, he was an ardent devotee of astrology. In keeping with the latter enterprise, he collected horoscopes. On 18 July 1602 Kepler sent his own horoscope to Fabricius.[1] Dissatisfied with the lack of detail in what he had received, Fabricius pressed Kepler for additional information. In particular he wanted to know "on what day in the carnival season in the year [15]91 a fever attacked" Kepler.[2]

Kepler's reply is preserved only in a copy prepared by a hired scribe. This copyist did not always understand what Kepler had written in his draft of the letter. In some cases the copyist made mistakes; in other cases he simply omitted what he could not read. Thus, where Kepler answered Fabricius's question about the fever in 1591, the surviving copy says:

> In the year 1591 on the Friday [1 March] following Ash Wednesday [27 February] a headache marked the beginning of a very acute fever that lasted 8 days and nearly killed me. If I remember correctly, the sun was 90° from Mars. After the preceding Christmas holidays [in 1590], as I was leaving the church and the services I suffered[3] very much from the extremely bitter cold. Hence, from my illness during the previous autumn[4] [of 1590] there had been remnants, which erupted during the carnival [in 1591]. Shortly before that time there was a remission of my skin ailment, to which had been added an agitation of body and mind due to the excitement of the play in which I performed the part of Mariamme.[5]

253

This (lost) Latin play by an unidentified dramatist highlighted the beheading of John the Baptist. Actresses being forbidden to perform, female parts were played by male students of slender build, like Kepler, who portrayed Mariamme. When Herod Antipas the Tetrarch celebrated his birthday, the voluptuous dancer of the veils demanded as her reward the head of John the Baptist. Our ancient sources do not specify the presence at this gruesome feast of anybody called Mariamme. But this was a favorite name among the women belonging to the family of Herod.[6] Which one of the many Mariammes was inserted in the cast of this play is known today no better than is the name of its author. But at least Kepler was not expected to execute the dance of the veils. This performance was staged out of doors in the marketplace of Tübingen on 17 February 1591, under wintry conditions that Kepler was not robust enough to withstand.[7]

Some such view may be our present-day understanding of this medical situation, but it was not Kepler's. As he wrote to a leading astrologer, "from the planetary configurations he cannot find the reasons why he suffered such an acute and intense fever in the carnival of the year 1591."[8] Kepler admitted his inability to find an astrological cause of his fever in a letter addressed to Helisaeus Roeslin (1544–1616). Like Kepler's letter a decade later to Fabricius, his letter to Roeslin has not been preserved. But whereas his lost letter to Fabricius can be dated 18 July 1602, his lost letter to Roeslin can be dated only conjecturally.

In this lost letter to Roeslin, Kepler withheld his name. But he supplied his horoscope, timing his birth at 2:30 a.m. on 27 December 1571.[9] He also described himself as a master of arts, having received that degree from the University of Tübingen on 11/21 August 1591.[10] In his own handwriting Roeslin replied on 17 October 1592,[11] explaining at the close that he "wrote these things with a very rapid pen, to satisfy your request to some extent, without reading them over."[12] It would therefore seem that Roeslin did not keep his correspondent waiting long. Between Kepler's attainment of the master's degree on 11/21 August 1591 and Roeslin's reply on 17 October 1592, some fourteen months elapsed. With due allowance for the time required to transmit a letter from Kepler in Tübingen to Haguenau in Alsace, where Roeslin then resided, Kepler may have consulted that eminent astrologer in the summer of 1592. This conjectural conclusion is supported by Kepler's comment in this lost letter to Roeslin that "those close encounters [between heavenly bodies; *occursus*] according to computation do not occur when the individual is passing through his 21st year."[13] Kepler completed his twentieth year and began to pass through his twenty-first year on 27 December 1591, so that the summer of 1592 seems a likely time for his admission of astrological puzzlement and consultation of Roeslin, who was about a quarter-century older and a widely recognized authority on astrology.

Kepler's reference to the twenty-first year elicited from Roeslin the following response:

> I reply that what I have learned in astrology precludes us from being able to restrict such things to years, not to mention days, especially since we are not absolutely certain about the minute in the hour [of the nativity]. And if the master [of arts] is assumed to have been born only 20 minutes before 19°30' Gemini rises, the nativity will occur with Mars in the aspect of quadrature while the master is in his 21st year, a configuration which surely could cause such a burning fever.[14]

According to the horoscope supplied to Roeslin by Kepler, he was born as 24° Gemini[15] was rising, with Mars close to, but not exactly in, the aspect of quadrature. Kepler made his astrological attitude toward Mars quite clear in a letter to Fabricius of 2 December 1602: "Regard this as certain, that Mars never crosses my path without involving me in disputes and putting me myself in a quarrelsome mood."[16] Mars had always been regarded from ancient times as a maleficent planet.[17]

In his reply of 17 October 1592 Roeslin further advised Kepler:

> I have learned that when two configurations come so close together, it happens that they sometimes diffuse their effect, one of them by delaying it, the other by advancing it. Anybody would be thoroughly mistaken who wants to restrict the effects emanating from the configurations to a particular year, let alone month and day. It is certain that the stars exert their effect, especially those outstanding configurations, as in this case the quadrature of Mars with the nativity. But the matter is not so certain that we can assign it to a definite time. For many details occur which conflict with such general rules of the heaven, so that the effect is either advanced or postponed. In addition, the motions of the heavenly bodies are not understood well enough, so that whole degrees will be missing, not to mention minutes. But one degree [in the nativity] corresponds to a whole year in the configurations. In like manner, a quarter of an hour in the nativity corresponds to four whole years. It is therefore safest for the astrologer making predictions to stick to generalities. Let him say: around this age a burning fever would come, and this person would be in danger of losing his life, that is to say, around these or those years, and this may well happen earlier or later.[18]

Roeslin felt that the current astronomical predictions of planetary positions were not absolutely precise, and therefore the astrology based
on them should be satisfied with approximations. His attitude recalls
the view of Claudius Ptolemy, the most influential astronomer of antiquity, who in his *Mathematical Syntaxis* undertook to explain how
to understand and predict the motions of the heavenly bodies endlessly
pursuing their changeless courses. This strictly astronomical treatise,
long miscalled the *Almagest*, was the propaedeutic to his astrological
work in four books – *Tetrabiblos* or *Quadripartitum* – setting forth the
more difficult, because more uncertain, method of foretelling changes
on earth. The eminent French empiricist Pierre Gassendi (1592–1655),
who assailed astrology in his posthumous *Syntagma philosophicum*,
first published in 1658, three years after his death, disputed the authenticity of the *Tetrabiblos:*

> Is there anybody who would be convinced that the *Tetra-*
> *biblos* is not spurious? That work was ascribed long ago to
> Ptolemy, undoubtedly because he was very famous on ac-
> count of his knowledge of heavenly phenomena . . . In the
> preface to his *Syntaxis* he declared that he undertook to ex-
> pound mathematics because it is a rational and unchallenged
> science. On the other hand, he ignored theology and phys-
> ical speculation, because they may both be labeled conjec-
> ture rather than established science, the former on account
> of the incomprehensible nature of divinity, and the latter on
> account of the variable condition of the subject, with the re-
> sult that because the material is not understood, philos-
> ophers never agree about it. Since Ptolemy felt this way, I
> say, could he later downgrade himself to thinking that he
> should embrace a study far more uncertain than theology or
> physics?[19]

Gassendi, an omnivorous reader who was thoroughly familiar with Kepler's publications, surely knew that Kepler regarded Ptolemy as the
author of the *Tetrabiblos*. As Kepler pointed out, it is addressed to a
certain Syrus,[20] to whom the *Syntaxis* and other genuine works of
Ptolemy are also addressed. In the main, modern critical scholarship
has sided with Kepler, as against Gassendi, on the ground that the two
works share a common vocabulary, style, and conceptual basis.[21]

Although Kepler accepted the *Tetrabiblos* as an authentic work of
Ptolemy, he did not agree with everything in it, just as he rejected parts
of the genuinely Ptolemaic *Syntaxis*. Traditional astrology clung to certain practices that Kepler tried to trim away as deleterious blemishes.
But he refused to align himself with those who sought to destroy astrology outright. In the battle swirling around him for and against astrology, he called himself the "third man in the middle." *Tertius in-*

terveniens was the Latin title he bestowed on the discussion he wrote in German and published at Frankfurt/Main in 1610. On the title page itself he referred to "star-gazing superstition [*sternguckerischer Aberglaube*]" as a conspicuous indication of where he stood in the controversy. On the other hand, he "warned . . . theologians, physicians, and philosophers . . . against throwing out the baby with the bath, and thereby maltreating their profession."[22]

As the "third man in the middle," Kepler enunciated 140 theses. In Thesis 39 he declared:

> The astrologers are accustomed to cast the nativity of every year, just as though it were another person being born . . . Now I cannot deny that this is a ridiculous fantasy, especially because a person is born in one moment with skin and hair. On the other hand, the year is not such a complete being. For when spring is in season, summer is not yet here; and when it comes, spring has already passed. A person is an earthly individual being, affected by heaven. The year is nothing but the heavenly motion itself, of which its supposed nativity, that is, the first day in spring, is a part. Consequently one day has no power to govern another day or alter it, but they must all together pass by according to the divinely established pristine order, each in its own special way.[23]

Just as Kepler denied that astrologers could foretell the character of the coming year from its first day, so he held that a person's future could not be predicted from the horoscope:

> Thus it is also not credible that it can be seen from the horoscope how things will work out for anybody with certainty. In general, everybody is the master of his fate, as may be indicated on the whole. Yet there are many more accidental causes than merely the heaven or the individual's feelings and habits, each of which by itself can produce a conflict in the person and lead him astray.[24]

A playwright born seven years before Kepler put in the mouth of his crafty character Cassius the famous lines:

> The fault, dear Brutus, is not in our stars,
> But in ourselves, that we are underlings.[25]

About a decade after Shakespeare's Cassius rejected astrological determinism, Kepler quoted with approval the "weighty saying: the stars incite, they do not control."[26] Convinced that a horoscope portended rather than coerced, Kepler spurned the

> completely worthless, gratuitous, superstitious, soothsaying predictions that the newlyborn's wife would be born in this or that country, have a hidden defect in her body, would not

remain faithful to her husband, have so or so many children, and the newlyborn would have two, three, or more wives. Since this is true, so is its application. With the lord of the seventh [house] in the tenth, if he is beneficent, if he is Jupiter, if he is in his own house, that signifies a rich wife; Venus in Saturn's house, an old wife; in the eighth, a widow; Mars in Venus' house and in trine aspect with the moon, a promiscuous wife; Venus in the rays [within 15° of the sun], a sick wife. Concerning these and similar lords of the houses, and the worldly happiness or misery deduced from them without the man's nature intervening, I say bluntly that I have no regard for them. In my opinion, this embellishment was devised to brag about ingenuity to people. For since they ask many questions, the astrologer thinks of a way to give many answers, God grant whether he finds it in nature or not.[27]

In trying to preserve what he believed was sound in his nondeterministic astrology, Kepler declared: "I have no intention to defend the predictions of individual future events insofar as they depend on a person's free will."[28] Although he practiced bloodletting on himself, Kepler repudiated the traditional astrologers' system of apportioning to the zodiacal signs the human limbs from which blood was to be drawn:

I do not hereby wish to have defended those fantasies . . . about assigning a person's limbs to the twelve [zodiacal] signs, scheduling blood-letting according to these assignments, dividing the twelve signs among the planets, and the recurring signs. For these childish observations have nothing in common with my thinking.[29]

By the same token Kepler ridiculed the traditional astrologers for attributing to the heavens the superiority of one country's products over another's:

The astrologers can indeed be fools, since they want to squeeze out of their art the reason why one country produces something better than another, that is to say, when they look for the relevant reasons in Terrestrial Triangles and Planetary Dominations.[30]

By contrast, Kepler aligned himself with the natural philosophers who arrive at the causes to some extent and find that these are disposed in accordance with the sun and its heat. In Italy there is good, spirited wine, since the countryside faces the noonday sun. Along the Rhine there is also much wine, but gentler, because the countryside faces north, and yet has deep valleys to retain the heat. Along the upper Danube there is no wine, because the countryside is not protected against the harsh air currents from the snowy mountains. But down below in Austria and Hungary there is good, strong wine because the land faces west and south, and begins to become deep down between very high mountains. The Elbe produces little wine, because the countryside faces north and is flatter than other regions.[31]

To explain why certain events took place on earth, traditional astrologers linked them with heavenly phenomena, seeking to demonstrate a relation of cause and effect. Such would-be demonstrations were condemned by Kepler as fallacious. As an example he chose

the conjunction of Saturn and the moon as the purported cause of a Jew cheating someone. For if this conjunction happens on Saturday, in Prague [where Kepler was then living] nobody is cheated by any Jew. On the other hand, several hundred Christians are cheated daily by Jews and vice versa, yet the moon runs below Saturn only once in a month.[32]

The traditional astrologers were condemned by Kepler because they "claim complete right for themselves to imagine, lie, deceive, and say whatever they want about the heaven, which is blameless."[33] Yet he himself blamed Mars for involving him in disputes and making him quarrelsome.[34] He renounced the horoscope as disobedience to religious teaching ("there shall not be found among you anyone . . . that useth divination")[35] and as disregard of reason and nature:

Suppose someone came to me and asked me to tell him whether his friend in a distant land were alive or dead, or whether his sick [friend] would recover or die. If I cast this questioner's horoscope, and told him yes or no, then I would be a soothsayer and he would disobey God's commandment about superstition, not only on account of the questioner's purpose and belief, but because the means I used in this instance would be absolutely irrational and unnatural.[36]

Yet when his own health was in question, Kepler had full faith in his horoscope, as we saw above in his letters to Fabricius and Roeslin.[37] For professional purposes he compiled a stock of about eight hundred horoscopes, on which he entered the dates and the planetary configurations with his own hand.[38] The most famous product of this horoscope factory was the prediction that Kepler prepared and later revised for Albrecht von Wallenstein (1583–1634), the military adventurer who was Kepler's last patron. The patron who had first appointed him imperial mathematician was Rudolph II (1552–1612), for whom Kepler prepared a report in 1611 concerning the assassination of King Henry IV of France on 14 May 1610:

> In the horoscope of the assassinated French king, for last May nothing is found, but there is something two years earlier, according to my calculation, or two years later, according to the calculation of Dr. Camerarius in Esslingen.[39] Yet it is true that on 14 May there was a conjunction of Mars and Venus 90° from Saturn. According to the significance of this aspect, one might write about sorcery and poison.
>
> I had often most humbly advised your Imperial Majesty that heaven alone can accomplish nothing. To this I now add that obviating a lot of trouble requires the help and good will of many people, without whom nothing happens. Nobody but yourself is concerned to relieve Your Majesty's struggle.
>
> Of course, in three successive years heaven has had evil configurations, and now Saturn is moving into opposition to Jupiter, which ruled the realm. Moreover in October, December, and June strong aspects are coming, which are related to your Majesty's birth. On the other hand, things should be much better because Jupiter is moving westward.
>
> The trends which emerged three years ago in connection with your Majesty [in June 1608 Rudolph II had been compelled to cede control of Austria, Hungary, and Moravia to his younger brother, Matthias], after the appearance of the comet at the end of the year 1607, happened under very evil configurations and are accordingly so extremely hostile to your Majesty. Because these trends will not die out by themselves, it is to be feared that your Imperial Majesty will not suppress them by force. For the evil configurations have entwined themselves with you and shaped you. Even though your Majesty faces several good configurations again, yet these are not as strong in their beneficence as the preceding configurations in maleficence. In part the consequence is that as often as your Imperial Majesty lays hands on the aforementioned trends with the intention of getting rid of them, you only harm yourself thereby, and attract discouragement and sickness therewith, while also vainly striving for the happiness portended by the present and future configurations.
>
> Consequently, if I were your Imperial Majesty's confidential adviser, and knew your Imperial Majesty's thoughts, feelings, and wishes, as well as the condition of everything, I would want to consider whether perhaps your Imperial

Majesty might not start something completely different, having nothing in common with the past agreements, whether for them or against them, and yet turning out to be best for your imperial administration and government. In such a situation, which is quite possible, you would make better use of your favorable and lucky configurations, and might accomplish a more fruitful result. This would proceed more splendidly if your Imperial Majesty decided to dismiss all these past vexatious agreements from your mind, and await the time when you yourself (as usually happens through God's vengeance) will go on the offensive and be a destroyer.[40]

Also in the year 1611, on Easter Sunday, 3 April, Kepler drew up a strictly confidential memorandum addressed to a court official, whose name he prudently withheld, as he also withheld his own. Just below the date, he wrote:

Never mind the salutations and titles, which are obligatory, but betray what is secret. I trust that you will recognize a man of German dependability. I draw the emperor's pay, and am not corrupted by the Bohemians and Austrians. After one or two contacts I deliberately refrain from talking to them. I am writing more freely to you, who are on the emperor's side, not only because your reputation but also my eyes and ears tell me about your good services.

Among the other things in yesterday's conversation, I said without wasting a word that "Astrology inflicts severe damage on monarchs if some cunning astrologer wants to fool around with people's gullibility." I think I must make an effort to stop this from happening to our emperor. The emperor is gullible. If he hears about that Frenchman's prediction,[41] he will give him great credit. Hence it is up to you, who are the emperor's adviser, to find out whether this is what the emperor is doing. For I believe you see that, if the foundations of sound management are missing, all confidence is empty and harmful. I now regard it as practically certain that the rumor about the French prediction has reached the emperor's ears.

Popular astrology, believe me, is a technique, and with a slight effort it can be induced to say what pleases both sides. For my part, I am absolutely convinced that not only popular astrology but also that astrology which I understand agrees with the nature of things should, according to my deepest conviction, be kept apart from discussions as difficult as these are. Of course, I do not offer this advice as though you needed it in official meetings, where I know it is customary to propound no arguments on this basis. But this little fox insinuates itself much more furtively, at home in the bedroom, outside on the street, inside in the mind, and sometimes it lets drop what somebody corrupted by it may introduce in the council, while concealing the source.

I was asked about the decrees of the stars by the side which I know is opposed to the emperor. I answered with what I think is not of any importance in itself, but with what impresses the gullible, to wit, the emperor's advanced age [Rudolph II, born in 1552, was then nearly sixty years old], and the absence of evil configurations. There were, of course, evil conformations and eclipses, but these had already occurred two or three years ago. On the other hand, Matthias [the emperor's younger brother, who was trying to push him off the throne] is threatened by disorders because Saturn is approaching the sun, and there will be a great opposition of Saturn and Jupiter in the sun's place. I say these things to the emperor's enemies because, even if they are not frightened thereby, they are certainly not made confident. To the emperor himself I would not want to say these things because they are not important enough to be relied on, in my opinion. On the other hand, I am afraid that they may strengthen the emperor's imprudent disregard of the ordinary channels which can perhaps lead to the intervention of loyal princes. In this way astrology might push him into much greater misfortune than he now faces.

To you, on the other hand, because you are loyal to the emperor, I shall plainly say, as I never shall to Matthias and the Bohemians, what I seriously think on the basis of the sounder astrology about the cooperation of the stars in these disturbances. In the meantime, however, I would not want anybody to rely on the stars while paying less attention to impending developments of the situation and to travelers on earth.

Matthias [1557–1619] has already passed through several very difficult configurations, by Jove, corresponding to his career: in 1566, the moon in quadrature with Saturn; in 1595, the sun's relation to Saturn. For in 1594 (the match is precise enough, with no more to be expected from the stars) he fared very badly at Esztergom and Györ, on the island.[42] This was like 1589, with mid-heaven opposite Saturn, and like 1605, 1606, with the sun in opposition to Mars, when there were uprisings in Hungary, and the archdukes coming to Prague and recommending Matthias to the emperor made the emperor much more antagonistic to him. From that time on, the configurations and recurrences were favorable: in 1606, the moon's trine aspect with Mars; in 1607, the moon's 60° aspect with Jupiter; and now in the current year [1611], the alignment of mid-heaven with the [heavenly] body Mars, a configuration which is stormy but potent, as it is in the forefront. Next year follows, with mid-heaven 60° from Jupiter, then the rising sign 60° from Mars (a feverish but favorable configuration for the time being), and finally the rising sign with the [heavenly] body Jupiter. Here I (astrologically, at any rate) believe that all objective events will turn out to be happy and honorific, and the emperor's destiny will pass to him. For, each of them has 60° between Jupiter and Mars, and of course the emperor passed through the same configurations on his way to the throne. The one and only enemy very harmful to him, in my opinion, will be the coming liquefactions. But even though Saturn is approaching the sun and a great opposition is taking shape in the place of the sun, nevertheless the same configurations occurred to the emperor too in 1593, 1594. Therefore, just as for the emperor a war began at that time [the Long War of 1593–1606 against the Turks], which was of course horrible, yet it turned out to be fortunate, in that he emerged in a strong position from that war, so Matthias can also hope for the same outcome, since Jupiter's approach to the rising point portends everything favorable.

On the other hand, the emperor has unfavorable configurations, with mid-heaven opposite the rays coming from Venus and Mercury, whereas Matthias has the moon in the rising point, very nearly 90° away from the rays of the sun, which is opposite Mars in Matthias's horoscope.

If any astrologer saw these configurations and took them into account, and if it were up to him to advise each of the rivals at the same time, he would naturally make Matthias extremely confident, but the emperor fearful. For my part, as I said, nothing is to be built up, in my opinion. Indeed, I have written and analyzed everything with the thought that you would derive from it an idea of how much credit should be given to the French prognostication: absolutely nothing at all, of course.

In short, I declare that astrology should vanish not only from the council but also from the very minds of those who today want to urge the best course on the emperor, and in

the same manner it should be kept completely out of the
emperor's sight.[43]

Kepler urged his highly placed friend to keep astrological predictions
away from Rudolph II, who was "gullible," as Kepler bluntly said. If
Rudolph II were ever convinced (as, for instance, by the French prognostication) that the stars were against him, he would fail to resort to
the nonastrological measures needed to save his throne from his enemies, who were seeking to oust him. Under such circumstances astrology could become a powerful weapon in political controversies.
The potency of the weapon depended upon the reputation of the astrologers. Nowadays the stockbroker who correctly forecasts the upward or downward movement of the prices of securities attracts the
greatest number of clients. In such enterprises accurate "prediction
after the event" helps to build a reputation among the gullible.

Kepler's acceptance of a sound astrology unintentionally left one of
his publications vulnerable to such distortion when he published a work
dealing mainly with the comets visible in 1618. Book III of this work,
On the Significance of the Comets, was finished and sent to the printer
on 17 May 1619,[44] almost exactly two months after the death of Rudolph
II's successor, Emperor Matthias, on 20 March 1619. His decease elicited from Kepler the following astrological reflections:

> Emperor Matthias, of most sacred memory, was born as
> Scorpion was rising. From that very same place the [third]
> comet [of 1618] arose. Originally southern, it became north-
> ern, that is, by running transversely along the ecliptic and,
> as it were, cutting this natural thread of the life of the living
> (on account of the sun's involvement with the comet). On
> astrological grounds Emperor Matthias could probably be
> predicted by me (he understands, whose business it was to
> understand), the emperor being the very person signified by
> the comet according to me. But the question whether he was
> going to die, even though this could not be rigorously de-
> duced from the comet, was nevertheless answered by his in-
> firm old age and his health, seriously undermined for two
> years. Consequently there was no need for the comet to
> forecast what could be foreseen as about to happen soon in
> accordance with nature's laws. But as regards what was
> going to happen as a consequence of his death, I am abso-
> lutely convinced that we are summoned by the comet (if we
> are summoned to any particular result) to analyze these de-
> velopments with very attentive minds and intense interest.[45]

As the imperial mathematician, Kepler believed that the third comet
of 1618 was a heavenly sign concerning Emperor Matthias. But Kep-

ler's treatise on the comets said quite plainly that the date of Matthias's death could not be deduced from the comet. As a natural philosopher, Kepler attached great weight to Matthias's advanced age and ill health. But even these considerations did not authorize Kepler to go beyond saying that Matthias would die "soon."

Yet the biography of Kepler, upon which all subsequent biographies of the astronomer-astrologer were based, went far beyond attributing to Kepler the simple statement that Matthias would die soon. Michael Gottlieb Hansch (1683–1749), a professor in the University of Leipzig, acquired the Kepler papers in 1707 and published a selection of nearly five hundred letters, for which his biography of Kepler served as an introduction.[46] There Hansch said of Kepler in 1618 that "in this year he had predicted the death . . . of Emperor Matthias in six M's: *Monarcha Mundi Matthias Mense Martio morietur.*"[47] In keeping with his usual practice, Hansch cited Letter 328 on page 520, right-hand column. But this letter says nothing about Matthias's death. Nor does another work, also cited by Hansch. This lack of sound documentation did not deter a popular anecdotal historian of the courts of the various German princes since the Reformation from expanding Hansch's six Ms to seven: "A great sensation was caused by the fulfillment of Keppler's prognostic of seven M's, drawn for the year 1619: *Magnus Monarcha Mundi Medio Mense Martio Morietur* (the great monarch of the world will die in the middle of the month of March)."[48] In our own time, when the popular acceptance of astrology has surpassed the level it attained in the Renaissance, the author of *The Hapsburgs* repeated what the translator of Vehse into English had said: "People were astonished and pleased to learn that Matthias's death had perfectly fulfilled the astrological prediction of the seven M's which Kepler had made for 1619."[49] No attempt to verify whether Kepler ever made "the astrological prediction of the seven M's" was undertaken by this recent work, which also cited the prediction from a page in the English translation of Vehse where it cannot be found.

Having seen how Kepler's repudiation of the claim that an individual's future could be foretold from the heavens was turned into a sensational confirmation of that claim, let us examine Kepler's understanding of the relation between astronomy and astrology. While he was the mathematician for the Estates of Styria, on 14 September 1599 he explained to a wealthy supporter:

> From time to time I write horoscopes and calendars. This is, by God, a most annoying servitude, but it is necessary, lest I be free for a short while but more shamefully obligated later. Therefore, to defend my annual salary, my title and position, I must humor uneducated curiosity.[50]

In the same vein, writing in Latin about the 1604 nova in the winter months of 1605–6, Kepler asked the supercilious observer in his spotless ivory tower:

> Why do you snarl, O dainty philosopher, if [astronomy,] the mother who is very wise but poor, is supported and nourished by the ditties of [astrology,] her foolish daughter, as she seems to you, if the mother does not find her appropriate place among the extremely stupid throngs of people otherwise than by the interventions of this lack of sophistication? For if someone else had not previously been so naive as to hope that he would foretell the future from the heavens, you would never have been so clever as to think that astronomy (since it was unknown) should be learned by itself. Were it not for wisdom, we are not conducted to philosophy, we shall never be conducted to it. In every state of wonderment, and in every state of desire, while it is undeveloped, there is a great deal of unreality. But on the road leading to philosophy, this unreality steers those it meets to the right place.[51]

Returning to this theme four years later in his *Tertius interveniens*, which he dedicated on 3 January 1610, in Thesis 7 Kepler said in German:

> We plainly see that this inquisitiveness [about the future] benefits the study of astronomy, which nobody condemns, but is highly praised and justly so. This astrology is indeed a foolish little daughter . . . but, dear God, where would her mother, the strictly rational astronomy, be if she did not have her foolish daughter? Yet the world is much more foolish, and so foolish that this reasonable old mother, astronomy, in herself honest, must just be bamboozled and deceived by the tomfoolery of her daughter, particularly because she too has a mirror.
>
> And yet otherwise the salaries of astronomers are so rare and so low that the mother certainly must suffer from hunger if the daughter earned nothing.[52] If previously nobody had been so silly as to conceive the hope of learning future developments from the heavens, then you too, O astronomer, would never have become so clever as to have thought of investigating the heavenly motions to honor God. Yes, you would have known nothing about the motions in the heavens.
>
> In fact, you learned to distinguish the five planets from other heavenly bodies not from Holy Scripture but from the superstitious books of the Babylonians.

If we could not have achieved the knowledge of nature otherwise than through pure understanding and wisdom, we would indeed never have approached[53] it at all.

All curiosity and all wondering is, then, in its first stage nothing but simple foolishness. Yet this foolishness plucks us by the ears and leads us on the road which proceeds directly to philosophy.[54]

What led Kepler to his achievements in science, or philosophy as he called it? As he wrote to the gifted English scientist Thomas Harriot (1560–1621)[55] on 2 October 1606: "Ten years ago I rejected the division [of the heavens] into 12 equal parts, the houses, the dominations, the triplicities etc., all of that, keeping only the aspects and transferring astrology to the science of harmonics."[56] In his treatise, *Harmonics of the Universe*, often misunderstood as the "Harmony of the Universe," which he dedicated to King James I of England on 13 February 1619, Kepler said:

But if I now speak of the outcome of my studies, what, may I ask, do I find far off in heaven that even remotely refers to it? No inconsiderable parts of science, according to the experts, have been either freshly constructed by me, or corrected, or completely finished. But in this regard my stars were not Mercury rising[57] in the corner of the seventh house 90° from Mars, but Copernicus[58] and Tycho Brahe. Without the latter's volumes of observations everything which has now been brought by me into the clearest light would lie buried in darkness. Not Saturn as lord of Mercury, but their Imperial Majesties, Rudolph and Matthias, were my lords. Not Capricorn with Saturn was my planetary abode, but Upper Austria, the heritage of the emperor,[59] and the extraordinary generosity of his officials, extended to me at my request. Here, not at the setting point of the horoscope, is that corner of the earth, to which with the approval of the emperor, my lord, I withdrew from his excessively restless court, and in which during these current years, leading to the close of my life, I have been working on the *Harmonics* and whatever else I have in my hands. An astrologer will search in vain in my horoscope for the reasons why in 1596 I discovered the relationship between the heavenly spheres; in 1604, the process of vision; in the current year 1618 the reasons why every planet has a particular eccentricity, neither smaller nor greater; and in the intervening years the reasons for the explanation of celestial physics and the ways in which the heavenly bodies are moved, and their true motions.[60]

We have heard Kepler in 1592 asking Roeslin about his fever, and in 1618 denying that any astrologer could find in his horoscope the reasons for his scientific discoveries. In 1596, as he told Harriot, he turned his back on most of the traditional astrology. What was left was his belief that the planetary configurations imparted to the individual at the moment of birth a lifelong temperamental influence.

The eminent historian of science, William Whewell (1794–1866) believed that

> the *mystical* parts of Kepler's opinions, as his belief in astrology, his persuasion that the earth was an animal . . . do not appear to have interfered with his discovery, but rather to have stimulated his invention, and animated his exertions. Indeed, where there are clear scientific ideas on one subject in the mind, it does not appear that mysticism on others is at all unfavourable to the successful prosecution of research.[61]

Whereas Whewell regarded some of Kepler's ideas as mystical, Kepler distanced himself from the mystical philosophy: "But whoever wants to nourish his mind on the mystical philosophy . . . will not find in my book what he is looking for."[62] Kepler's wealthy supporter once asked him: "Why just seven planets, and not fewer?"[63] Kepler referred to his supporter's thinking about the "occult excellence of the number 7, and surmise about the same number of planets. But I wish to prove nothing from the mystique of numbers, and I believe nothing can be proved."[64] Yet an essay published under the auspices of the U.S. History of Science Society in a tercentenary commemoration of Kepler's life and work remarked that "the belief in the magic power of certain numbers was so enduring that over three centuries later [than Roger Bacon] we find Kepler struggling through long years to apply these principles in determining the planetary orbits."[65] When Kepler determined the planetary orbits, he did not struggle in a tall building, as I do, where the floor immediately above the twelfth floor is the fourteenth, seven centuries after Roger Bacon.

Select bibliography

Beer, Arthur, and Beer, Peter (eds.), *Vistas in Astronomy*, vol. 18, *Kepler: Four Hundred Years* (Oxford, 1975), pp. 397–469. Includes articles by A. Beer, J. O. Fleckenstein, G. Simon, M. List, G. Colombo, A. Postl, J. A. Hynek, and K. Figala.

Brackenridge, J. Bruce. "Kepler, Elliptical Orbits, and Celestial Circularity," *Annals of Science*, 39 (1982), pp. 117–43.

Kepler, Johannes. *Opera omnia*, ed. Christian Frisch (Frankfurt, 1858–71; repr. Hildesheim, 1971–7). Cited in notes as *F*.

Kepler, Johannes. *Gesammelte Werke*, ed. Max Caspar *et al.* (Munich, 1937–). Cited in notes as *GW*.

Shumaker, Wayne. *The Occult Sciences in the Renaissance* (Berkeley, 1972).
Simon, Gérard. *Kepler astronome astrologue* (Paris, 1979).
Strauss, Heinz Artur, and Strauss-Kloebe, Sigrid. *Die Astrologie des Johannes Kepler* (Munich and Berlin, 1926).

Notes

1 This horoscope has not been preserved because the letter transmitting it was lost. Kepler's next letter to Fabricius survives only because Kepler had a copy made of it. At the end of the copy, Kepler recalled: "I think [I wrote this] on 1 October 1602. I wrote the previous one on 18 July 1602" (*GW*, XIV, 280/687).
2 *GW*, XIV, 243/168–9 (1, 5 August 1602); omitted in *F*, I, 310/22.
3 Where the copyist wrote only the initial letter *s*, *F*, I, 310/8 ↑, VIII, 676/23 printed *sudavi* (I sweated). Would not *sufferebam* be more appropriate? (I use the vertical arrow to denote a line counted upward from the bottom of the page.)
4 *F*, I, 310/8 ↑ omitted *autumnalis praecedentis*.
5 *GW*, XIV, 275/473–9; *F*, I, 310/12 ↑ –5 ↑.
6 Arnold H. M. Jones, *The Herods of Judaea*, 2nd ed. (Oxford, 1967), table facing p. 266.
7 The professor of Greek at Tübingen, Martin Crusius (1526–1607), kept a diary, which has been published beginning with 1596: *Diarium Martini Crusii* (Tübingen, 1927–61). For the unpublished entry concerning 1591, see Edmund Reitlinger et al., *Johannes Kepler* (Stuttgart, 1868), p. 94.
8 *GW*, XIX, 320/last 3 lines; *F*, VIII, 294/15–16.
9 *GW*, XIX, 320/7–12/1–2; *F*, VIII, 294/3–4 (without the horoscope). Besides the horoscopes Kepler sent to Roeslin and Fabricius, he wrote out a third horoscope with his own hand in 1597, when he was twenty-six years old (*GW*, XIX, 331/15 ↑; *F*, V, 479/5). This 1597 horoscope, differing in some details from the other two and containing many more entries, served him as the basis of an elaborate self-analysis, which he left unfinished. Never sent out to anyone else, it remained among his papers and was preserved in Pulkovo before being transferred to Leningrad. It times his birth at 1:30 a.m., but says "the hour was 2:30" (*GW*, XIX, 329). This second entry (which was omitted in *F*, V, 476) is consistent with Kepler's investigation of the moment of his conception (16 May 1571, 16 hours, 37 minutes) and the duration of his mother's pregnancy (224 days, 10 hours; *F*, VIII, 672/24 ↑ – 23 ↑, 19 ↑ –18 ↑). The earlier hour of birth (1:30) was inserted here parenthetically by Frisch. But it is inconsistent with Kepler's investigation, which places his birth at (about) 2:30.
10 *GW*, XIV, 276/502–3, 491, on 226/503; *GW*, XIX, 319–20.
11 1593 (*F*, VIII, 294/2, 295/27) is either a misreading or a mistranscription, followed by *GW*, IV, 434/10 (in 434/12 the footnote number should be 2).
12 *GW*, XIX, 321/5 ↑ –4 ↑; *F*, VIII, 295/25–6.
13 *GW*, XIX, 321/3–5; *F*, VIII, 294/20–1.
14 *GW*, XIX, 321/5–10; *F*, VIII, 294/21–6.
15 *GW*, XIX, 320. Kepler shifted to 25° Gemini in his letter of 9 April 1597 to his former teacher, Michael Maestlin (*GW*, XIII, 119/262); in his 1597 horoscope (*GW*, XIX, 329); and in his *Harmonics of the Universe*, IV, 7, published in 1619 (*GW*, VI, 279/9).
16 *GW*, XIV, 328/426–8; *F*, I, 319/9–11.

17 Claudius Ptolemy, *Tetrabiblos*, ed. and trans. F. E. Robbins, Loeb Classical Library, with Manetho (Cambridge, Mass., 1940, 1971), p. 39.

18 *GW*, XIX, 321/12–26; *F*, VIII, 294/6 ↑ –295/9.

19 Pierre Gassendi, *Opera omnia* (Lyons, 1658; repr. Stuttgart-Bad Cannstatt, 1964), I, 741 rt. col./23 ↑ –6 ↑ ; (Florence, 1727), I, 647 lt. col./21–39. Gassendi's *Syntagma philosophicum*, pt. 2 (physics), sec. 2, bk. 6, chap. 5, was anonymously translated into English under the title *The Vanity of Judiciary Astrology* (London, 1659).

20 *GW*, X, 38/22–4.

21 Franz Boll, "Studien über Claudius Ptolemäus," *Jahrbücher für classische Philologie*, Suppl. 21 (1894), pp. 124–7, 168–79.

22 *GW*, IV, 147; *F*, I, 547.

23 *GW*, IV, 183/19–31; *F*, I, 581/19 ↑ –5 ↑ .

24 *GW*, IV, 231/21–6; *F*, I, 626/11–17.

25 William Shakespeare, *Julius Caesar*, 1.2.

26 *GW*, IV, 243/5; *F*, I, 636–7. In *astra incitant, non necessitant*, the last word is postclassical. Kepler's ultimate source might be an antideterminist astrologer in agreement with Thomas Aquinas and Dante (*Purgatorio*, XVI, 67–9):

> Voi che vivete ogne cagion recate
> pur suso al cielo, pur come se tutto
> movesse seco di necessitate.

> You who are living refer every
> cause to heaven, as if it moved
> everything with it by necessity.

27 *GW*, IV, 232/37–233/6; *F*, I, 627/27–41.

28 *GW*, IV, 198/5–7; *F*, I, 595/27–9.

29 *GW*, IV, 226/20–4; *F*, I, 621/16 ↑ –11 ↑ .

30 *GW*, IV, 235/7–10; *F*, I, 629/18 ↑ –15 ↑ .

31 *GW*, IV, 235/14–24; *F*, I, 629/10 ↑ –630/2.

32 *GW*, IV, 163/14–18; *F*, I 562/13 ↑ –8 ↑ .

33 *GW*, IV, 230/9–10; *F*, I, 625/4–6.

34 See text above at note 16.

35 Deuteronomy 18:10.

36 *GW*, IV, 238/34–9; *F*, I, 633/3 ↑ –634/4.

37 See text above at notes 5, 8, and 13.

38 Martha List, "Das Wallenstein-Horoskop von Johannes Kepler," in *Johannes Kepler Werk und Leistung*, ed. Gerold Maar (Linz, 1971), p. 129 rt. col./IV.

39 Johann Rudolph Camerarius, *Horarum natalium centuria una* (Frankfurt, 1607), p. 95/12 ↑ –11 ↑ , expressed "fear that Henry IV would be subject to no mean danger to his life at the age of 59 years, 9 months" in 1612. A photocopy of the relevant pages of Camerarius was kindly sent to me by Barbara Shailor of the Beinecke Rare Book and Manuscript Library, Yale University.

40 *GW*, XVI, 466/4–37.

41 This Frenchman has not yet been identified.

42 In the Long War (1593–1606) between the Holy Roman Empire and the Turks, the commander-in-chief of the imperial forces, Matthias, in 1594 failed at Esztergom (Strigonium, Gran) and lost the important fortress Györ (Jaurinum, Raab), situated at the confluence of the Danube and the Rába.

Miklós Istvánffy's detailed account of these struggles in Book 28 of his
Historia de rebus ungaricis, 1490–1616 (Cologne, 1622; 2nd ed., Vienna,
1758) does not mention that Matthias was wounded or that his brother was
killed, imaginary casualties pointed out by the Turkish historian Mustafa
Naima (1652–1716), *Annals of the Turkish Empire*, trans. Charles Fraser
(1832; repr. New York, 1975), p. 31.

43 This document was discovered among the Kepler papers by Otto Struve
(1819–1905), who published it in his "Beitrag zur Feststellung des
Verhältnisses von Keppler zu Wallenstein," *Mémoires de l'Académie
impériale des sciences de St. Pétersbourg*, ser. 7, tome 2, no. 4 (1860), pp.
11–12. In 1871 the document was printed a second time in *F*, VIII, 343–5.
GW, XVI, 373–5 reprinted Struve's text because the original in Kepler's
handwriting was no longer present in the Pulkovo manuscripts.

44 *GW*, VIII, 262/13–14.

45 *GW*, VIII, 259/20–32.

46 Martha List, *Der handschriftliche Nachlass der Astronomen Johannes
Kepler und Tycho Brahe*, Bayerische Akademie der Wissenschaften,
Deutsche geodätische Kommission, [Veröffentlichungen], ser. E, Heft 2,
1961, pp. 24–7.

47 M. G. Hansch, *Joannis Keppleri aliorumque epistolae mutuae* (Leipzig,
1718), p. xxvii/15–18.

48 (Karl) Eduard Vehse (1802–70), *Geschichte der deutschen Höfe seit der
Reformation*, 48 vols. (Hamburg, 1851–8); vol. IX, *Geschichte des
östreichischen Hofs und Adels und der östreichischen Diplomatie*, III, 123
(Hamburg, 1851); trans. Franz Demmler as *Memoirs of the Court,
Aristocracy, and Diplomacy of Austria* (London, 1856), I, 282.

49 Dorothy Gies McGuigan, *The Hapsburgs* (Garden City, N.Y., 1966), p. 164.

50 *GW*, XIV, 63/14–17; *F*, I, 71/17–20.

51 *GW*, I, 211/6–16,; *F*, II, 657/4–14.

52 For his *Rudolphine Tables* (Ulm, 1627) Kepler wrote the preface toward the
end of his life, and there he still asked in a marginal note: "How does
astronomy support itself?" His answer was that "the silly daughter,
astrology, in an enterprise not uniformly approved by everybody, nourishes
and sustains astronomy, her mother, who is very knowledgeable but
dreadfully impoverished" (*GW*, X, 40/27–8; *F*, VI, 670/23 ↑).

53 Reading *gelangen* (1610 ed., sig. A4ᵛ/16; *F*, I, 561/15).

54 *GW*, IV, 161/9–31; *F*, I, 560/8 ↑ –561/19.

55 *Thomas Harriot, Renaissance Scientist*, ed. John W. Shirley (Oxford, 1974).

56 *GW*, XV, 394/74–7; *F*, II, 68/13 ↑ –11 ↑ .

57 Not *untergeht*, as in Wilhelm Harburger, ed. and trans., *Johannes Keplers
kosmische Harmonie* (Leipzig, 1925), p. 203, in the series *Bücher deutscher
Mystik* (Books by German Mystics).

58 "Pico della Mirandola . . . pointed out that the astronomical basis of
astrology would be shattered when astronomers adopted the Copernican
system, as he believed they would," according to J. Bruce Brackenridge
and Mary Ann Rossi, "Johannes Kepler's *On the More Certain
Fundamentals of Astrology*," *Proceedings of the American Philosophical
Society*, 123 (1979), p. 106. Pico died in 1494, before the Copernican system
was initiated.

59 Not "Upper Austria formed the house of the emperor," as in Max Caspar,
Kepler, trans. and ed. C. Doris Hellman (London and New York, 1959), p.
279/14 ↑ ; 1962, p. 290/14–15.

60 *GW*, VI, 280/5–24; *F*, V, 262/23 ↑ –7 ↑ .

61 William Whewell, *History of the Inductive Sciences*, 3rd ed. (London, 1857; repr. London, 1967), pt. I, pp. 319–20.

62 *GW*, VI, 397/5–7; *F*, V, 423/8 ↑ –6 ↑ .

63 *GW*, XIV, 60/61, 61/71.

64 *GW*, IV, 75/507–9.

65 E. H. Johnson, "Kepler and Mysticism," in *Johann Kepler 1571–1630*, ed. F. E. Brasch (Baltimore, 1931), p. 63.

8

Kepler's rejection of numerology

JUDITH V. FIELD

In the Copernican description of the planetary system there are six planets instead of the Ptolemaic seven, the moon having become a subsidiary body of a type new to astronomers – and for which Kepler was to invent the term satellite in 1611.[1] In 1540, Rheticus felt the need to defend this new number of the planets:

> Who could have chosen a more suitable and more appropriate number than six? By what number could anyone more easily have persuaded mankind that the whole universe was divided into spheres by God the Author and Creator of the world? For the number six is honoured above all others in the sacred prophecies of God and by the Pythagoreans and the other philosophers. What is more agreeable to God's handiwork than that this first and most perfect work should be summed up in this first and most perfect number?[2]

As Rosen remarks in his note on his translation of this passage, Rheticus's numerological argument finds no parallel in the work of Copernicus himself. It does, however, find an answer in Kepler's defense of Copernicanism in the *Mysterium cosmographicum*.

Kepler's own explanation of the number of the planets is geometrical: There are exactly six orbs because there are exactly five regular solids to define the spaces between them. As Kepler points out, the fact that there are exactly five such solids is proved in a scholium to the last proposition of *Elements*, Book XIII. Kepler had, however, considered the possibility of a numerical explanation of the structure of the planetary system, in connection with his earliest attempts to find a pattern in the ratios of the dimensions of the planetary orbs. It seems, nevertheless, that even had these attempts succeeded he would not have been willing to accept a purely numerical explanation, for when re-

273

flecting on their failure, in the Preface to the *Mysterium cosmographicum*, he says:

> Nor could I conjecture from the nobility of any number why instead of an infinite number there should be so few moving spheres. Nor is it likely that Rheticus is correct in what he says in his *Narratio* when he argues from the sanctity of the number six for there being six moving spheres in the heavens. For in discussing the formation of the world itself [*de ipsius mundi conditu*] one should not draw reasons from those numbers which have taken on some dignity from things which came into being later than the world [*ex rebus mundi posterioribus*].[3]

A further indication of Kepler's opinions at this time is to be found in a letter he wrote to Maestlin on 3 October 1595:

> We see that God created the bodies of the world in a definite number. Now number is an accidental property of quantity, number in the world, I mean. For before the world [was created] there was no number, apart from the Trinity, which God himself is. Therefore if the world is constructed according to numerical measure it is according to the measure of quantities.[4]

It is clear that Kepler is distinguishing two kinds of numbers – abstract or undimensioned numbers, and numbers derived from measurement – and allowing only the latter to play a part in describing the plan according to which the universe was created. In the Appendix to *Harmonice mundi*, Book V (Linz, 1619), he was to refer to these two kinds of number as "counting numbers" (*numeri numerantes*) and "counted numbers" (*numeri numerati*).[5] He does not use these terms in the *Mysterium cosmographicum*, though the first appears in a note he added to the work in the second edition (Frankfurt, 1621), referring to the passage I have just quoted from the Preface: "See, even then I rejected counting numbers, as they call them [*numeros numerantes, ut appellant*]."[6] The *ut appellant* suggests that the term was well known, and the distinction between the two kinds of number is, indeed, pointed out by Aristotle, in connection with his theory of time. However, the application of this distinction in constructing mathematical models of natural phenomena appears to be original to Kepler.[7]

In the *Mysterium cosmographicum* the plan according to which the universe was created is repeatedly referred to as an Idea in the mind of the Creator,[8] and Kepler's recourse to geometry in order to describe it thus seems to be entirely Platonic in its inspiration. Indeed, Kepler's later works abound in references to geometrical figures as Ideas coeternal with the Creator. However, in the *Mysterium cosmographicum* there is a relic of the opinions of Aristotle in Kepler's comment on his unsuccessful attempt to describe God's plan in terms of regular polygons (which led to the successful version involving the regular polyhedra): "The figures [regular polygons] pleased me, as being quantities, and things prior to the heavens. For quantity was created at first, with body, the heavens were created the next day."[9] Kepler is taking it for granted that God's geometrical Idea for the design of the universe could not become operative until after matter ("the Heavens and the Earth" of Genesis 1.1), and hence space, had been brought into being on the first day. Nevertheless, the geometrical entities that form the basis for God's plan, being Ideas, are regarded as real, and the numbers derived from them are regarded as *numeri numerati* equally with the numbers derived from material entities by observation. It appears to have been unimportant that in the former case the numbers must have been counted by God whereas in the latter they were the responsibility of man.

I do not propose to concern myself further with the historical roots or philosophical consistency of Kepler's position in regard to numbers. The use of mathematics in the *Mysterium cosmographicum* has clear general affinities, as well as many detailed parallels, with Plato's use of mathematics in *Timaeus*. Moreover, many years later, in *Harmonice mundi*, Book IV, Kepler was to refer to *Timaeus* as being "beyond any possible doubt a commentary on the book of Genesis,"[10] though it should be noted that Plato's numerical series 1, 2, 4, 8 and 1, 3, 9, 27 (34b–36d) are dismissed in the introduction to *Harmonice mundi*, Book III, as examples of unacceptable Pythagorean numerology (with no mention of the name of Plato).[11] However, Kepler, like many another reader, believed that *Timaeus* was not to be taken entirely at face value (it is, after all, hardly overtly Copernican) so, despite this partial rejection, it seems possible that he believed he was reviving the "true" Platonic theory of number. To discuss Kepler's theory would thus certainly involve considering Plato's theory of number, which is a matter of such dispute among scholars[12] that it seems wise to follow the example set by T. L. Heath, in 1949, in declining to add to the secondary literature.[13]

There is, however, no doubt that Kepler's position in regard to numbers is tenable in mathematical terms. Euclid's *Elements* are constructed in such a way that it is evident that arithmetic may be considered as a subset of geometry: Arithmetic, which handles only integers, is clearly presented as no more than a part of geometry, which handles magnitudes in general. Mathematically speaking, one creates no insurmountable difficulties by demanding that all arithmetical results shall be deduced from geometrical ones. Kepler does not make this demand explicitly in the *Mysterium cosmographicum*, but, as we shall see below, it is made in *Harmonices mundi libri V* (Linz, 1619).

One of Robert Fludd's most perceptive comments on this work is his complaint that Kepler is "so addicted to geometrical proofs that he has forgotten about truly physical and formal Units determined by no dimensions."[14] Kepler himself had said much the same thing in a letter to Christopher Heydon in 1605, when he asserted that the archetype of the world "lies in Geometry, and specifically in the work of Euclid, the thrice-greatest philosopher [*et nominatim in Euclide philosopho ter maximo*]."[15]

Music theory

It was in the theory of music that Pythagorean numerology had its most lasting success. The decline in prestige of its ancient form in the later sixteenth and early seventeenth centuries, the period that concerns us here, was apparently connected with the fact that there was a close relationship between theoreticians and practitioners. Thus the inadequacy of the traditional Pythagorean intonation (based on the integers 1 to 4) for performing the polyphonic music being written by musicians such as Orlando di Lasso (1531–94) seems to have led to a demand for a system that would admit thirds and sixths as consonances. The system described by Gioseffo Zarlino in his *Istitutioni harmoniche* (Venice, 1558) admits these consonances by using two further integers, 5 and 6. The spirit is still purely arithmetical, though Zarlino provides a handsome geometrical diagram to illustrate the arithmetical properties of his *senario* (see Figure 1).

Eventually, Zarlino's system also began to seem inadequate to practicing musicians. For example, in response to an attack on him for his use of dissonance, by Giovanni Maria Artusi (1545–1613), Claudio Monteverdi (1567–1643) appended a few lines of Italian prose to the fourteen pages of music for the *basso continuo* parts to his fifth book of madrigals for five voices (Venice, 1605):

> I have written a reply to make it clear that I do not compose my works at random, and as soon as it is rewritten it will be published under the title Second System [*Seconda Prattica*], or Perfection of Modern Music, at which perhaps some may be surprised, not believing that there is any other system than that described by Zarlino; but be assured that in connection with consonances and dissonances there is another explanation [*consideratione*], different from that already given, which, in accordance with reason and the evidence of the senses, defends the modern style of composition.[16]

It is clear from what Monteverdi says that Zarlino's system still represented musical orthodoxy – whatever dissonances Giovanni Gabrieli (1557–1612) might be sending echoing round the gold mosaics of Saint Mark's. Monteverdi's *Seconda Prattica* was never published,

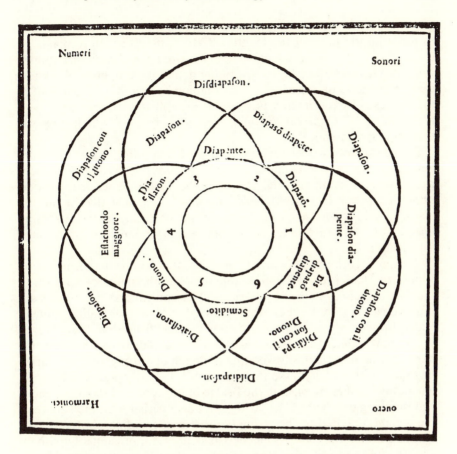

Figure 1. The six sonorous numbers. (Gioseffo Zarlino, *Istitutioni harmoniche* [Venice, 1558], p. 25)

but it seems clear from the way he handles dissonances in his music (for example in *Orfeo*, first performed in 1607) that when he refers to "the evidence of the senses" he is appealing to the way the ear recognizes a smooth gradation from consonance to dissonance. This was, of course, not a new observation, and it had already attracted the attention of theoreticians, including that of the distinguished mathematician Giovanni Battista Benedetti (1530–90). Benedetti's account of an alternative to Zarlino's theory that will explain this gradation is contained in a letter to his friend the composer Cipriano de Rore (1516–65), which was probably written about 1562, but was not published until 1585.[17] Benedetti's theory is based on the coincidences in the vibrations of two or more sound waves. In the particular example that is discussed, the waves are generated by vibrating parts of the string of a monochord. Benedetti's description allows consonances to be graded according to the lowest common multiple of the numbers used to describe the ratio of the lengths of string involved. In principle he is working with measured lengths, that is, with *numeri numerati*, though it is not clear whether he is describing actual experiments. We know that he carried out experiments with falling bodies, but his account of these musical experiments is introduced with the words "imagine a monochord [*Concipiatur in mente monochordus*]," which suggests thought experiments.[18]

Benedetti draws no conclusions from his results beyond the conventional one that consonances soothe the ears whereas dissonances cause distress by their roughness. His theory has depended upon arithmetical properties of numbers, and it would appear that he did not distinguish his own use of number from that of Zarlino in the *Istitutioni harmoniche*, to which there is a complimentary reference earlier in the same letter.[19]

Zarlino's work was, however, violently attacked by his former pupil Vincenzo Galilei (ca. 1520–91), apparently in the name of a return to the pure ancient system of Pythagoras. This controversy has been beautifully described by Walker, who shows that Vincenzo Galilei actually ended up with a system barely distinguishable from that of Zarlino.[20]

The new element in the work of Vincenzo Galilei was his carrying out experiments which showed that the pitch of a string not only varied inversely with its length (so that one could, for example, obtain a note one octave higher by placing the bridge in the center of the monochord and plucking one half of the string), but also varied with the square of the weight attached to the string (so that one could obtain the octave by quadrupling the load). Vincenzo also asserted, presumably without experiment, since the result is erroneous, that the note emitted by an organ pipe depends upon its volume, giving a ratio $1:8$ for the octave. On the basis of these experimental results, Vincenzo Galilei constructed a mathematical scheme designed to prove that Zarlino's was incorrect. He set up his integers 1 to 8 as an alternative to Zarlino's 1 to 6.[21] What might have been a weapon against the whole numerological basis of Zarlino's theory was used merely to attack its details. Some of Vincenzo's later work remained in manuscript, but there is general agreement that it is likely that these experiments were known at the time to Vincenzo's son Galileo Galilei, who later used some of them in the First Day of his *Discourses Concerning Two New Sciences* (Leiden, 1638).

Galileo's work describes the experiments with weighted strings, together with some further experiments that he may have carried out for himself and a few that will not work as described (Walker refers to them as "so-called 'thought experiments' about which he [Galileo] had not thought quite enough"). The results of Vincenzo's experiments are used by Galileo to cast doubt upon the standard numerological explanation of consonances, for if the tensions are in the ratio 1:4 when the lengths, to give the same interval, are in the ratio 1:2, there is clearly no reason why only the second ratio should be seen as determining the fact that the interval of an octave is a consonance.[22] Galileo then uses the results of his own experiments to construct a musical theory that is novel in not being dominated by consonances and the grading of consonances. As Walker shows, the theory is not finally self-consistent,[23] but whatever its shortcomings it is significant for our present purposes that Galileo, like his father, has departed from the time-hallowed practice of attributing the properties of musical intervals to the properties of the *numeri numerantes* used to express the corresponding ratios of lengths of strings. The numbers to which Galileo appeals are *numeri numerati*, the numbers of vibrations that strings make in certain times. However, as is his wont, he gives no explicit refutation of numerology as such, contenting himself with a rhetorical paragraph referring to the experimental results obtained by Vincenzo.

Kepler, who made a serious study of music theory and who seems to have been rather well read in both ancient and modern works on the subject,[24] does not refer to Vincenzo Galilei's *Discorso intorno alle opere de Gioseffo Zarlino . . .* (Venice, 1589), which contains the description of the experiments later used by Galileo, though Vincenzo's *Dialogo della musica antica et della moderna* (Florence, 1581) is the work he cites most often in his own most substantial contribution to musical theory, namely, Book III of *Harmonices mundi libri V* (Linz, 1619). It should be remarked, however, that Kepler does not cite Vincenzo Galilei for Vincenzo's own theories: Kepler approved of Zarlino's system and apparently failed to anticipate Walker in cutting through the forest of polemical barbed wire to the conclusion that Vincenzo's system is substantially the same as Zarlino's. Kepler's references to Vincenzo Galilei's work in *Harmonice mundi*, Book III, are almost entirely for its clear account of ancient theories. Perhaps it was in the reasonable expectation that the brief *Discorso* would merely repeat the polemical points already made in the much longer *Dialogo* that Kepler neglected to read the work? It seems very unlikely that if he had read it he would have failed to comment on its appeals to experiment, for his own avowed reason for accepting Zarlino's 1 to 6 instead of the traditional Pythagorean 1 to 4 system was the empirical fact that thirds and sixths sounded consonant to the ear,[25] and he makes many other references to musical experience.

It will be noted that in defending Zarlino's system Kepler was, for once, on the side of orthodoxy rather than standing up to be counted as a partisan of the avant garde. However, he was unorthodox in seeking an archetypal explanation for the properties of the consonant ratios in geometry (giving rise to *numeri numerati*) rather than in arithmetic (appealing to the properties of *numeri numerantes*).

Kepler's concern with music theory

It is in connection with cosmology, in the *Mysterium cosmographicum*, that we find the earliest indication of Kepler's concern with music theory, and it seems not unlikely that it was indeed in this connection that his concern first arose. Since the Pythagoreans had been wise enough to recognize that the sun was the center of the planetary system (as Kepler believed they had), it was natural for him to take a serious interest in their belief in *musica mundana* and hence in the theory of music. Moreover, Ptolemy had related musical ratios to astrological "aspects," and Kepler was also interested in astrology, which he nevertheless believed to be, like astronomy, in need of reform.[26] (Two heavenly bodies were considered to be "at aspect" to one another if their angular separation – usually measured merely as the difference of their ecliptic longitudes – took some special value. In the work of Ptolemy, these special values were 0°, 180°, 90°, 120°, and 60°, as shown in Figure 2.)

Music theory makes only a brief appearance in the *Mysterium cosmographicum*. In Chapter X, Kepler identifies the "noble" numbers involved in consonances and aspects as *numeri numerati* associated with the five Platonic solids. The effect is somewhat like a geometrical version of "Green Grow the Rushes, O" since most of the chapter consists of a list:

Unus est cubus, Una pyramis . . .
Duo corpora secundaria . . .
Tres anguli basium in pyramide, Icosaedro, Octaedro . . .
Quatuor anguli & latera basis in Cubo . . .[27]

and so on via 5, 6, 8, 12, 20, and 30 to 60 (the number of plane angles in the dodecahedron and icosahedron).

The following chapter is concerned with the origin of the zodiac. Music and astrology return, together, in Chapter XII, in which Kepler relates aspects to consonances and both to the Platonic solids and regular polygons inscribed in circles. Even at the time he wrote this chapter, Kepler seems to have regarded it as somewhat unsatisfactory, for he comments that "because we do not know the causes of this relationship it is difficult to associate particular harmonic ratios with particular solids."[28] By May 1599 he had entirely rejected this account of aspects and suggested, in a letter to Herwart von Hohenburg, that they should be derived from musical ratios among the arcs into which the circle of the zodiac is divided by bodies that are at aspect to one another. The accompanying group of diagrams shows the zodiac opened out to resemble the string of the traditional monochord (see Figure 3).[29]

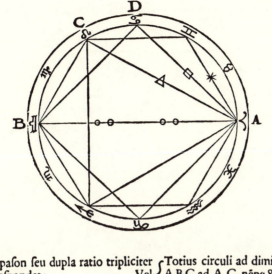

Diapafon feu dupla ratio tripliciter ⎧Totius circuli ad dimidium
   refpondet: Vel⎨A B C ad A C, nêpe 8 ad 4.
                              ⎩A CB ad A D, fex ad tria.

Diapente feu fefquialtera, item tri- ⎧Totius circuli, feu 1 2. 8
   pliciter: Vel⎨D A B, ideft, 9 ad 6
                              ⎩A B, ideft, 6 ad A C, ideft 4.

Diateffaron feu fefquitertia , item ⎧Totius circuli ad A B C D,
   tripliciter: Vel⎨ feu 1 2 ad 4 in A B D.
                              ⎪A B C ad A B, ideft 8 ad 6.
Diapafon & Diapente, item tripli- ⎩A C ad A D, 4 ad 3.
   citer, Bifdiapafon uero dupliciter
   Tonus femel.

Figure 2. Astrological aspects. (Claudius Ptolemy, *Harmonica*, trans. Antonio Gogava [Venice, 1562], p. 144)

These attempts are clearly related to the system described by Ptolemy in his *Harmonica* and displayed in the diagram supplied by Antonio Gogava for his Latin translation of the work (Venice, 1562) (see Figure 2), though Kepler's scheme involves three aspects not used by Ptolemy (see Figure 3). It seems that in 1599 Kepler knew Ptolemy's work only by report, though he was later to obtain a copy of Gogava's translation and a Greek manuscript of the work.[30]

In another letter to Herwart, written later in 1599, Kepler gives a sketchy description of a possible derivation of musical ratios from geometrical figures, commenting: "For in this matter nothing can come of Arithmetic, since whatever fitness numbers have arises from Geometry and the things that are numbered."[31]

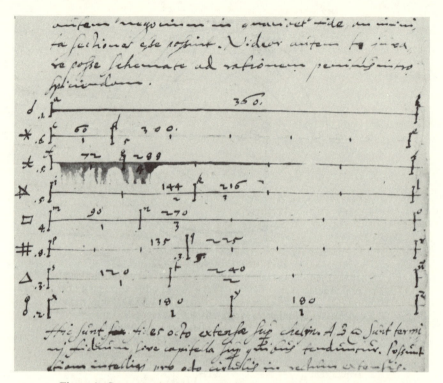

Figure 3. Correspondence of astrological aspects and harmonic ratios.
Ptolemaic aspects: AB, conjunction (0°, undivided string) – used as an
aspect by Ptolemy, though he does not define it as such; CDE, sextile (60°,
1:5); MNO, quadrature (90°, 1:3); STU, trine (120°, 1:2); XYZ, opposition
(180°, 1:1). *New aspects*: FGH, quintile (72°, 1:4); IKL, biquintile (144°,
2:3); PQR, sesquiquadrature (135°, 3:5). (From a letter written by Kepler to
Herwart von Hohenburg, 30 May 1599)

In fact, between them these two letters contain almost all the elements that were to be assembled, in a much more orderly manner, to
form the geometrical explanations of consonances and aspects given
nearly twenty years later in *Harmonice mundi*, Books III and IV. Two
further letters written in 1599 indeed supply sketches of a book Kepler
was planning to write, entitled *De harmonia mundi* in one letter and
De harmonice mundi in the other.[32]

Harmonices mundi libri V
The titles of the "short book" planned in 1599, and the eventual
title of the work that was published twenty years later as a folio of
about 320 pages, consign it unambiguously to the Pythagorean tradition
of *musica mundana*.

Of the five books that make up Kepler's long-meditated work, the first two deal with geometry and the third with musical theory. The two final books contain applications of the mathematical and musical theorems to construct explanations of the efficacy of aspects, in Book IV, and the structure of the solar system, in Book V (it being taken as an accepted fact that the five Platonic solids explain the number of the planets and the approximate sizes of the spaces between their orbits).

The order in which the books are presented may at first seem to have been dictated by the necessity of proving theorems before using them, but closer scrutiny reveals that this is not entirely the case: The astrological book, Book IV, uses the geometrical results directly, not in the musical form they have been given in Book III, as Kepler himself remarks,[33] and the musical results of Book III are not applied until Book V. While it is possible that the central position of the book concerned with music merely seemed appropriate in a work whose title proclaimed it to be concerned with harmony, it should be noted that throughout the work Kepler, like other authors of works on *musica mundana*, uses the word "harmony" and its cognates in a much wider sense than the purely musical one. For example, the title of Book IV refers to "harmonious configurations of stellar rays," when, as we have seen, the harmony is, according to Kepler's explicit statement, not a musical one. Elucidation of this usage is provided by the discussion of the Greek term $\dot{\alpha}\rho\mu o\nu i\alpha$ and its cognate verb in the Introduction to Book II.[34] It seems that the positioning of the musical book immediately after the geometrical ones was in fact designed to emphasize that the "musical" ratios, long seen as arithmetical in origin, had now been given a basis in geometry.

Moreover, this geometrical basis is mentioned explicitly in the Introduction to Book I, which appears to be intended as an introduction to the whole work, since it ranges very widely and only its final paragraph, with the marginal note "the purpose of this first book," serves as a specific introduction to the geometrical work of Book I.[35]

The first sentence of this general introduction reads:

> Since today, to judge by the books that are published, there is a total neglect of the intellectual distinctions to be made among geometrical entities, I thought fit to state at the outset that it is from the divisions of the circle into equal aliquot parts, by means of geometrical construction [i.e., using straight edge and compasses], that is, from the constructible Regular plane figures, that we should seek the causes of Harmonic proportions.[36]

The geometrical work of Book I, which serves to construct the musical ratios of Book III, is largely based on Euclid's classification of surds in Book X of the *Elements* (long famous as the most difficult part of Euclid's work). Whereas Euclid is concerned with classifying magnitudes according to their commensurability with one another, or their commensurability in the square, and so on, Kepler recasts the work so that he is classifying regular polygons in terms of the degree of commensurability of the side of a regular polygon with the diameter of the circle in which it is constructed. This corresponds to the number of operations one has to carry out to construct the side in the circle by means of compasses and straight edge.

While *Harmonice mundi*, Book I, is mainly derivative, solid, and hard going, Book II is both highly original and very easy. It contains the first systematic treatment of the problem of fitting regular polygons together to form either a polyhedron or a pattern that entirely covers the plane (a tessellation).[37] The geometrical results of Book II are applied in Book IV to explain the efficacy of certain configurations of heavenly bodies, that is, to account for astrological aspects. By this time, Kepler had long abandoned attempts to explain aspects in terms of musical ratios, having decided that there were some aspects which did not divide the zodiac in the appropriate manner.[38]

There is, of course, no doubt in any historian's mind as to Kepler's competence and originality as a mathematician, but the weight of the geometrical work in *Harmonice mundi*, Books I and II, marks it as nontrivial even by Kepler's standards and must be seen as indicating that he took very seriously his endeavor to prove that God was a Platonic geometer rather than a Pythagorean numerologist. There are, indeed, arithmetical passages in *Harmonices mundi libri V* – for example, in Book III (Chapter II, Section XIX, and Chapters IV and XI)[39] – but they are all directly proposed as arithmetical corollaries to geometrical theorems or clearly indicated as concerned with *numeri numerati*, derived from geometrical figures or from observation. The rejection of numerology in *Harmonices mundi libri V* is not something that has to be proved by a selection of revealing quotations (though Kepler is conveniently given to laying his opinions on the line); it is expressed in the very structure of the work.

Kepler's harmonies and those of Fludd

I am not certain that historians should be unequivocally grateful to those unnamed friends of Kepler who persuaded him that in the Appendix to *Harmonice mundi*, Book V, he

> should not omit to mention Robert Fludd, a doctor of medicine from Oxford, who filled his book, published a year ago, on the Microcosm and the Macrocosm, with meditations on harmonies [*Harmonicis contemplationibus*], but should briefly show the reader on which matters he and I are in agreement and on which we differ.[40]

This discussion of Fludd's work takes up about three sides, folio. The length of the first two parts of *Utriusque cosmi . . . historia* (Oppenheim, 1617, 1618), with which the comparison is made, is 992 pages, folio.

To put it even more briefly than Kepler did: For the most part they differ.

Fludd replied to this with a work of 54 folio pages: *Veritatis proscenium . . . seu demonstratio quadam analytica . . . in appendice quoadam a Joanne Keplero, nuper in fine harmoniae suae mundanae edita . . .* (Frankfurt, 1621). He criticized Kepler's summary of the few sections of *Utriusque cosmi . . . historia* that were in question, but most of his book was concerned to point out that where he and Kepler disagreed, for example about Copernicanism, he, Fludd, was in the right.

Kepler replied to this with a work of 50 folio pages: *Pro suo opere harmonices mundi apologia. Adversus demonstrationem analyticam Cl.V.D.Roberti de Fluctibus medici oxoniensis* (Frankfurt, 1622).

Fludd replied to this with a work of 83 quarto pages: *Monochordum mundi symphoniacum seu replicatio Roberti Flud . . . ad apologiam . . .* (Frankfurt, 1622). Most of this last work (pp. 19–75) is taken up with a refutation of Copernicanism, a rejection of Kepler's reply to Fludd's comments on one particular part of the Appendix (Text XII in Fludd's *Demonstratio*). In fact, throughout the exchanges, Kepler and Fludd mainly confine themselves to discussing, in order, the particular passages of Kepler's Appendix that Fludd cited in his *Demonstratio*, so that one can read the succession of rival opinions as forming a series of individual dialogues concerning particular texts.

The clearest message of this exchange of open letters is that relations between reviewed and reviewer never did run smooth. Each accuses the other of not understanding his book, and claims to be unable to understand the other's book, and so on. There are, nevertheless, some revealing remarks. For example, Kepler complains of Fludd's Hermetic analogies, which, he says, "are dragged in by the hair."[41] In similarly nonreportorial style Fludd replies that Kepler heaps up definitions, axioms, and propositions.[42] Apart from their tone, both these comments appear to be entirely justified and revealing of the very different preferences of the authors as to what constitutes a convincing style of argument.

One of Fludd's comments in the *Demonstratio* is luridly inappropriate to his ostensible interlocutor: He notes that Kepler is arguing that Ptolemy's description of the planetary system must be abandoned, and replies at some length that he disagrees, finally asserting that the Ptolemaic theory gives rise to tables of proven accuracy and most exact ephemerides that predict astronomical events to the very hour and minute.[43] In his reply, Kepler asks to be forgiven for not helping Fludd to understand his arguments against Ptolemy (on the grounds that Fludd has not said exactly which bit he has not been able to follow), and adds that the remark about the accuracy of predictions from the Ptolemaic theory is the clearest proof that Fludd is not well versed in astronomy.[44] Connoisseurs of controversy must have regretted that Tycho Brahe was dead.

For our present purposes, it is of interest that one of the contrasts Kepler points out between his own work and that of Fludd is that Fludd's harmonies ignore actual units and use abstract numerical relationships, whereas Kepler finds musical ratios among quantities measured in the same units, such as the extreme angular speeds of planets as seen from the sun.[45] Fludd picks this up as Text XVII and launches into a defense of numerology:

> In this passage I see that the author is entirely ignorant of the true numbers of natural Harmony: . . . Yet he describes Pythagoras' triangular number [the tetractys, 1 to 4] on page 4 of his Book III [Has Fludd not noticed the refutation that follows Kepler's translation of Camerarius' commentary on the Pythagorean *Carmina aurea*?][46] . . . He tries to avoid abstract numbers in his harmony; yet it seems that without using abstract numbers nothing can genuinely [*sincere*] be expressed in numbers, for no less abstract are the Mathematical numbers from lines, surfaces and bodies, or roots, squares and cubes, than are those found in common Algorithmic Arithmetic. The wiser philosophers, Themistius, Boëthius, Averroes, Pythagoras and Plato. . .[47]

According to Fludd, numerological explanation is applicable not only to music but even at the highest level:

> Further, all kinds of natural things, and those which are supernatural, are bound together by particular formal numbers. The mystery of these occult numbers is best known to those who are most versed in this science, who attribute the Monad or unity to God the artificer, the Dyad or duality to Aqueous Matter, and then the Triad to the Form or light and soul of the universe, which they call virgin.[48]

The numerological creed that Fludd advances here seems to be in perfect accord with what we find in *Utriusque cosmi . . . historia*. I have quoted it in preference only for its greater concision. Fludd's numerology is, however, less radical than Kepler's commitment to geometry: He is prepared to use geometrical methods in the same way that he uses arithmetical ones. For example, he explains the fact that the sun's orbit is midway between the two boundaries of the celestial region by appealing to two intersecting pyramids (as he calls them),

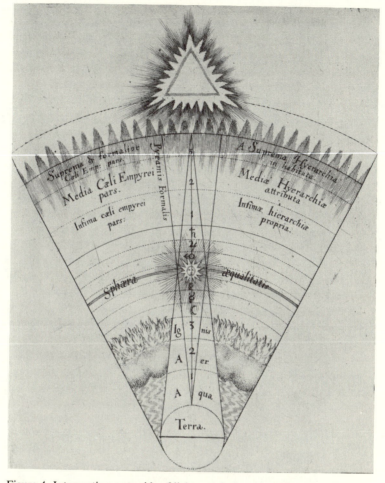

Figure 4. Intersecting pyramids of light and darkness. (Robert Fludd, *Utrius-que cosmi . . . historia, tractatus I* [Oppenheim, 1617], p. 89).

one of light radiating from an aureoled triangle that represents the Trinity, and another of darkness whose base lies in a plane through the center of the body of the earth (Figure 4).[49] However, the monochord that defines the more detailed structure of the celestial region is purely arithmetical (Figure 5). Indeed, it is very similar in its import to the diagram Zarlino gave in his *Istitutioni harmoniche* to illustrate the numerological cosmology of the ancients (Figure 6).

In fact, the overall plan of Fludd's work commits him to giving a very simple account of *musica mundana*, for he chooses to deal with it before describing what he calls "artificial music," that is, music as

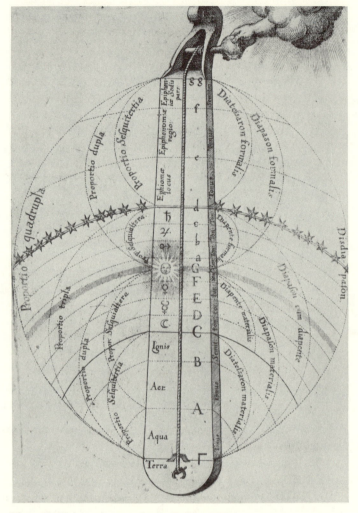

Figure 5. Universal monochord. (Robert Fludd, *Utriusque cosmi . . .
historia, tractatus I* [Oppenheim, 1617], p. 90)

a microcosmic phenomenon, including musical theory. Thus there is
no more to Fludd's *musica mundana* than the system of ratios displayed
in the diagrams of the monochord. The later account of microcosmic
music takes the form of a condensed version of a musical textbook,
going into much practical detail. This contrasts with Kepler's treatment, in which musical theory is discussed in some detail, in *Harmonice mundi*, Book III, before it is shown, in Book V, that there
exists an observable celestial counterpart to human music.

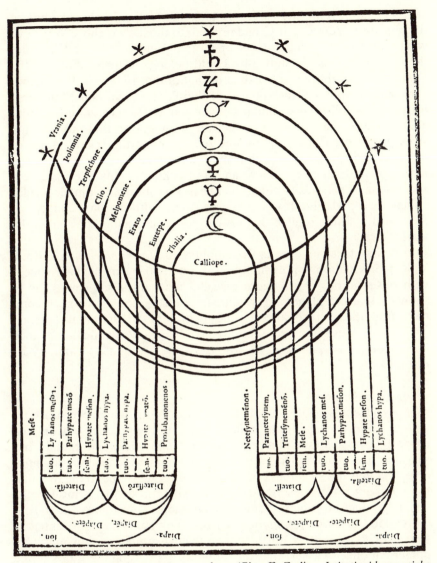

Figure 6. Ancient *musica mundana*. (Gioseffo Zarlino, *Istitutioni harmoniche* [Venice, 1558], p. 102)

Faced with Fludd's simple ratios and Kepler's elaborate polyphony, one may well feel that the first looks a better proposition, scientifically speaking, than the second. Indeed, many a modern astronomer has been known to balk at the suggestion that nature shares Kepler's admiration for Orlando di Lasso. However, the objections to Fludd's monochord are by no means only such as may appeal to the twentieth century. Even if we set aside the fact that Galileo's observations of the phases of Venus in 1613 had proved (to those who chose to believe what Galileo saw through telescopes) that Venus must orbit the sun, the system of spheres shown by Fludd belongs to the tradition of the *Sphere* of Sacrobosco rather than to that of the *Almagest*. It was conventional to show all the spheres of the planets as of equal thickness (Copernicus followed a similar convention in *De revolutionibus*), but although Ptolemaic astronomy did not allow one to find the absolute sizes of the orbs, it did allow one to determine the relative sizes of deferent and epicycle for each planet, and hence the ratio between the radius of the planetary sphere and its thickness. This did not give a succession of spheres of equal thicknesses as shown by Fludd. Kepler complained that Fludd was concerned only with his own concept of the world.[50] Fludd replied that his harmonies existed in the soul of the world.[51]

In contrast, if one checks the numbers in *Harmonice mundi*, Book V, Kepler's elaborate polyphony turns out to be in excellent agreement with observation; and modern celestial mechanics has got as far as showing that if a planetary system starts off by containing such commensurabilities (''resonances'') they will persist indefinitely.[52] However, there does not as yet seem to have been any significant advance upon Kepler's explanation of how the system originally came by its resonances.

Conclusions

It seems legitimate, for the period with which we are concerned, to restrict the term "numerology" to the practice of using the properties of abstract numbers, *numeri numerantes*, to explain observable phenomena. Thus we are using a numerological argument in asserting that the sweetness of a musical interval is due to the fact that the numbers which describe the observed ratio of the string lengths which produce it (which, being measurements, are *numeri numerati*) correspond to *numeri numerantes* which belong to the set of integers 1 to 4 or 1 to 6. It is not a numerological argument, but merely a numerical one, to assert that the numbers expressing the ratios of the lengths have a low common multiple, so that sound waves corresponding to the two strings coincide frequently, producing a pleasant effect on the ear.

Kepler, as we have seen, makes this distinction quite clearly, but it appears not to have played any significant part in the natural philosophy of either Vincenzo Galilei or Benedetti, despite the fact that their musical theories appeal to *numeri numerati*. It seems that the eventual drift away from numerology in musical theory was not connected with philosophical objections of the kind that occurred to Kepler, but was caused by the pressure of observation and experiment and the activities of practicing musicians. The concern was not with the fundamental causes of consonance and dissonance, but rather with the practical and technical problem of finding a stable system of intonation: procedures rather than metaphysics.[53]

Kepler's musical theory was not numerological in Kepler's terms, since it depended upon numbers drawn from geometry, which he regarded as *numeri numerati*. As we have seen, this subtle philosophical point seems to have been wasted on Robert Fludd, who roundly asserted that numbers derived from geometry were just as abstract as those used in common algorithmic arithmetic.[54] To Fludd, therefore, Kepler was a numerologist, and many have been the historians who have agreed with him. To Kepler, however, the distinction seems to have been of some importance, and it serves as an indication of the tradition to which Kepler himself believed his work belonged.

The most obvious sources of influence upon Kepler's work are some of the most widely influential books of all time: the Bible, *Timaeus*, and the *Elements*. The mediator between these three is Proclus. In the Introduction to *Harmonice mundi*, Book I, Kepler even goes so far as to regret that Proclus has not left a commentary on Book X of the *Elements*, for if he had then he, Kepler, would not have needed to write the present work.[55] From this tetractys of authorities, together with Kepler's professed admiration for the Pythagoreans, one might expect a style of thought which has something in common with that of Fludd. Fludd clearly did. Since Kepler spent some of his most productive years in Prague at the court of Rudolph II, he had presumably had occasion to say to other people what he wrote about reliance upon authorities in his reply to Fludd in 1622: "Why should you follow Trismegistus if you forbid me to associate myself with Plato? Why are you allowed to call upon Iamblichus and Porphyry, enemies to Christian belief, while I am not allowed to call on Proclus or Aristarchus?"[56]

Kepler's philosophical rejection of numerology, a characteristic element in the complex of beliefs associated with the neo-Platonic, neo-Pythagorean, and cabalistic tradition of the Renaissance, to which Fludd belongs, is a mark of his conviction that his own natural philosophy is not indebted to this tradition. Historians are, of course, under no compulsion to agree with him, but it seems that, in concentrating attention on the tradition from which Kepler distances himself, too little attention has been paid to the one with which he associates himself. While the 1599 title *De harmonia mundi* may stir memories of the cabalistic work of the same name by Francesco Giorgi (1466–1540),[57] the final title, *Harmonices mundi libri V*, should certainly be seen as a reference to that of the *Harmonica* of Claudius Ptolemy (fl. 125–141). In the Introduction to *Harmonice mundi*, Book V, Kepler

Figure 7. The creation of light. (Robert Fludd, *Utriusque cosmi . . . historia, tractatus I* [Oppenheim, 1617], p. 49.)

refers to the failure of Ptolemy's attempt to describe the *musica mundana*[58] and then contrasts it with his own success: "I have stolen the golden vessels of the Egyptians to build with them a tabernacle for my God far from the confines of the land of Egypt."[59]

Kepler originally intended that the Appendix to his work should contain a translation of Ptolemy's *Harmonica*.[60] He gives only vague reasons why the proposed translation did not appear, including a line and a half of Horace to the effect that "the plan was to make an amphora but as the wheel turned a pitcher emerged instead."[61] In fact, it seems likely that by 1618 Ptolemy's work no longer seemed very important to Kepler, except historically. Its purely musical part had been developed by Zarlino and others;[62] and the application of the musical results to astronomy and astrology was indissolubly wedded to the Ptolemaic description of the planetary system and the astrological ideas of the *Tetrabiblos*. Kepler nevertheless gives a brief summary of the work,[63] and it is clear that he recognized it as having exerted considerable influence on his own, which deals with essentially the same problems, though with a heliocentric planetary system and different astrological beliefs. A further point of difference is that whereas Ptolemy gives a numerological explanation of consonances and then uses consonances to account for astrological aspects, Kepler finds separate explanations in geometry for both consonances and aspects. As we have seen, however, his explanation of consonances is very similar to the relationship Ptolemy makes between consonances and aspects, both depending on the division of a circle by the inscription of regular polygons.

Despite the differences between their works, I think Kepler sees himself as carrying on a mathematical tradition derived from Plato, Euclid, and Ptolemy. *Harmonices mundi libri V* is related to the *Harmonica* in much the same way as the *Astronomia nova* is related to the *Almagest*, covering the same ground and designed to produce answers that are in accord with the author's notions of physics and in adequately accurate agreement with observation. The most important difference in scope between the cosmological works is that Ptolemy's is merely cosmological, whereas Kepler's describes God's archetype for the Creation and is thus also concerned with cosmogony. As Kepler told Heydon in 1605: "Ptolemy had not realised that there was a creator of the world: so it was not for him to consider the world's archetype."[64] The Bible, the only influence that truly unites Kepler with Fludd, is the source of the only crucial divergence in outlook between Kepler's work and that of Ptolemy.

Acknowledgments

I am grateful to Professor A. R. Hall and Stephen Pumfrey, who read and criticized an earlier draft of this chapter.

For permission to reproduce illustrations, I wish to thank the following: for Figures 1, 3, and 6, the British Library; for Figures 4, 5, and 7, the Science Museum, London; for Figure 2, the Bayerische Staatsbibliothek, Munich.

Notes

1 Johannes Kepler, *Narratio de observatis a se quatuor Iovis satellitibus erronibus* (Frankfurt, 1611) in Kepler, *Gesammelte Werke*, ed. W. von Dyck, M. Caspar, F. Hammer, et al. (Munich, 1938–), IV. Hereafter cited as *GW*.

2 Rosen, *Three Copernican Treatises* (New York, 1971), p. 147, trans. from Rheticus, *Narratio prima* (Danzig, 1540). Rheticus's work is reprinted in *GW*, I. The passage in question is on p. 105, ll. 34–9.

3 Johannes Kepler, *Mysterium cosmographicum* (Tübingen, 1596), preface, p. 7; *GW*, I, 10, ll. 32–7.

4 Kepler to Maestlin, 3 October 1595, *GW*, XIII, 35, letter 33, ll. 54ff.

5 *GW*, VI, 370, ll. 19–25.

6 *GW*, VIII, 28, ll. 39–40. Both types of number are mentioned in his following note.

7 Aristotle, *Physics*, IV. xi (219b3–9). Aristotle's account of Pythagorean number theory in the *Metaphysics*, and Proclus's reply to it in the Introduction to his commentary on Euclid, merely discuss the ontological status of numbers, apparently without regard to the possibility of there being more than one kind of number. I cannot see that Kepler's application of this distinction would be likely to appeal to an Aristotelian.

8 For example, Kepler, *Mysterium cosmographicum*, chap. II; *GW*, I, 24, ll. 1ff.

9 Ibid., preface, p. 8; *GW*, I, 12, ll. 12.

10 Kepler, *Harmonice mundi*, bk IV, chap. I; *GW*, VI, 221.

11 *GW*, VI, 100.

12 See, for example, R. D. Mohr, "The Number Theory in Plato's *Republic VII* and *Philebus*," *Isis*, 72 (1981), pp. 620–7.

13 T. L. Heath, *Mathematics in Aristotle* (Oxford, 1949), p. 220.

14 Robert Fludd, *Monochordum mundi symphoniacum seu replicatio Roberti Flud . . .* (Frankfurt, 1622), p. 10, ll. 3–4.

15 Kepler to Heydon, October 1605, letter 357, ll. 166–7; *GW*, XV, 235.

16 Claudio Monteverdi, *Basso continuo del quinto libro de le madrigali a cinque* (Venice, 1605), p. 21.

17 See C. Palisca, "Scientific Empiricism in Musical Thought," in *Seventeenth Century Science and the Arts*, ed. H. H. Rhys (Princeton, 1961), pp. 91–137.

18 G. B. Benedetti, *Diversarum speculationum mathematicarum et physicarum liber* (Turin, 1585), p. 283, ll. 3ff. Benedetti's description is reprinted with an English translation by Palisca, though with the omission of the first three paragraphs, which read:

> Nec alienum mihi videtur a proposito instituto, speculari modum generationis ipsarum simplicium consonantiarum; qui quidem modus fit ex quadam aequatione percussionum, seu aequali concursu undarum aeris, vel conterminatione earum.
>
> Nam, nulli dubium est, quin unisonus sit prima principalis audituque amicissima, nec non magis propria consonantia; & si intelligatur, ut punctus in linea, vel unitas in numero, quam immediate sequitur diapason, ei simillima, post hanc vero diapente, caeteraeque. Videamus igitur ordinem concursus percussionum terminorum, seu undarum aeris, unde sonus generatur.
>
> Concipiatur in mente monochordus, hoc est chorda distenta, quae cum divisa fuerit in duas aequales partes a ponticulo, tunc unaquaeque pars eundem sonum proferet, & ambae formabunt unisonum, quia eodem tempore, tot percussiones in aere faciat una partium illius chordae, quot et altera: ita ut undae aeris simul eant, & aequaliter concurrant, absque ulla intersectione, vel fractione illarum invicem.

19 Ibid., p. 281, ll. 23–5.

20 D. P. Walker, *Studies in Musical Science in the Late Renaissance* (London, 1978).

21 Ibid., p. 24.

22 G. Galilei, *Discorsi*, Day 1, in *Two New Sciences*, trans. S. Drake (Madison, Wis., 1974), pp. 101–2; *Opere*, Edizione Nazionale (Florence, 1964–6), VIII, 143–4.

23 Walker, pp. 27–33.

24 See M. Dickreiter, *Der Musiktheoretiker Johannes Kepler* (Bern, 1973).

25 Kepler, *Harmonice mundi*, bk. III, *Proemium*; *GW*, VI, 99.

26 See G. Simon, "Kepler's Astrology: The Direction of a Reform," *Vistas in Astronomy*, 18 (1975), pp. 439–48, and *Kepler astronome astrologue* (Paris, 1979).

27 Kepler, *Mysterium cosmographicum*, chap. X; *GW*, I, 37.

28 Ibid., chap. XII; *GW* I, 41, ll. 12ff.
29 Kepler to Herwart, 30 May 1599, letter 123, ll. 359–416; *GW*, XIII, 348–50.
30 See U. Klein, "Johannes Kepler's Bemühungen um die Harmonieschriften des Ptolemaios und Porphyrios," in *Johannes Kepler Werk und Leistung* ed. Gerold Maar (Linz, 1971), pp. 51–60.
31 Kepler to Herwart, 6 August 1599, letter 130, ll. 339–41; *GW*, XIV, 29.
32 Kepler to Maestlin, 29 August 1599, letter 132, ll. 136ff.; *GW*, XIV, 46; Kepler to Herwart, 14 December 1599, letter 148, ll. 11ff.; *GW*, XIV, 100.
33 Kepler, *Harmonice mundi*, bk. IV, chap. IV; *GW*, VI, 234, l. 32.
34 Ibid., bk II, *Proemium*; *GW*, VI, 69; trans. in J. V. Field, "Kepler's Star Polyhedra," *Vistas in Astronomy*, 23 (1979), pp. 109–41.
35 Kepler, *Harmonice mundi*, bk. I. *Proemium*; *GW*, VI, 19, l. 21.
36 Ibid.; *GW*, VI, 15.
37 It is in the course of this work that Kepler gave his first published description of the two regular star polyhedra he had discovered. See Field, "Kepler's Star Polyhedra."
38 See J. V. Field, "Kepler's Geometrical Cosmology," unpublished Ph.D. thesis, University of London, 1981.
39 *GW*, VI, 118, 130, and 155.
40 Kepler, *Harmonice mundi*, bk. V, *Appendix*; *GW*, VI, 373, ll. 13ff.
41 Ibid.; *GW*, VI, 375, ll. 2–6.
42 Robert Fludd, *Veritatis proscenium . . . seu demonstratio quadam analytica . . . in appendice quadam a Joanne Keplero, nuper in fine harmoniae suae mundanae edita . . .* (Frankfurt, 1621), text XV, p. 25. Cited hereafter as *Demonstratio*.
43 Ibid., text XII, p. 19, ll. 17–40; ll. 33–5 for comment on accuracy. In fact, though tables showed the predicted positions to minutes and seconds of arc, observation proved that these values were grossly inaccurate. Errors of several degrees were not uncommon. Even Reinhold's *Prutenic Tables* (Tübingen, 1551), which used relatively new observations and Copernicus's models of planetary orbits, sometimes showed similarly large errors – prompting Tycho Brahe to spend twenty years making astronomical observations, from which Kepler laboriously deduced the elliptical orbit of Mars (*Astronomia nova* [Heidelberg, 1609]) and then calculated elliptical orbits for all the other planets. Kepler's *Rudolphine Tables*, based on the new orbits, were published only in 1627, and at the time he read Fludd's *Demonstratio* he was still engaged in trying to remedy the deficiencies of which Fludd seems to have been so sublimely unaware.
44 Kepler, *Apologia, ad analysin XII*; *GW*, VI, 419, ll. 28ff.
45 Kepler, *Harmonice mundi*, bk. V, *Appendix*; *GW*, VI, 375, ll. 25ff.
46 Ibid., bk. III, *Proemium*; *GW*, VI, 99–101; Camerarius, *Libellus scholasticus* (Basel, 1551), pp. 205–8.
47 Fludd, *Demonstratio*, text XVII, p. 25, l. −3 to p. 26, l. 5.
48 Ibid., p. 26, ll. 17ff.
49 In fact, this plane cannot pass exactly through the center of the earth because the surface of the "pyramid" appears to be intended to be tangential to the surface of the earth. The "pyramid" would thus have to become an infinite cylinder, rather than a finite half-cone as shown, if its base were to pass through the center. Close inspection of Fludd's plates reveals that the illustrator has done a certain amount of cheating to overcome this problem.
50 Kepler, *Harmonice mundi*, bk. V, *Appendix*; *GW*, VI, 377, ll. 1ff.

51 Fludd, *Demonstratio*, text XXV, p. 35, last line.
52 See J. V. Field, "Kepler's Cosmological Theories: Their Agreement with Observation," *Quarterly Journal of the Royal Astonomical Society*, 23 (1982), pp. 556–68, and."Kepler's Geometrical Cosmology" for references to the work in celestial mechanics.
53 See E. W. Strong, *Procedures and Metaphysics* (Berkeley, 1936).
54 See passage referenced in note 47 above.
55 Kepler, *Harmonice mundi*, bk. I, *Proemium*; *GW*, VI, 15, ll. 16ff.
56 Kepler, *Apologia*, p. 47; *GW*, VI, 451, ll. 30ff.
57 F. Giorgi, *De harmonia mundi totius cantica tria* (Venice, 1525).
58 Kepler, *Harmonice mundi*, bk. V, *Proemium*; *GW*, VI, 289, ll. 19ff.
59 Ibid.; *GW*, VI, 290, ll. 4–5.
60 Gogava's translation, published in 1562, had been made from a corrupt manuscript, and Kepler felt he could improve on it. See Klein, "Johannes Kepler's Bemühungen."
61 Kepler, *Harmonice mundi*, bk. V, *Appendix*; *GW*, VI, 369; Horace, *Ars Poetica*, 21. Kepler is misquoting, having omitted an interrogative.
62 In fact, Kepler continually refers to Zarlino's system as Ptolemy's, with the result that Zarlino's name occurs only once in *Harmonice mundi*, bk. III.
63 *GW*, VI, 369–73. This summary is described as excellent by I. Düring, *Ptolemaios und Porphyrios über die Musik*, German trans. of the *Harmonica* of Ptolemy with notes based on the commentary of Porphyry (Göteborg, 1934), Göteborgs Högskolas årsskrift, bd. 40, no. 1.
64 Kepler to Heydon, October 1605, letter 357, ll. 164ff.; *GW*, XV, 235. This passage is followed immediately by the lines referred to in note 15 above.

9

Francis Bacon's biological ideas: a new manuscript source

GRAHAM REES

In the past few years a couple of unprecedented events have taken place in the normally rather quiet field of Bacon studies. Several hitherto unknown Bacon manuscripts, the most substantial ones to have come to light since the seventeenth century, have been identified. In addition to that, a whole new branch of Bacon's philosophy, the branch I have called the "speculative philosophy," has been discovered and put together again. These two developments turn out to be mutually reinforcing. The new manuscripts tell us a lot about the speculative philosophy, and what we already know about the speculative philosophy from the printed sources helps us to make sense of the manuscript materials – materials that promise to give us new insights into the growth, scope, and character of the speculative philosophy itself.

Until recently it was generally believed that the canon of Bacon's work had been substantially established by the great Victorian editors Spedding, Ellis, and Heath.[1] But in 1978 an unpublished natural-philosophical manuscript was found in the British Library (Additional Manuscripts 38,693, fols. 29^r–52^v). A transcription of and commentary on this piece was published in 1981.[2] However, that discovery was quite overshadowed by findings made by Dr. Peter Beal in the course of his researches for the monumental *Index of English Literary Manuscripts*.[3] Beal discovered a manuscript copy of an unknown fragment, *Historia et inquisitio de animato et inanimato*, a copy (possibly complete) of the *Abecedarium novum naturae*, and a 13,500-word Latin manuscript on biological topics.[4] In my view, the most important of these discoveries is the last and, in due course, I hope to publish this text with translation and full commentary. What follows is a preliminary report on this exciting find.

The manuscript, designated Hardwick 72A, is lodged at Chatsworth House, the unassuming country *palazzo* of the Duke of Devonshire.

The manuscript runs to some thirty leaves. The first half was written out by an amanuensis and subsequently revised and corrected by Bacon (fols. 1^v–15a^r).[5] The second half (fols. 16^r–30^v) is in Bacon's hand alone and consists of a series of rough drafts (and rough drafts of rough drafts),[6] which were meant to be spliced into the material in the first half.[7] By dint of beta-radiographical analysis and interpretation of internal clues, it is possible to establish that the amanuensis's copy was probably made about 1612–13 and that Bacon's revisions and additions were probably drafted over a period ending not later than 1618 or 1619, when the manuscript was abandoned unfinished.[8] If these dates are about right, the Hardwick manuscript embodies Bacon's first extant attempt at a proper treatment of several topics that were to become crucial in the final years of his philosophical career.

The manuscript is entitled *De viis mortis et de senectute retardanda, atque instaurandis viribus*, and as the title suggests, the work is concerned with the processes of aging and with ways of slowing them down. Bacon, like the alchemists, regarded the prolongation of life as a protosalvation, as a soteriology for this life.[9] Yet, if Bacon adopted an alchemical aim, he did not subscribe to alchemical means.[10] And consideration of means gives us access to the largely unreconstructed realm of his biological ideas – ideas that constitute a distinct, yet integral, subdepartment of the speculative philosophy.

Now it is still not widely understood that Bacon was the architect not of one but of two bodies of philosophy. Most people know, of course, that he put together a method and program for the regeneration of the sciences by inductive means. But the contents of the Hardwick manuscript have precious little to do with the method and program. Instead, the manuscript expresses aspects of a quite different body of philosophy, a body that is nothing less than a systematic, deductive model of the phenomena of nature. This speculative system, this highly integrated and wide-ranging set of *explanations*, permeates Bacon's writings.[11] In fact, Bacon's philosophical work resembles one of those perspective drawings of the *Gestalt* psychologists. Looked at in one way, the method and program flash upon the eye; but if one looks a bit more intently one begins to discern the outlines of the speculative philosophy, a philosophy that coexists and intersects with the method and program.

Since the Hardwick manuscript develops aspects of the speculative philosophy, it is in order to say a little about the principal features of that philosophy here, though what follows is a mere summary of a rather complicated structure. Bacon visualized the universe as a finite, geocentric plenum, a plenum divided into three regions. The central region is the interior of the earth, which is the abode of the extremely dense, passive, and immobile *tangible* matter. The celestial heavens,

Table 1. *The structure of Bacon's matter theory*

		Sulfur quaternion	Intermediates	Mercury quaternion
Tangible substances (with attached spirits)	{	Sulfur (subterranean)	Salt(s) (subterranean and inorganic beings)	Mercury (subterranean)
		Oil and oily, inflammable substances (terrestrial)	Juices of animals and plants	Water and "crude" nonflammable substances (terrestrial)
Pneumatic substances	{	Terrestrial fire (sublunar)	"Attached" animate and inanimate spirits (in tangible bodies)	Air (sublunar)
		Sidereal fire (planets)	Heaven of the fixed stars	Ether (medium of the planets)

by contrast, consist entirely of spirits or *pneumatic* substances, and these are weightless, highly active, but thoroughly corporeal. Between the earth's interior and the heavens lies the realm of mutability, a frontier zone where tangible and pneumatic matter mix and associate together.[12] The distinctions between these three regions and between tangible and pneumatic matter are the hinges on which Bacon's entire speculative philosophy turns. They form the basis for further distinctions that constitute Bacon's matter theory.

Some twelve different manifestations of matter lie at the heart of this theory (see Table 1).[13] Eight of these belong to one or the other of two rival families or "quaternions," each quaternion consisting of four qualitatively related substances. The sulfur quaternion comprises sulfur, oil, terrestrial fire, and the substance of the planets, sidereal fire. The mercury quaternion consists of mercury, water, air, and ether – the last of which fills the interplanetary spaces. These two mutually hostile quaternions are the keys to Bacon's hybrid, semi-Paracelsian cosmology and astrophysics.[14] He invoked them to explain the structure and motion of the heavens, wind and tidal motion, and even the directional tendency (or verticity) latent in the earth's crust.[15] In short, the two quaternions are the main pillars of Bacon's speculative edifice.

Structurally dependent on the theory of the two quaternions is the theory of intermediates or substances that combine or embody qualities inherent in one member of a quaternion with qualities inherent in its opposite number in the other quaternion (see Table 1). Just as there are four members in each quaternion, so there are four kinds of intermediates.[16] From the point of view of Bacon's reflections on biological topics, the most important intermediates are the air–flame ones. These are the "attached" spirits: "attached" because (unlike air, ether, terrestrial fire, and sidereal fire) they are imprisoned in tangible matter. The "attached" spirits are either animate or inanimate. The inanimate spirits are diffused through all tangible bodies, living and nonliving, that dwell within the mutable frontier region between the heavens and the bowels of the earth. These spirits are distributed through tangible matter in discrete portions (rather like bubbles in ice), and their activity within tangible bodies is responsible for most of the changes observable in the terrestrial realm.[17]

The animate or vital spirits are present in living bodies only, where they coexist with the inanimate spirits. Unlike the inanimate spirits, the vital ones have more fire than air in their constitution and are organized not in discrete portions but in continuous channels.[18] The two sorts of "attached" spirits are the principal agents of biological change and, through the theory of intermediates, the engines of biological change are integrated into the structure of the speculative philosophy. In fact, Bacon's biological theories absolutely depend, from a logical point of view, on a more comprehensive, intellectual framework – a framework that embraces the whole universe and rests on the biquaternion theory and, ultimately, on the tangible–pneumatic distinction.

While Bacon was revising and adding to the amanuensis's draft, he seems to have been particularly interested in exposing the wider theoretical setting of his biological ideas. He alluded to the two quaternions, some of the intermediates (fol. 17^{r-v}),[19] and the special character of the earth's interior (fol. 16^r). He also went to some trouble to formulate his ideas about the zone of mutability (fols. 22^r, 25^r).[20] That is not surprising, for that zone is the home of the "attached" spirits, and the Hardwick manuscript was the first in which ideas about these spirits were systematically developed. Until the manuscript was discovered, one might have been forgiven for supposing that Bacon had done very little to develop a philosophy of terrestrial change before 1620, when the first parts of the *Instauratio magna* were published. Before 1620 one finds few references to the inanimate spirits and fewer still in connection with biological topics.[21] Only in the Hardwick manuscript were they used extensively in relation to biology, and for the first time. As for the vital spirits, they simply do not figure in any work written before the Hardwick manuscript. In short, it used to be thought that the theory of "attached" spirits was a late development. But it now seems that the theory probably surfaced not long after 1611–12, at the same time as, or not long after, the great upsurge of speculative activity that marked the appearance in plenary form of the cosmological and astronomical aspects of the speculative system.[22] In fact, the second decade of the seventeenth century may be seen as a crucial one for the system, as the one in which the cosmological dimension of the speculative philosophy was elaborated and the one in which the frontiers of the speculative philosophy were pushed into the realm of biology. The development of the speculative philosophy was not, as it once seemed, intermittent, but was a concerted process, and a process that was well under way before 1620.

What does the Hardwick manuscript actually say about the vital and inanimate spirits? For Bacon, staying young and beautiful was largely a matter of preserving the integrity of the vital spirits against the assaults of the inanimate spirits whose natural tendency was to destroy the conditions necessary for the persistence of the vital ones. Bacon was very proud of this idea. In fact, he distinguished himself from earlier writers on aging and death by insisting that answers to the problem of aging should be sought as much in what makes living beings like inorganic ones as in what makes them different (fols. 1^v–3^r, 21^r, 29^r).[23] The main thing that living beings have in common with inorganic bodies is inanimate spirit, and it is to the doings of the inanimate spirit that much of the Hardwick manuscript is devoted. The manuscript tells us very little about vital spirit though what it does tell us is important.

Let us consider the vital spirit first. Possession of vital spirit is what makes a living being different from an inanimate one, and the specific quality of the vital spirit is, ultimately, what makes one living being different from another. All beings belong to a hierarchy of forms, for Bacon organized the daunting variety of living things in terms of a thoroughly traditional, commonplace regulative belief: the belief in the chain of being. His particular version of the hierarchy can be reconstituted in considerable and rather attractive detail, though such a reconstruction would have to be based on evidence drawn from texts other than the Hardwick manuscript.[24] Bacon eventually assimilated the hierarchy to his biological theory by associating it with the notion of vital spirit. He annexed an old concept to a new philosophy or, conversely, justified an aspect of his matter theory in terms of its capacity to explain the fact of a natural hierarchy. He did this by asserting that the higher the organic form in the chain of being, the warmer would be its vital spirit, the greater the quantity of the vital spirit relative to the body, and the more complex the spatial layout of the spirit.[25] He began this task in the Hardwick manuscript and started by establishing a fundamental trichotomy – a trichotomy that was to reappear in the works of his final years.[26] According to the manuscript, spirit is arranged or disposed in three distinct ways: in discrete portions, in branching channels, and in branching channels connected to a cellular or ventricular concentration of spirit. This trichotomy marks out three main sections of the chain of being. Inorganic bodies possess only the first kind of spirit: the inanimate spirit. Plants have inanimate spirit but they also contain the ramifying, channeled vital spirit. Animals possess the two former sorts of spirits but the ramifying vital spirit is rooted in a concentration of vital spirit located in the head (fols. 16^r, 26^v, 28^r).[27]

Plants possess only vital spirit arranged in branching channels, and the vital spirit, coursing through the channels, superintends the vegetative functions: nutrition, maintenance, growth, and reproduction. In animals, on the other hand, the vital spirit does not just keep the involuntary functions going; it also mediates centrifugal motor functions and centripetal sensory ones. Animals are capable of sensation and voluntary motion because the vital spirit, channeled throughout the body in the nerves and sinews, communicates with a ventricular concentration of spirit. The spirit, as an intermediate, embodies the properties of air and fire. The fiery component is the source of motion; the airy, the source of sensation. Sensations are propagated through the spirit in the nerves and communicated to the ventricular concentration. Conversely, motions of the spirit in the ventricles of the brain travel outward through the nerves, and the force so produced causes voluntary motion (fol. 26^v).[28]

For the sake of completeness it should be added that Bacon carried his consideration of the functions of the vital spirits further in works belonging to the *Instauratio magna*. We learn from these works that the spirit in the ventricles of the brain is the material substrate of higher mental functions. Animals, even insects, have memory, and memory seems to be an aggregation of qualitative changes induced in the ventricular spirit by sensory inputs. These changes, preserved in the spirit, can be passed over to the imagination (itself a state of the vital spirit in the appropriate ventricles) to produce voluntary motion. Memory and imagination are therefore precursors of voluntary motion in animals.[29] Whether that is true in the case of man, and whether human memory and imagination are functions of the spirit, are moot points. The evidence is contradictory, though on balance Bacon does appear to assign some of the higher faculties in man to the activity of the vital spirit rather than to man's unique attribute – the immortal, incorporeal soul, which is the seat of the rational faculty.[30] At all events, Bacon's view of the higher faculties in human beings needs more discussion, though I do not propose to take the matter any further here as the Hardwick manuscript says nothing about it.

So much for the functions of the vital spirit. What of the conditions for its persistence? Here again the works of the *Instauratio* tell us a great deal more than the Hardwick manuscript. The manuscript merely points out that living bodies are hot and so need respiration to cool them and that the vital spirits consume the juices of the body so that the body requires "alimentation [*alimentatio*]" (fol.29ᵛ). The later works develop these points and add a further one: that the spirits need sleep and motion for their preservation. Sleep seems to be necessary because it reduces the activity of the spirits and their demand for food.[31] Motion is intrinsic to the vital spirit, and loss of mobility causes the spirit to perish. Thus a severe blow to the head will be fatal if the blow constricts the spirit in the cerebral ventricles.[32] The spirit needs respiration to cool it, for such is the nature of its chemistry that without respiration the spirit would destroy itself in its own heat.[33] As for alimentation, Bacon's views are somewhat mysterious. In the *Historia vitae* he declares that "the living spirit subsists in identity, and not by succession or renovation [*in identitate, non per successionem aut renovationem*]." But he also says that no living being can go without food for long – a fact which shows that consumption is the work of the living spirit, which either repairs itself or makes it necessary for the parts of the body to repair themselves or both.[34] In other words, Bacon comes close to the contradiction of representing the spirit as a self-subsisting entity requiring nourishment – a contradiction he never entirely resolved. These then are the conditions for the persistence of the vital spirit, conditions that hold only if the organs of the body function efficiently. Destruction of vital organs leads to destruction of the vital spirit, and what destroys the vital organs in the process of aging is the other class of "attached" spirit, namely, the inanimate. But before I say anything more about inanimate spirit, I will add a few words about the origins of Bacon's concept of vital spirit.

The concept of vital spirit, as it emerges in the Hardwick manuscript and as it is developed in the works of the *Instauratio*, has fairly obvious affinities with the Galenic theory of the cerebral pneuma (*pneuma psychikon*) – the substance secreted from the arterial blood of the choroid plexus or *rete mirabile*.[35] Indeed, Bacon sounds very much the Galenist when he remarks that the vital spirit is "repaired from the fresh and lively blood of the small arteries which are inserted into the base of the brain."[36] It is also true that, as far as one can see, the concept of vital spirit was integrated into a broadly Galenic view of the functions of the organs of the body, though it would be proper to point out that Bacon, with his emphasis on invisible spirits as the chief agents of biological change, and his correlative scepticism regarding the efficacy of anatomical researches (fol. 28ʳ), said very little in the Hardwick manuscript or anywhere else about the functions of the organs.[37]

Nevertheless, it would be a mistake to think of the vital spirits simply as offspring of Galenism coopted by Bacon in mid-career to plug theoretical gaps in his emergent system of biological ideas. Apart from anything else, the concept of vital spirit has affinities with Neoplatonic and Paracelsian notions of the astral body – a notion admirably explicated by Owen Hannaway in his recent study of Oswald Croll, whose *Basilica chymica* (1609) was certainly well known to Bacon.[38] Nor should one overlook Bacon's acknowledgment, in the *De augmentis scientiarum* (1623), of links between the theory of vital spirit and the speculations of Bernardino Telesio and Agostino Doni on the nature of the human soul.[39] Above all, it should not be forgotten that Bacon's idea of vital spirit is locked, together with the idea of inanimate spirit, into a theory of matter rather different from anything devised by his predecessors or contemporaries. Taken in isolation, the doctrine of vital spirits may resemble the Galenic cerebral pneuma, the Paracelsian astral body, or whatever, but the doctrine's theoretical setting is unique. The view that Bacon was an "unoriginal" philosopher often seems to stem from an unfortunate tendency among scholars to look at particular ideas of his but to forget the relationship of those ideas to their wider intellectual context.

Turning now to the inanimate spirits, these are (as I have said) imprisoned within the tangible matter of all living and nonliving things in our environment. Trapped in discontinuous portions in tangible matter, like yeast in dough, they are the principal agents of change in the terrestrial realm. The greater part of the Hardwick manuscript is given over to a grinding struggle to formulate basic ideas about these spirits, a struggle that manifests itself in a degree of repetitiousness, the lengthy revisions inserted by Bacon in the margins of the amanuensis's draft (fols. 5^r, 8^v, 10^v, 11^r, 11^v, 12^v), and complete rethinking and rewriting of an entire section of the amanuensis's copy (fols. 5^v–8^r, 16^r–18^v).[40] The manuscript deals with the different varieties of inanimate spirits (fols. 5^v, 16^v, 23^r), the effects of different distributions of spirits within bodies (fols. 6^r, 9^v–10^v, 16^r, 24^r), the effects on the spirits of the various kinds of tangible substances that imprison them (fols. 6^{r-v}, 17^{r-v}), and the effects of ambient bodies on the relationship between the tangible body and its imprisoned spirit (fols. 7^{r-v}, 18^{r-v}). But chiefly the manuscript is concerned with the principal operations of the spirit and the impulses that give rise to them.

The inanimate spirits have three fundamental impulses: to move about, to multiply themselves, and to unite with kindred substances. They detest close confinement in the alien environment of tangible matter, so they become restless and predatory. They move about and search for weak points in the tangible matter surrounding them. This desire is intensified by their conspiratorial relationship with the ambient air, a relationship that arises from the chemical constitution of the spirits. The spirits are compounds of air and fire, but the airy component is the dominant partner; consequently, they long to unite with external air. In their efforts to accomplish their union with the air, they eventually wreck the fabric of the objects containing them. If the spirits cannot at first escape from their tangible prisons, they attack the substance of the tangible bodies and turn susceptible parts of the bodies into more spirit. By multiplying themselves in this way, they weaken the structure of the tangible body – eventually to the extent that they are able to decamp into the surrounding air (fols. 7^v, 17^r, 27^{r-v}).[41]

These impulses give rise to the three principal operations of the spirit: attenuation (*attenuatio*), escape (*evolatio*), and contraction (*contractio*). Attenuation takes place when the spirits attack and convert the susceptible parts of a tangible body. Escape occurs when the body is sufficiently undermined for the spirits to leave it. That leaves the body dry and fissured, and the fissuring is intensified when the remaining tangible parts contract to avoid a vacuum by filling up the spaces vacated by the spirits and close ranks ("*ut fit in bello*") to resist further alteration of their nature (fols. 3^v–4^r, 7^r–8^r, 12^r, 18^r).[42] Thus wood dries and cracks with age, kernels shrivel in their shells, and human beings suffer wrinkling, loss of skin tone, and progressive loss of flexibility and efficiency in those internal organs that provide for the persistence of the vital spirits (fols. 4^r, 9^v). In fact, Bacon viewed death largely as a species of terminal desiccation, a condition not confined to scholars and similar undesirables, but a fate shared by all living things.

Since aging results in large measure from progressive desiccation caused by the depredations of the spirits, protracted youthfulness is to be gained by stilling the spirits' motions, insulating them from the seductions of the external air, and preventing their flight from the body. When revising the amanuensis's draft, Bacon systematically added practical recommendations for achieving these aims. He advised exercise, astringent medicines, the consumption of oily foods, massage, and strict avoidance of *balnea voluptaria* (fols. 10^v–12^v). He also commended the ancient Britons for their custom of wearing little more than woad, and the American Indians for smearing themselves with paint (fol. 13^v), though Bacon no doubt recognized that neither of these methods would be quite *comme il faut* at Gray's Inn or at Gorhambury. All the same, he did urge (in another work) that discreetly oiled underclothes might help to keep the air at bay. It is not known whether he actually took his own advice.[43]

In spite of everything said so far, it should not be imagined that Bacon regarded the inanimate spirits solely as agents of disintegration. In the works of his last years, two processes, vivification and putrefaction, stand out as reciprocal creative processes of the spirit. According to the *Sylva sylvarum* (1626), these operations are "as nature's two terms or boundaries; and the guides to life and death."[44] However, the Hard-

wick manuscript has relatively little to say about vivification and practically nothing to say about putrefaction. The manuscript tells us that
vivification, the process that turns a nonliving into a living thing, does
not occur in substances from which the inanimate spirits can easily
escape. In such substances, desiccation is the usual outcome. Nor does
vivification occur in hard bodies where the spirits are firmly imprisoned. It happens only in special kinds of matter where the spirits are
neither firmly suppressed nor able to find ready escape routes. In other
words, a balance has to be struck between detention and discharge of
spirit – and that is a characteristically Baconian intellectual reflex,
typical of his vision of cardinal biological phenomena as manifestations
of opposite tendencies held in equilibrium, or (in the case of vital spirit
and organic juices) as fusions of entities embodying contrary qualities.
The ideal matter for vivification is compliant and sticky, matter that
hinders the spirit but nevertheless allows it to move about and shape
the matter itself. Such matter is found, according to the works of the
Instauratio, in eggs, seminal fluid, "all menstruous substance," and
rotting flesh.[45] In such substances, according to the manuscript, the
spirit "fashions members, an organised body and things of that kind"
by a "simple and gentle thrusting out" – as can be seen especially in
putrefaction, which gives birth to vegetable and animal productions
like moss and worms. Indeed, it is even possible to see the movement
of incipient worms in putrefaction before their formation is complete
(fol. 15^v).[46]

At some point during the formation of a living body, the vital spirit
must come into being. In the works of his last years Bacon said nothing
about the origin of the vital spirit save, in the *De augmentis* (1623),
that it sprang from the "wombs of the elements [*e Matricibus Elementorum*]."[47] Only from the Hardwick manuscript and from no other
text do we learn that vital spirit is elaborated from inanimate. The
motions of the inanimate spirits cause discrete portions of the spirit to
join up in a network of channels and thereby to become vital spirit (fol.
26^v).[48] This organizational change is presumably accompanied by a
qualitative one, for the vital are more fiery than the inanimate spirits.
But nowhere did Bacon actually discuss the qualitative change and its
causes. All we know is that, in general, he believed that larger quantities of any substance could take on qualities other than those apparent
in smaller volumes of the same substance.[49] Perhaps he believed that
the mere joining together of pockets of inanimate spirit would induce
the gradual qualitative change that would convert it into vital spirit,
the substance that sparks inert matter into life.

It seems then that the formation and destruction of living bodies may
be seen as a cyclic process. The activity of the inanimate spirits in
suitable substances organizes the particles of gross matter into a com-

plex structure, and some of the inanimate spirits are converted into vital spirit that animates the developing, incipient organism. But with the passage of time, the residue of inanimate spirit in the tissues follows its usual course, attacks and undermines the tangible matter around it, and eventually renders the body inefficient, desiccated, and, ultimately, unable to sustain the vital spirit. The result of this is the disintegration of the body and putrefaction. The nauseating smell of rotting bodies is nothing less than the escape of the inanimate spirits into the air (fol. 4^r),[50] though putrefaction itself provides suitable conditions for new life. The pullulating deliquescence of corrupting flesh provides exactly the right kind of matter for the generation of imperfect creatures – worms, maggots, and flies – creatures whose formal instability results from the fact that they have sprung not from the highly specific matter of egg or seed but from the unsavory and indeterminate concoctions of putrefaction.[51] That is why Bacon described putrefaction as the "bastard brother" of vivification.[52] Death and putrefaction, themselves consequences of the action of inanimate spirits, furnish new matter on which the inanimate spirits can exercise the creative aspect of their nature.

These then are the most important doctrines adumbrated in the Hardwick manuscript, doctrines often (though not always) explicated more fully in the *Novum organum*, *Historia vitae et mortis*, and *Sylva sylvarum*. Let me now try, in my concluding paragraphs, to indicate why the manuscript is important for our understanding of Bacon's philosophical work as a whole. The manuscript tells us new things about the development of his biological ideas. It allows us to date their emergence to the years before the publication of the *Novum organum*. It enables us to identify the period 1611–19 as the period when the speculative philosophy underwent its most rapid growth – growth not simply in its cosmological ramifications, but in its biological ones as well. The manuscript seems, in fact, to represent a stage between Bacon's early allusions in the first decade of the seventeenth century to the aim of prolonging life[53] and his full-blown treatment of the subject in various parts of the *Instauratio magna* – notably, of course, in the *Historia vitae*.

But perhaps the most important fact about the manuscript is simply that it is concerned with an aspect of the speculative philosophy, a philosophy greatly indebted to sixteenth-century naturalism, to chemical, magical, and occult traditions. It is significant that all the recently discovered manuscripts have much to say about the speculative philosophy but very little to say about the much better known preoccupation with the inductive method and its accompanying program. The manuscripts, taken together with the huge volume of speculative material in the printed works, cannot but alter our understanding of the *balance* between major themes of Bacon's philosophical enterprise. The manuscripts lend their formidable weight to the lively suspicion that the speculative philosophy was in many ways just as important to Bacon as the method itself.

Were there any truth in this suspicion, would it not become all the more necessary to explain why Bacon, advocate of inductive routines and enemy of premature theorizing, should have poured so much effort in the speculative philosophy, into a philosophy that apparently violates his most cherished methodological principles? I used to think that was the right way of putting the question. But in light of recent developments, I begin to think that it is the existence of the method rather than the speculative philosophy that needs explaining. Putting the matter in a rather extreme form, why did Bacon bother to write the *Novum organum* at all if he believed he already possessed the makings of a creditable and credible body of positive science? However, this (like other questions about the relationship between Bacon's empirical, experimental method and his speculative philosophy with its occult, chemical, and magical antecedents) has so far proved singularly resistant to treatment. Perhaps the acute shortage of ready answers merely indicates that the question is wrong or inappropriate; at present I do not know.

All the same, there may be a couple of ways forward. The first has to do with the sort of trust Bacon extended to his speculative system. Put briefly, he must have believed that the system was (1) true, (2) false, or (3) possible or probable. If he thought his system was true, he would have had no reason for constructing the method. If he thought it was false, what possible motive could he have had for constructing it? In fact, it seems likely that he viewed most of the explanations embodied in the system as possible or probable, though one should be alive to the possibility that his view of the reliability of this or that part of the system may have varied as time passed. These modalities need exploring so that we can get a firm idea of what status Bacon accorded what parts of the system at what times. For the present, my view is that, on the whole, he reposed more confidence in his explanations of terrestrial phenomena than in ones relating to the cosmological domain.[54] This impression certainly accords with the Hardwick manuscript, where Bacon writes as if he were in possession of the truth or something like it. Given that degree of certainty and the speculative route by which he attained it, perhaps we ought to look again at the status Bacon accorded the inductive method. Perhaps, after all, he did not regard it as the unique, exclusive, omnicompetent method that some of his followers took it for. Certainly he believed that progress might be made without thorough application of the principles and routines of the method – so long as students of nature made a general commitment to look more closely than they had in the past at the facts of nature.[55]

But, leaving aside the possibility that Bacon may have been prepared to dilute his claims for the method quite considerably, there may be another approach to questions about the relationships between the speculative philosophy and the method, an approach by way of the earliest stages of Bacon's career. I am thinking here of the much-neglected documents of the 1590s,[56] inspection of which suggests that the vast and influential method and program may have originated, in part, from an attempt to establish general criteria for assessing the merits of existing natural philosophical systems – with a view to demonstrating, ultimately, the superiority of his own speculative system. Bacon seems to have put together elements of the speculative philosophy *before* he thought, in any concrete way, about formulating a new method; so why was he not satisfied simply to elaborate a theoretical system, without then going on to expound a new way of doing natural philosophy? Perhaps in the 1590s he could not be sure that he could persuade others that his system was a better bet than the ones of Telesio, Doni, Paracelsus, or anyone else. Perhaps his preoccupation with method may have originated in attempts to deal with a problem of this kind.[57] In other words, the celebrated (and currently reviled) method may, in part, have been a product of his early reflections on the obscure, forgotten, speculative system of which the Hardwick manuscript was a later expression. It might also be worth considering the possibility that the two branches of Bacon's philosophical endeavor may thereafter have grown up in some sort of dialectical relationship. If Bacon's speculative doubts and certainties stimulated his methodological thought (and vice versa), then our understanding of his natural philosophical enterprise as a whole will have to be revised substantially.

Acknowledgment

I would like to thank Dr. Christopher Upton for the invaluable help he has given in checking the transcription and in revising and correcting the translation of MS. Hardwick 72A.

Notes

1 *The Works of Francis Bacon*, ed. J. Spedding, R. L. Ellis, and D. D. Heath, 7 vols. (London, 1859–74), cited hereafter as *Works*; J. Spedding, *The Letters and Life of Francis Bacon*, 7 vols. (London, 1861–74); *A Conference of Pleasure, Composed for Some Festive Occasion About the Year 1592 by Francis Bacon*, ed. J. Spedding (London, 1870). Brian Vickers listed MSS of Bacon discovered subsequently to Spedding in his *Essential Articles on Francis Bacon* (Hamden, Conn., 1968), pp. xix, xxi–xxiii (n. 15).

2 Graham Rees, "An Unpublished Manuscript by Francis Bacon: *Sylva Sylvarum* Drafts and Other Working Notes," *Annals of Science*, 38 (1981), pp. 377–412.

3 Peter Beal, *Index of English Literary Manuscripts*, vol. 1, *1450–1625*, pts. 1 and 2 (London, 1980), items BcF 286, BcF 296. William Rawley, Bacon's secretary and biographer, wrote in 1657 that the "*Abcedarium*" (*sic*) was "lost" (see *Works*, I, 9). However, Tenison published a fragment of it in *Baconiana* (London, 1679). This fragment was reprinted by Spedding (see *Works*, II, 85–8). It is not absolutely certain that the manuscript discovered by Beal is complete. The manuscript appears in the same volume and is written in the same hand as copies of the *Historia . . . de animato* and *Historia densi et rari* (BcF 295), Bibliothèque Nationale, fonds français n°. 4745, fols. 39^r–62^r, 5^r–8^r, 9^r–38^v, respectively. I have inspected a photocopy of this material, and comparison of the *Historia densi* with the printed text (*Works*, II, 243–305) shows that the copyist worked from a version written before the one on which the printed text was based. However, the printed portion of the *Abecedarium* seems to me to have been based on a manuscript fragment antedating the one from which the Paris manuscript was copied. Nevertheless, one cannot be certain that the draft represented by the Paris manuscript was Bacon's final one. As the *Abecedarium* and *Historia densi* are both late works (ca. 1622 and ca. 1624, respectively), it is possible that *Historia . . . de animato* belongs to the same period of Bacon's career.

4 Beal treats the manuscript as if it comprised two distinct items (BcF 287, BcF 294). In fact, the manuscript materials are parts of a single, unfinished work.

5 Beal's numbering of items BcF 287 and BcF 294 omits the leaf after folio 15. I have designated the omitted leaf (the bottom half of which has been torn off) "folio 15a."

6 For instance, the material on folio 25^r is an early version of the draft entitled "Aphorismus 1" on folio 22^r. The passage beginning "At animatorum . . ." on folio 28^r appears in a revised form on folios 28^v–29^v. The final section of the revised form was revised again at the bottom of folio 29^v.

7 The manuscript is a horrendous palimpsest of shifting intentions, changes of mind, of canceled and reinstated drafts, and successive rearrangements of materials. I think I have managed to sort out the final order in which the drafts were meant to appear and the stages by which Bacon came to settle on the final order. But I will leave discussion of these matters until a later date lest I burden this relatively short account with detail more appropriate to a monograph.

8 Proper discussion of the dating of the manuscript would require a gargantuan note. For the present, suffice it to say that the material which concerns us begins on folio 1^v of the manuscript. At some point in the history of the manuscript folio 1^r had an extra leaf gummed onto it. That leaf bears a copy (in an unidentified seventeenth-century hand) of a panegyric on Bacon by George Herbert, *In honorem illustrissimi D. D. Verulamii*. It is possible to read folio 1^r through the gummed leaf if the leaf is held up to the light. It turns out that folio 1^r carries the closing 150 words of Bacon's *De fluxu et refluxu maris* – words drafted by the *same* amanuensis who drafted the material on folios 1^v–15a^r. It is very probable, therefore, that the Hardwick manuscript once contained a complete copy of

the *De fluxu*. It is certain that folio 1ʳ bears the only known fragment of the *De fluxu* actually copied under Bacon's supervision. At all events, it is obvious that the amanuensis must have drafted the material on folios 1ᵛ–15aʳ at some time after the completion of the *De fluxu*. The *De fluxu* was not, as many modern scholars seem to think, written ca. 1615–16. It was written ca. 1611–12. Bacon mentions it in the *Thema coeli* (1612); see *Works*, III, 776.

It is likely that the manuscript was abandoned not later than 1618–19. William Rawley seems not to have seen or had access to it. Had he had access to it, he would probably have published it after Bacon's death. Rawley did not enter Bacon's service until 1618 and does not mention this manuscript in the chronological list of the works of the last five years of Bacon's life (*Works*, I, 8–10). It is also the case that much of the manuscript material was superseded by material in the *Novum organum* (1620) and *Historia vitae et mortis* (1623). Bacon may have been planning the *Historia vitae* as early as 1620; see *Works*, I, 708.

9 Roger Bacon, for instance, viewed the prolongation of life in this way; see Edmund Brehm, "Roger Bacon's Place in the History of Alchemy," *Ambix*, 23 (1976), pp. 53–7, 56.

10 For Bacon's criticism of iatrochemical means of prolonging life, see *Works*, I, 574, 599; II, 105.

11 There is scarcely a work untouched by it; see Graham Rees, "Francis Bacon's Semi-Paracelsian Cosmology and the *Great Instauration*," *Ambix*, 22 (1975), pp. 161–73; Rees, "Unpublished Manuscript," pp. 381–93.

12 For the origins and significance of the three-zone theory, see Graham Rees, "Francis Bacon on Verticity and the Bowels of the Earth," *Ambix*, 26 (1979), pp. 202–11.

13 For a full reconstruction of the theory, see Graham Rees, "Matter Theory: A Unifying Factor in Bacon's Natural Philosophy?" *Ambix*, 24 (1977), pp. 110–25.

14 See Graham Rees, "Francis Bacon's Semi-Paracelsian Cosmology," *Ambix*, 22 (1975), pp. 81–101.

15 Ibid., pp. 91–101; Rees, "Bacon on Verticity," pp. 202–11.

16 For the theory of intermediates, see Rees, "Matter Theory," pp. 116–18.

17 See *Works*, I, 310–11; II, 213, 451, 616.

18 Ibid., II, 214–15, 528.

19 "At de crudo et pingui, sunt illa duo primae corporum texturae et maximae rerum familiae . . . eorumque judicium proprie capitur ex illo quod inflammatur vel non. Nam tertium illud, quod satis ignoranter addunt chimistae de sale, compositum quiddam plane ex duobus reliquis. Omnis enim sal habet partes quae ignem concipiunt, et partes quae ignem exhorrent et fugiunt unde etiam rudimentum vitae est. Omnis enim aura vitae consistit ex spiritu aereo et igneo." For later and similar formulations, see *Works*, II. 82–3, 351–2, 543.

20 "In superficie terrae et incrustatione illa exteriore, quae certe non multum producitur in profundum . . . omne ens tangibile . . . habet . . . portiones pneumaticas . . . nullum prorsus reperiatur corpus ex crassis et tangibilibus sincerum sed quod habeat inclusum et commistum aliquid notabile ex pneumatico . . . non intelligimus virtutes, aut energias, aut facultates corporis aspectabilis et tangibilis, sed plane aliud corpus, corpore illo crassiore obductum et obsessum" (fol. 22ʳ). These words were later echoed in the *Historia vitae et mortis* (1623); see *Works*, II, 213.

21 The inanimate spirits were invoked in, for instance, the *Cogitationes de natura rerum* (ca. 1604) and the *De principiis atque originibus* (161-?) to explain various physical phenomena (*Works*, III, 24–5, 32, 34, 89, 109). In works written before the Hardwick manuscript there are only two brief references to the biological functions of inanimate spirits. The references, in *The Advancement of Learning* (1605) and the *De sapientia veterum* (1609), occur in connection with the aim of prolonging life (*Works*, III, 362; VI, 760–1).

22 This upsurge manifests itself in the *De fluxu et refluxu maris* (1611?), *Thema coeli* (1612), and *Descriptio globi intellectualis* (1612).

23 "At animatorum natura est partim communis cum inanimatis, partim propria. Omnia enim quae diximus insunt etiam animatis neque superaddita natura vitalis ea extinguit, sed in ordinem redigit. Latent vero illae operationes sub actionibus vitalibus ad tempus . . . At in decursu aetatis praegravant operationes substantiae super actiones vitales nisi accuratis remediis altera natura confortetur, altera immutetur. Itaque omnia viventia patiuntur et subeunt tormenti illud genus Mezentii, ut viva in complexu mortuorum pereant . . . Nam ut nunc sunt res, consueverunt medici, et maxime illi qui in anatomia diligentiam suam ostentant . . . actiones vitales solum et per se contemplari atque omnia ad illas referri" (fol. 29r). For later echoes or versions of these views, see *Historia vitae et mortis* and *Sylva sylvarum*, *Works*, II, 106–7, 364.

24 Bacon's version of the hierarchy has never been examined in appropriate detail. For some of the materials from which a reconstruction of that version might be made, see *Works*, I, 231, 278–9, 283, 525–6, 543–4, 604–7; II, 208, 262–3, 340, 453, 474, 506–8, 517, 529, 531, 547, 557, 560, 592–3, 630–1, 638, 639. The seminal study of the chain in Western intellectual history is A. O. Lovejoy's *The Great Chain of Being* (New York, 1936); see also William F. Bynum, "The Great Chain of Being After Forty Years; An Appraisal," *History of Science*, 8 (1975), pp. 1–28.

25 *Works*, II, 208, 214–15, 474, 528, 530. The manuscript (fols. 26v, 29v) also suggests that higher beings have warmer vital spirits in greater concentrations than lower ones.

26 *Works*, I, 311; II, 214–15, 528.

27 "Spiritus entis aut intermistus est aut ramosus, aut cellulatus cum universitate. Spiritus intermistus ille est qui a se per partes rei crassiores penitus abscissus est. Atque iste spiritus invenitur in omni ente tangibili inanimato, et in mole et partibus tangibilibus omnis entis viventis. Spiritus ramosus sibi continuus est per poros et meatus suos, sed ista continuatio datur tantum per lineas exiles et canales minutos, qualis est spiritus omnis vegetabilis. At spiritus cellulatus et ipse scilicet ramosus est, sed habet cellam, id est arta loca et spatia, cavaque in re, ubi spiritus congregatur purus et per se, in quanto pro ratione rei notabili et bene magno ad quem rivuli illi spiritus ramosi se referunt tanquam ad universitatem. Atque huiusmodi est spiritus omnis sensibilis" (fol. 16r).

28 See Table 1, above. For references to sensation and motion in the printed sources, see *Works*, I, 278, 328, 609–10; II, 351–2.

29 See, for instance, *Works*, I, 649; II, 559.

30 K. R. Wallace says that Bacon did not commit himself on the incorporeality of the immortal soul; see *Francis Bacon on the Nature of Man* (Urbana and London, 1967), pp. 14–15. In fact Bacon was quite certain that the soul was incorporeal; see *Works*, II, 225. In the *De augmentis* he attributed man's

higher faculties to this soul (ibid., I, 494), but later in the same work we learn that the study of voluntary motion and imagination belongs unequivocally to the research field concerned with the human vital spirit (ibid., I, 609–10). The whole question of Bacon's view of the human faculties needs to be looked at again.

31 *Works*, II, 205–6, 363.
32 Ibid., II, 204.
33 Ibid., II, 205.
34 Ibid., II, 206; cf. V, 314.
35 See Rudolph E. Siegel, *Galen on Psychology, Psychopathology, and the Function and Diseases of the Nervous System*, (Basel, 1973), pp. 37–9, 61–4; E. Ruth Harvey, *The Inward Wits: Psychological Theory in the Middle Ages and the Renaissance* (London, 1975), pp. 4–7.
36 *Works*, II, 226; cf. V, 335.
37 Bacon believed that the stomach converted food into "chylus," which was in turn elaborated into blood in the liver; that urine, "the whey of blood," was drawn off by the kidneys; that the venous blood created in the liver nourished the tissues, and that the blood of the veins supplied the blood of the arteries, which in turn supplied the vital spirits (ibid., II, 130, 180, 207, 358, 362, 613). For Bacon's scepticism of anatomical researches and its association with the theory of spirits, see ibid., I, 232–4; also see Graham Rees, "Atomism and 'Subtlety' in Francis Bacon's Philosophy," *Annals of Science*, 37 (1980), pp. 549–71, 567–9.
38 See *Works*, II, 671; O. Hannaway, *The Chemists and the Word: The Didactic Origins of Chemistry* (Baltimore and London, 1975).
39 *Works*, I, 606. For Bacon's debt to Telesio and Doni, see D. P. Walker, *Spiritual and Demonic Magic from Ficino to Campanella* (Liechtenstein, 1969), pp. 199–201, and "Francis Bacon and Spiritus," in *Science, Medicine and Society in the Renaissance*, ed. Allen G. Debus, 2 vols. (London, 1972), II, 121–30. I suspect that further comparative study of the spirit theories of Doni and Bacon might prove illuminating for our understanding of the latter. Doni's physiology of spirits seems, at points, remarkably like Bacon's; see Luigi De Franco, *L'eretico Agostino Doni, medico et filosofo cosentino del '500. In appendice: A. Doni – De natura hominis – con traduzione a fronte* (Cosenza, 1973), pp. 308–12, 326–32, 348–52.
40 The copy and rewriting account for about 20 percent of the manuscript.
41 "Quod vero ad ipsa desideria spiritus . . . illa tria omnino esse reperiuntur. Omnis spiritus triplicem habet appetitum, et secundum eum perfungitur et operatur: primus est agitationis et motus et fruendi natura sua, secundus multiplicandi sui super aliud, tertius exeundi sive conjungendi se cum cognatis. Itaque spiritus . . . corpus illud crassum convellit et fodicat et subruit . . . et in hunc modum se multiplicat" (fol. 17ʳ). "Itaque evolat spiritus non solum ob desiderium suum exeundi, verum etiam plane sollicitatur et evocatur ab aere tanquam inita conspiratione" (fol. 27ʳ).
42 "Atque universus iste processus nihil aliud est quam Actio triplex, videlicet Attenuatio, et subinde partis attenuatae Evolatio, partis vero manentis et non attenuatae Contractio . . . Sed spiritus ille innatus et praeinexistens primo depraedatur aliquid ex substantia crassiore, illudque confecit, et in spiritum vertit atque una secum vehit, eaque simul evolant, unde fit diminutio ponderis" (fol. 3ᵛ). "Atque actio illa Contractionis . . . Postquam enim tenuior pars inclusa tolli et rapi caeperit . . . partes crassiores se cogant in angustius et spatium desertum impleant" (fol. 7ᵛ).

43 *Works*, II, 178, 180.

44 Ibid., II, 451.

45 Ibid., I, 316; II, 451, 557–8, 638.

46 "Videtur enim omnis vivificatio esse quiddam medium inter detentionem et emigrationem spiritus. Ubi enim spiritus . . . incidit in materiam obedientem et sequacem . . . ita tamen ut spiritus dilatet se localiter, et vias ad exeundum tentet . . . sequitur vivificatio, et membrificatio, et corpus organicum, et huiusmodi. Etenim simplex illa et mollis protrusio . . . est procul dubio rerum rudimentum . . . et principium ipsius vivificationis . . . Itaque plane cernitur quandoque muscus paulo arctior devenire herbidus, et formatus, et instar pusillae plantae. Putredo autem facile transit in vermiculos, etiam motu se manifestante antequam efformatio sit absoluta" (fol. 15ᵛ).

47 *Works*, I, 604.

48 "Quod si detur copia se sibi continuandi, et per hoc natura sua utendi et fruendi, tum demum se incendit, et se gerit pro potestate sua, unde primo corpus ad integrale figurat et determinat . . . Quod si non tantum diffundere se spiritus possit per canales illos et ramos, sed etiam sedem aliquam et cellam sibi parare ubi in quanto aliquo notabili congregari possit, tum vero sequuntur effecta multa nobilia . . . ex regimine spiritus in cella, spiritus in canalibus se comprimit et dilatat unde sequitur pulsus et motus localis."

49 See *Works*, I, 329.

50 Also see ibid., II, 120–1.

51 Ibid., II, 359, 507, 557–8.

52 Ibid., II, 452.

53 Ibid., III, 362; VI, 760–1.

54 See Rees, "Matter Theory," pp. 118–21.

55 See *Works*, I, 223.

56 See "Letter to Burghley" (1592?), in Spedding, *Letters and Life*, I, 108–9; "Mr. Bacon in Praise of Knowledge (1592?), ibid., I, 123–6; "Gesta Grayorum" (1594), ibid., I, 334–7; "A device to celebrate Queen's Day" (1595), ibid., I, 379–85. "Mr. Bacon in Praise of Knowledge" is only one section of a longer text, the whole of which was published in a record-type transcription by Spedding: see *A Conference of Pleasure*. The manuscript from which Spedding worked is lodged at Alnwick Castle, MS. 525 (safe 4), fols. 3–25.

57 These are large issues that cannot be examined fully here. The documents of the 1590s are difficult to interpret: They are allusive and ambiguous. There are also very great risks that an interpreter may unwittingly allow knowledge of Bacon's later writings to impose upon a reading of the early sources. However, it is certain that Bacon had clear speculative commitments in the early 1590s (see Spedding, *Letters and Life*, I, 124–5). No *specific* methodological commitment is evident in these early texts. There is a general commitment to the reform of knowledge for the material benefit of mankind, a reform to be accomplished by drawing the empirical and rational faculties into a new relationship (ibid., I, 108–9, 123–4), but no reference to induction, the natural-historical program, etc. The earliest reference to the speculative philosophy is coupled with a profoundly sceptical attitude to established authority in astronomy and cosmology. The scepticism spills over into a generalized call for the reform of knowledge (ibid., I, 124–6). In other words, the evidence of allegiance to specific speculative doctrines is linked to general programmatic declarations by criticism of views rivaling the speculative doctrines.

10

Newton and alchemy

RICHARD S. WESTFALL

On the whole, Newton preferred not to publicize his involvement in alchemy. Unlike his other major pursuits, nothing of his alchemy, or at least nothing explicitly labeled as alchemy, appeared in print during his lifetime or in the years immediately following his death. A few people did know about it. A fascinating correspondence between Newton and John Locke following the death of Robert Boyle reveals that the three men, possibly the last three men from Restoration England whom one would have expected, only a generation ago, to find so engaged, exchanged alchemical secrets and pledged each other to silence.[1] John Conduitt, the husband of Newton's niece, who gathered material about his life, knew of his experiments in Cambridge and reported that his furnace there remained an item of curiosity shown to visitors. Nevertheless, the adjective Conduitt used was "chymical," not "alchymical,"[2] and in a similar manner knowledge of Newton's interest in the art quickly sank from view. When David Brewster found alchemical manuscripts in Newton's own hand among his papers, he was appalled and quickly dismissed them as a curious relic of an earlier age.[3] It waited until the twentieth century for the record to become public, with the auction of the papers still in the hands of the Portsmouth family, and for scholars to come to grips with it. Lord Keynes purchased some of the alchemical papers at the auction and insisted forcefully on their importance,[4] but only in our own generation have scholars ready to take the papers seriously systematically studied the entire corpus, or rather that part – well over 90 percent – of the corpus known to exist that is available to the public. Betty Jo Dobbs and Karin Figala have been the leaders of this investigation.[5] As a result of their outstanding work, we probably know more today about Newton's endeavors in alchemy than anyone, including even his confidants in the art, Locke and Boyle, ever has.

The record is subject, of course, to varying interpretations. Newton was the single most important figure in establishing modern science with its unique view of reality and of the proper procedures to study it. Alchemy was one of the enterprises that modern science put out of business. Indeed, as David Brewster's references to "the most contemptible alchemical poetry," and, in regard to another paper, "the obvious production of a fool and a knave" make manifest, it appears to many as the quintessential embodiment of all that modern science opposes.[6] Not surprisingly then, some scholars, some very considerable scholars, reject the suggestion that alchemy played a significant role in Newton's intellectual life. Despite the manuscripts – and it should be obvious, as they contend, that the existence of the manuscripts does not of itself establish Newton's attitude toward their content – alchemy was in their view an activity peripheral to his central concerns. Those concerns manifested themselves in his *Principia*, his *Opticks*, and his fluxional calculus, the achievements that both shaped the modern scientific tradition and ensured their author's undying fame. Thus Bernard Cohen's recent *Newtonian Revolution* presents an analysis of the development of the *Principia* that focuses on problems internal to the science of dynamics and on Newton's transformation of received concepts of mechanics without saying more than a single word about alchemy. The single word is his emphatic rejection of the argument made by several scholars, including me, that Newton drew the concept of attraction out of the alchemical tradition.[7] Rupert Hall is uneasy that attention to Newton's alchemy will "cloud the clarity of reason and intellectual integrity . . . I would have regarded Newton as a founder of reason; so I think he wished to be regarded (for him reason included God, of course) not as flotsam on the weltering sea of the human unconscious. You must see that if you deny Freud in Manuel, you admit Jung with alchemy. *That* I am sorry about."[8] Cohen and Hall are names to be reckoned with in any discussion of Newton. A consideration of Newton and alchemy that proceeds by ignoring their opinions cannot hope to be taken seriously.

As there are those who reject the contention that alchemy was a central aspect of Newton's career, so there are others who make it the most central aspect. David Castillejo's recent *Expanding Force in Newton's Cosmos* presents the most fully developed expression this position has yet received. Significantly, the *Principia* scarcely appears in a work whose title proclaims the exact opposite of universal gravitation, and Newton's achievement in mathematics receives no mention at all. Castillejo opens, rather, with a chapter on alchemy, moves on to a chapter on the prophecies, and primarily from those two topics weaves a fabric that portrays not merely a Newton who let alchemy influence him, but a Newton whose entire intellectual life was thoroughly occult. In Castillejo's opinion, that intellectual life focused always on one investigation of which Newton's various studies were only specific facets, an investigation of two opposing forces, capable both of spiritual and material manifestations, the cyclical pattern of whose contentions has shaped both the universe and human history.[9] Castillejo does not enjoy the renown that Cohen and Hall command. Nevertheless, the book rests on very extensive research in the manuscripts, and it is written with insight and conviction. No serious discussion of Newton and alchemy can afford to ignore it any more than Cohen and Hall.

My goal in this chapter is to neglect neither of the two positions, represented by Cohen and Hall on the one hand and by Castillejo on the other, but also to agree with neither. I shall attempt rather to define and defend a position between them, one that asserts the significance of alchemy in Newton's scientific career while it refuses to equate him with the occult.

I begin by taking my stand on three empirically established facts. First, Newton left behind a corpus of papers about alchemy which testify that he took an interest, the nature of which requires definition, in the art. Second, as a natural philosopher Newton introduced a major revision in the prevailing mechanical philosophy by asserting the existence of forces, attractions and repulsions between particles of matter that are not in mutual contact. Third, there was a chronological nexus between the first two points, the interest in alchemy spanning the period that witnessed the revision of natural philosophy. My argument must, of course, include elaborations drawn from the nature of the alchemical papers, but it rests squarely on these three foundation stones and depends directly on their solidity.

As far as I can tell from the surviving manuscripts, alchemy was not among the topics that introduced Newton to natural philosophy while he was still an undergraduate in a university that, like all universities of the age, did not energetically promote anything we would call science. Chemical questions of any sort scarcely figured in his initial reading in natural philosophy. Not long after taking his bachelor's degree, however, Newton did discover chemistry, and according to his custom with any new study, he attempted to systematize what he was learning in a glossary of chemical terms.[10] The distinction between chemistry and alchemy in the seventeenth century, if indeed it is valid to speak of a distinction, is difficult to place with precision, but most people, I think, would incline without hesitation to place the glossary squarely on the side of chemistry. Robert Boyle was his primary authority at this time. His studies did not remain on the chemical side of the line for long, however. His accounts show that on a trip to London in 1669 he purchased *Theatrum chemicum*, the huge collection of alchemical writings in six quarto volumes. He also purchased two furnaces, glass equipment, and chemicals.[11] As we shall see, he quickly learned to put the equipment to work. For the moment note that he also did not allow *Theatrum chemicum* to lie idle. Notes from the essays it contains began to appear among his papers, and a few years later he compiled a list of its most important items.[12] Nor did he confine himself to the *Theatrum*. He ransacked other major collections, such as *Ars aurifera*, *Musaeum hermeticum*, and *Theatrum chemicum britannicum*. In collections, collected works of single authors, and individual books, he consulted all the major authorities of the long alchemical tradition: Morien, Rosinus, the *Turba philosophorum*, the *Scala*, the *Rosary*, Ripley, Michael Maier, Sendivogius, Eirenaeus Philalethes, and many others it would be pointless to list exhaustively. As he read, he developed criteria of judgment such that, for example, he canceled one passage of notes with a curt dismissal: "I believe that this author is in no way adept."[13] In the opinion of Professor Dobbs, Newton probed "the whole vast literature of the older [i.e., pre-seventeenth-century] alchemy as it has never been probed before and since."[14] A similar assessment of his reading in seventeenth-century alchemists from Sendivogius and Michael Maier to Eirenaeus Philalethes, Theodore Mundanus, and Didier does not seem excessive. Eventually he compiled a massive "Index chemicus," the likes of which alchemy has never seen, to guide him to relevant discussions – over 100 pages crammed with 879 separate headings and approximately 5,000 page references to more than 150 different works.[15] At the same time he began to assemble what must have been one of the great collections, in his day, of alchemical works, so that at his death, nearly thirty years after he had ceased to buy alchemical literature, alchemy still constituted more than 10 percent of his library.[16]

One interesting feature of Newton's alchemical papers, and one that helps to illuminate his interest in the art, is the appearance among them of copies, in Newton's own hand, of unpublished treatises. Some of them would later see publication. Thus he made extensive notes on Philalethes's *Ripley Reviv'd* about ten years before it appeared in print and copied out a version of his "Exposition upon Sir George Ripley's Epistle to King Edward IV" that differs from the published one.[17] Over a period of nearly thirty years, he appears to have had access to manuscripts that remain unpublished to this day: for example, an anonymous "Sendivogius Explained" and John DeMonte-Snyders's "Metamorphosis of the Planets."[18] A sheaf of unpublished treatises, in at least four different hands, among his papers and his own copies elsewhere of five of the treatises suggest what appears to me as the only plausible interpretation of these papers.[19] Someone lent him the collection to study and copy, and in this case, for reasons we cannot possibly know, he never returned the originals. Similarly, a treatise named "Manna," which is not in his hand, concludes with two pages of variant readings added by Newton together with the information that they were "collected out of a M.S. communicated to Mr F. by W. S. 1670, & by Mr F. to me 1675."[20] I do not see how to account for these copies of unpublished papers without admitting that Newton was in touch with the largely clandestine circle of English alchemists from whom he received manuscripts to copy and to whom, quite possibly, he himself communicated others. In 1683 one Fran. Meheux wrote to him about the progress of some unnamed third man in alchemical experimentation. In 1696, scarcely two weeks before his appointment as warden to oversee His Majesty's coinage in gold and silver, Newton received a visit from a Londoner who was a friend of Boyle and of Dr. Dickinson (a well-known alchemist of the day) who stayed for two days to discuss the work.[21] Mr. F., who lent copies of "Manna," was probably Ezekiel Foxcroft, a fellow of King's College.[22] W. S., Meheux, and the Londoner have all the solidity of shadows at this distance in time, but Newton knew them as sources of information on alchemy.

Newton did more than read. Almost from the beginning he experimented as well. When he moved into the chamber beside the great gate of Trinity in 1673, he set up a laboratory in the garden outside, and there he continued to experiment for more than twenty years.[23] At first glance, nothing could look less alchemical than his laboratory notes. They described severely quantitative experiments with specific substances, even if we cannot always identify the substances Newton's symbols represented; frequently, for example, he systematically varied the amount of a single ingredient (measured by weight) in order to determine the ideal proportions in a given compound.[24] Nevertheless, Professor Dobbs has succeeded in correlating some of the early experiments with the alchemical manuscripts and has shown that two substances he learned to produce, the star regulus of antimony and the net, were forms of the alchemical hermaphrodite, in which the sulfuric seed of iron (or Mars) was planted in a mercuric matrix, of antimony in the one case, of copper (or Venus) in the other.[25] Hence it appears impossible to avoid the conclusion that the early experiments were alchemical. No one has yet unraveled the later experiments, but it seems suggestive at least that Newton used materials such as the net and the oak, names drawn from the imagery of alchemy that appeared in his alchemical papers, and that he sometimes interrupted his notes with interpretive interjections couched in the imagery of alchemy. "I understood the trident." "I saw sophic sal ammoniac." "I made Jupiter fly on his eagle."[26]

The experimental notes aside, Newton's alchemical papers are sometimes said to consist solely of reading notes. This is simply incorrect.

Indeed, the concept of reading notes is itself less clear than one might think. Although some papers are certainly that, others reveal a typically Newtonian effort to organize information, to bind various authorities together into a systematic statement of the art. Thus one early paper drew up a list of forty-seven axioms with references to the authors on whom they were based.[27] He began to correlate the varied imagery he met.

> Concerning Magnesia or the green Lion [he wrote in a list of "Notae" which also treated other terms]. It is called prometheus & the Chameleon. Also Androgyne, and virgin verdant earth in which the Sun has never cast its rays although he is its father and the moon its mother: Also common mercury, dew of heaven which makes the earth fertile, nitre of the wise . . . It is the Saturnine stone.[28]

Some passages of this sort listed as many as fifty different images.[29] In a later paper, Newton distilled the work down to seven aphorisms. "This process," he stated, "I take to be ye work of the best authors, Hermes, Turba, Morien, Artephius, Abraham ye Jew & Flammel, Scala, Ripley, Maier, the great Rosary, Charnock, Trevisan. Philaletha. Despagnet."[30] He collected at least two sets of "Notable Opinions,"[31] and in his most extensive effort at synthesis he set out to compile a treatise in nine "works," for separate parts of which he left in one case seven, in another five, drafts.[32] Newton put these compilations together entirely from the writings of others. Nevertheless, to describe them as mere "reading notes" does not begin to suffice.

And finally, he also composed alchemical treatises himself. Professor Dobbs identified a paper from the late 1670s, entitled "Clavis," as Newton's own composition.[33] Although I find her argument, based on the paper's apparent use of Newton's own experimental results, wholly convincing, the identification has been challenged.[34] No one, I think, could challenge his authorship of another from the same period, entitled "Separatio elementorum," or his latter commentary on the "Tabula smaragdina."[35] Both papers are filled with emendations, Newton's typical habit with his own writing but one he never exercised on the writings of others. Undoubtedly his most important composition was an essay he finally called "Praxis," apparently composed in the summer of 1693.[36] It also is undoubtedly his own. We have four successive drafts of it,[37] and it cited Fatio's letter to Newton of May 1693.[38] At its climax, "Praxis" described a process that achieved multiplication, the ultimate goal of alchemy, in which the active essence of gold is set free to function.

> Thus you may multiply each stone [alchemical ferment] 4 times & no more for they will then become oyles shining in ye dark and fit for magicall uses. You may ferment it w^{th} ⊙

[gold] by keeping them in fusion for a day, & then project upon metalls. This is ye multiplication in quality. You may multiply it in quantity by the mercuries of wch you made it at first, amalgaming ye stone wth y^e ☿ [mercury] of 3 or more eagles [?] and adding their weight of ye water, & if you designe it for metalls you may melt every time 3 parts of ☉ wth one of ye stone. Every multiplication will encreas it's vertue ten times &, if you use y^e ☿ of ye 2^d or 3^d rotation w^{th}out ye spirit, perhaps a thousand times. Thus you may multiply to infinity.[39]

When Newton wrote this passage, he was in the state of acute tension that led to his breakdown in September 1693, and we must accordingly use it with caution. On grounds of scientific opinion, I cannot believe that Newton achieved multiplication. Because of his personal state when he wrote it, the passage does not convince me that he thought he had done so. I do accept it as valuable evidence of the extent of his immersion in the world of alchemy.

As another measure of the extent of his immersion, I propose the sheer quantity of the alchemical papers. Indications of their extent have appeared throughout my discussion, but we all know how readily one can contrive to inflate the impression of a small number of papers. Hence it has seemed important to me to arrive at a quantitative measure of these manuscripts by counting pages and words per page. There would be no point in estimating in a similar way the number of words Newton devoted to mathematics or dynamics or even theology, enterprises his commitment to which no one questions. The estimate is, of course, very crude; implicitly it equates the effort devoted to copying a page of a treatise with the effort given to composing a page of his own or to filling a page with experimental notes. Such a count serves only two purposes. It gives substance to the claim that the papers are very extensive, and when it is divided into chronological periods, it gives a rough measure of the intensity of his involvement with alchemy at different times. Restricting myself for the moment to the first, I note that Newton left behind about 1,200,000 words on alchemy. I see no way to dismiss it as an occasional interest. I think the other evidence I have brought forward indicates beyond reasonable doubt that the interest was sympathetic, the interest of a man who took the art seriously.

Meanwhile, alchemy did not exhaust the whole of Newton's intellectual life. As I suggested, he had found natural philosophy several years earlier. Specifically, about 1664, he had found the new natural philosophy that the seventeenth century called the mechanical philosophy, and in a notebook he recorded his initial contact with it under the heading "Quaestiones quaedam philosophicae."[40] For about three

years, as his earlier notes indicate, the university had been feeding him
on the dry bones of an Aristotelian philosophy desiccated beyond any
hope of renewal. The "Quaestiones quaedam" recorded a conversion
experience, not unlike the revelation we find in the pages of Galileo
and Descartes, that natural philosophy could be done in a different
way. Under his title Newton later returned to record a slogan: "Amicus
Plato amicus Aristoteles magis amica veritas." He had discovered the
world of the mechanical philosophy, his new friend Truth, for whom
he brusquely abandoned Plato and Aristotle.

If he never returned to the old academic philosophy, he did not long
remain entirely happy with his new friend either. About 1668 or 1669
he started a treatise with the title *De gravitatione et equipondio flui-
dorum*.[41] The Introduction, which was a discussion of the general questions of space, time, body, and motion, together with a couple of propositions, was all he completed. Only four or five years earlier,
Descartes had functioned as the guide who led Newton into the new
world of the mechanical philosophy. Nevertheless, *De gravitatione* was
not merely an anti-Cartesian treatise; it was a violently anti-Cartesian
one. The focus of his objection was the charge of atheism. Years later
Newton would tell John Craig that "the reason of his showing the errors
of Cartes's philosophy, was because he thought it was made on purpose
to be the foundation of infidelity."[42] Although Newton showed more
sympathy, both in *De gravitatione* and elsewhere, for Gassendi's alternative mechanical system, the weight of his objection to Descartes,
that he set up the material world as an autonomous order, did not fall
exclusively on the Cartesian version of the mechanical philosophy. Nor
did Newton confine himself to hurling the general charge of atheism.
The title of the piece suggested a work on fluid mechanics, and his
conflict with Descartes took the form of an argument on natural philosophy and on its subtopic, motion. From the time of the composition
of *De gravitatione*, Newton regarded the mechanical philosophy with
ambiguous feelings. He never made the slightest move to return to
academic Aristotelianism, which remained for him as dead as dead
could be. At the same time, he never ceased to believe that the mechanical philosophy of nature in its received form required fundamental
revision. I do not find it entirely accidental that the composition of *De
gravitatione* fell very close to the first recorded manifestations of Newton's interest in alchemy, which embodied a view of nature that gave
primacy to spiritual agents.

The ambiguity of his stance appeared in the "Hypothesis of Light"
which he sent to the Royal Society in 1675.[43] With its universally diffused ether that he employed in mechanistic explanations of the reflection and refraction of light and the descent of heavy bodies toward
the earth, the "Hypothesis" reads easily as a mechanical system of

nature. Other aspects of it fit that mold less readily. Indeed, it has been described as an alchemical cosmology, and one can see why.

> For nature is a perpetuall circulatory worker [Newton asserted], generating fluids out of solids, and solids out of fluids, fixed things out of volatile, & volatile out of fixed, subtile out of gross, & gross out of subtile, Some things to ascend & make the upper terrestriall juices, Rivers and the Atmosphere; & by consequence others to descend for a Requitall to the former.[44]

He ascribed a "principle of motion" to the corpuscles of light, and, in regard to chemical phenomena, he spoke of a "secret principle of unsociablenes," which kept certain substances from mixing together.[45] He specifically denied that the latter could be explained solely by the sizes of particles and pores, as mechanical philosophers tended to do.

About three years later, early in 1679, Newton wrote a long letter to Robert Boyle which was in some ways similar to the "Hypothesis of Light."[46] In discussing the cause of solubility, he again introduced his "secret principle in nature by wch liquors are sociable to some things & unsociable to others," and again he denied that the mere sizes of pores and particles could explain it. The question of volatility further drew upon the principle of unsociability, while the tendency of bodies to recede from each other gave the discussion a veneer of mechanical respectability by relating the causes of both phenomena to a universal ether. An unfinished treatise, *De aere et aethere*, from about this time appears to have been an effort to put the content of the letter to Boyle into a systematic form.[47] It began with a consideration of the tendency of air to expand and to avoid bodies, proceeded to note that in general bodies avoid each other, and concluded that air is composed of particles of bodies "torn away from contact, and repelling each other with a certain large force." Once again he apparently set out to explain the repulsion by means of an ether, but he abandoned the effort after only a few lines and never returned to it. Well he might have abandoned it, for his principle of unsociability and related ideas were moving steadily away from orthodox mechanical philosophy. It cannot have been long after *De aere et aethere* when Newton performed a carefully designed experiment with a pendulum, described in the *Principia*, that encouraged him to abandon belief in the very existence of an ether.[48] An ether, the invisible medium called upon as a causal agent for every apparently nonmechanical phenomenon, was the sine qua non of a workable mechanical philosophy of nature.

When we consider his constant probing of the mechanical philosophy over a period of nearly two decades, we are not surprised that Newton's masterpiece, the *Principia*, based celestial dynamics on a concept no ordinary mechanical philosopher would have considered, a principle of universal attraction. As we now know, Newton intended at one point to go further. In a drafted "Conclusio," he proposed a general revision, based on forces that act at a distance, of all natural philosophy. Nature, he noted, is simple and conformable to itself.

> Whatever reasoning holds for greater motions, should hold for lesser ones as well. The former depend upon the greater attractive forces of larger bodies, and I suspect that the latter depend upon the lesser forces, as yet unobserved, of insensible particles. For, from the forces of gravity, of magnetism and of electricity it is manifest that there are various kinds of natural forces, and that there may be still more kinds is not to be rashly denied. It is very well known that greater bodies act mutually upon each other by those forces, and I do not clearly see why lesser ones should not act on one another by similar forces.[49]

Newton was well aware that he was proposing a major philosophic innovation, and he tried to shield himself from expected criticism. When, in Book I, he came to Section XI and the mutual attraction of bodies, which suggested a more concrete notion of force than earlier abstract propositions had implied, he assured his readers that the demonstrations were purely mathematical. "I here use the word *attraction* in general for any endeavor whatever, made by bodies to approach to each other," he asserted, "whether that endeavor arise from the action of the bodies themselves, as tending to each other or agitating each other by spirits emitted; or whether it arises from the action of the ether or of the air, or of any medium whatever, whether corporeal or incorporeal, in any manner impelling bodies placed therein towards each other."[50] Similarly, some years later, in Query 31, he would declare once more that attractions could be performed by impulses.[51] He went on there to argue for the general necessity of "active Principles" since a purely mechanical universe would run down, and again he attempted to blunt expected objections. "These Principles I consider, not as occult Qualities, supposed to result from the specifick Forms of Things, but as general Laws of Nature, by which the Things themselves are form'd; their Truth appearing to us by Phaenomena, though their Causes be not yet discover'd. For these are manifest Qualities, and their Causes only are occult."[52] Since Book II of the *Principia* had demonstrated both the impossibility that the heavens can be filled with a material medium and the impossibility that a mechanical system can sustain itself without the constant addition of new motion, demonstrations he sought only to strengthen in subsequent editions, Newton had also made it evident to discerning readers that his vision of reality was even farther removed from orthodox mechanical philosophy than the mere concept of action at a distance implied.

Newton was not the only one who recognized that he was proposing a fundamental reordering of natural philosophy. For a generation, mechanical philosophers on the Continent, though they recognized the mathematical power of Newton's demonstrations, refused to have truck with a concept of attraction. Leibniz hinted that it was a return to the "enthusiastic philosophy" of Robert Fludd.[53] He was by no means alone, and more than one mechanical philosopher applied to it the very pejorative, "occult," that Newton had sought to avoid. For their part, Newtonians eventually seized on the concept of forces at a distance as the central characteristic of a new approach to the whole of natural philosophy. Not only British followers, such as Cotes, Pemberton, and McLauren, but early Continental Newtonians, such as Voltaire, 'sGravesande, and Algarotti, all grasped attractions and repulsions, not as mathematical abstractions, but as forces that really exist, and treated them as the foundation on which both a different picture of nature and a different form of scientific investigation rested. By the middle of the eighteenth century, there was no one who mattered left to argue with them.

My third premise is the close chronological correlation between the appearance of the Newtonian concept of force and his interest in alchemy. I shall assume that any further discussion of the chronology of the concept of force, which emerged fully with the *Principia*, is unnecessary. Newton's concern with alchemy, however, has not been public knowledge. In describing the papers, I mentioned some dates. Let me be explicit that for most of the papers dating rests solely on the hand in which they were written. Hence a degree of imprecision about their chronology appears unavoidable. The imprecision is less than the uninitiated might think, however. Newton's hand developed through a number of distinctive phases. To me it seems virtually impossible, for example, to confuse the tiny perpendicular hand of the 1660s with the large, sloping, careless hand of the 1690s or the medium-sized but shaky and crabbed hand of the old man. In a number of cases, some of which I mentioned, dates internal to the manuscripts support evidence drawn from the hand. The laboratory notes are sprinkled with dates that extend from 1678 to 1696. It is relevant to note that Newton performed one set of experiments in the spring of 1686, when the *Principia* was still under composition. Correspondence, such as the letter from Meheux and the exchange with Locke, inevitably carries dates, and Newton dated his memorandum about the Londoner who stayed two days discussing the work. His citation of Fatio's letter of May 1693 establishes the time before which "Praxis" could not have been written. In all, I feel complete confidence about the general period as long as one does not insist on precise years. Newton began serious study of alchemy in the late 1660s. I know of nothing that extends it back into his undergraduate career. Once aroused, his interest continued for nearly thirty years, well into the 1690s. Allow me to note that the alchemical papers come from the years of Newton's intellectual maturity, from the very time when, with his capacity at its highest pitch, he produced the book that has made him immortal. There are a few scraps about alchemy on papers associated with his early years at the Mint, but the manuscripts strongly imply that his active involvement with the art ended near the time when he moved to London.

My central question is implicit in the three premises of my argument. Given Newton's interest in alchemy, given his concept of forces that act between particles, and given the fact that the concept of forces appeared during the period when he was immersed in alchemy, can we establish a connection between the two? In my own view, my question is equivalent to asking whether Newton's alchemy was an activity isolated from the rest of his natural philosophy or whether it exerted an influence on his work in physics. Thus the question also implicitly asks if the structure of modern science embodies concepts that trace their lineage in part to alchemy.

In attempting to answer the question, we must plunge into the content of the alchemical papers. One of the earliest of them, a paper of Newton's own composition though it is not a single connected essay, which is known as "The Vegetation of Metals" from a phrase in the opening lines, probed the distinction between vegetation and purely mechanical changes. Rearrangements of particles effect mechanical changes; vegetation brings about more profound alterations.

> There is therefore besides ye sensible changes wrough in ye textures of ye grosser matter a more subtile secret & noble way of working in all vegetation which makes its products distinct from all others & ye immediate seate of thes operations is not ye whole bulk of matter, but rather an exceeding subtile & inimaginably small portion of matter diffused through the masse wch if it were seperated there would remain but a dead & inactive earth.[54]

As the concept of the vegetation of metals implies, Newton did not limit vegetation to the realm of plants, but treated it as a process present throughout nature. He sometimes called the principle of vegetable action a spirit, which he described as a "Powerfull agent"; sometimes he referred to it, in the plural, as seeds or seminal virtues, which are nature's "only agents, her fire, her soule, her life."[55] That is, what he found in the world of alchemy was the conviction that nature cannot be reduced to the arrangement of inert particles of matter. Nature contains foci of activity, agents whose spontaneous working produces results that cannot be accounted for by the mechanical philosophy's only category of explanation: particles of matter in motion.

The ultimate active agent of nature is what alchemists called the philosophers' stone, the goal of their search. They applied to it images of all sorts, all of them embodying a concept of activity that contrasted with the passivity of matter in the mechanical philosophy. Flammel called it "a most puissant invincible king"; Philalethes, the "miracle of the world" and "the subject of wonders." The author of *Elucidarius* proclaimed that "it is impossible to express [its] infinite virtues."[56] Sometimes activity took on the form of attraction, which was likened to a magnet. Whereas mechanical philosophers explained magnetic attraction away by imagining whirlpools of invisible particles, alchemists embraced it as a visible image of nature's mode of operation. "They call lead a magnet," Newton learned from Sendivogius, "because its mercury attracts the seed of Antimony as the magnet attracts the Chalybs." He also noted that "our water" is drawn out of lead "by the force of our Chalybs which is found in the belly of Ares [i.e., iron]."[57]

His laboratory experience constantly reinforced the message of the alchemical literature. Thus it is relevant to note the steady appearance of active verbs in his experimental notes. When he added spelter to a solution of aqua fortis and sal ammoniac, "ye menstruum [solvent] wrought upon ye spelter [zinc] continually till it had dissolved it." A solution often "fell a working wth a sudden violent fermentation." The spirit, he sometimes noted, "draws" or "extracts" the salts of metals, a usage similar to Sendivogius's magnetic image. When one substance combined with another, it "laid hold" on it; if the two sublimed, one "carried up" the other; if they failed to sublime, one "held" the other "down."[58] It citing these verbs, I seek only to record Newton's immediate perceptions of spontaneous activity in many chemical reactions. The alchemical concept of active agents directly expressed such perceptions. Mechanical philosophers argued that the perceptions were illusions and that the reality behind them consisted solely of inert particles in motion. One cannot infer a choice between two philosophies of nature from the verbs in Newton's experimental notes. They do suggest, however, how he would have been able to understand the images alchemy employed because he too had witnessed the activity the images expressed.

As he was completing the *Principia* in 1686, Newton composed a "Conclusio," from which I have already quoted, an essay that expanded the message of the book beyond universal gravitation into a manifesto of a new philosophy of nature based on forces that act at a distance. In the end he suppressed the "Conclusio," but twenty years later he expanded it into what we know as Query 31. Newton drew upon a number of sources for his assertion that a wide range of forces exists in nature – phenomena such as the expansion of gases, capillary action, surface tension, and the cohesion of bodies, which had seized his attention already in his undergraduate "Quaestiones" and had appeared in later speculations, such as the "Hypothesis of Light," that probed the limits of the mechanical philosophy. Above all, however, he drew upon chemical phenomena.

> Hitherto I have explained the System of this visible world [the "Conclusio" began], as far as concerns the greater motions which can easily be detected. There are however innumerable other local motions which on account of the minuteness of the moving particles cannot be detected, such as the motions of the particles in hot bodies, in fermenting bodies, in putrescent bodies, in growing bodies, in the organs of sensation and so forth. If any one shall have the good fortune to discover all these, I might almost say that he will have laid bare the whole nature of bodies so far as the mechanical causes of things are concerned.[59]

The chemical reactions that impressed Newton fell into two general types. Reactions that produce heat formed one of them.

> If spirit of vitriol (which consists of common water and an acid spirit) be mixed with Sal Alkali or with some suitable metallic powder, at once commotion and violent ebullition occur. And a great heat is often generated in such operations. That motion and the heat thence produced argue that there is a vehement rushing together of the acid particles and the other particles, whether metallic or of Sal Alkali; and the rushing together of the particles with violence could not happen unless the particles begin to approach one another before they touch one another . . . So also spirit of nitre (which is composed of water and an acid Spirit) violently unites with salt of tartar; then, although the spirit by itself can be distilled in a gently heated bath, nevertheless it cannot be separated from the salt of tartar except by a vehement fire.

The other type of reaction that he called upon displays selective affinities analogous to his secret principle of sociability and unsociability. Thus he argued that the ability of salt of tartar to precipitate bodies dissolved in acids stems from "the stronger attraction by which the salt of tartar draws those acid spirits from the dissolved bodies to itself. For if the spirit does not suffice to retain them both, it will cohere with that which attracts more strongly."[60]

Newton did not discover the reactions cited here. He could have found them all in the writings of mechanical chemists such as Boyle, with which he was certainly familiar. In Boyle, however, he could not have found the conclusion he derived from them: that particles of matter attract and repel each other. For that matter, he could not have found the conclusion, in the form stated above, in alchemical literature either. What he could have found there, as I have indicated, was a concept of active principles that bears a close resemblance to the manner in which Newton frequently expressed his concept of forces. It is also of some importance to my argument to insist that, without exception, all the chemical phenomena cited in the "Conclusio" had appeared in Newton's experimental notes during the previous decade.

It is further relevant to note that Newton composed a paper, "De natura acidorum," in which we can observe the transition from the alchemical concept of active principle to the Newtonian concept of attraction expressed in his own words. In Newton's alchemy, philosophic sulfur, the male principle, was the ultimate causal agent in nature. "De natura acidorum" argued that the activity of sulfur, perhaps common sulfur in this case, springs from the acid it conceals. "For what attracts and is attracted strongly, we call acid." Under the images of dragons and serpents that devoured uncounted kings and queens, acids were also active in the world of alchemy. The particles of acids, Newton asserted in a statement that grasps that world in one embrace with his own concept of force, "are endowed with a great attractive force and in this force their activity consists by which they dissolve bodies and affect and stimulate the organs of the senses."[61]

Newton composed "De natura acidorum" during the early 1690s, in the years immediately following the *Principia*. It was a period of almost manic intellectual activity in his life. Buoyed by the twin successes of the *Principia* and the Glorious Revolution, in which he had played a significant if minor role, he apparently set out to codify his philosophic legacy. He devoted extensive energy to revising the *Principia*. The book had taken shape, developing and expanding as Newton explored its topic, during a period of about thirty months that began in August 1684. There is every reason to think that he did not regard the form in which it appeared in 1687 as final. We have the manuscripts for important revisions both of the early demonstrations in Book I and of the opening propositions of Book III. The proposed new edition never saw publication in the form then planned, but the surviving manuscripts leave no doubt that Newton worked at it. The same years saw intense mathematical endeavor, including the composition of a definitive exposition of his fluxional calculus. He began to write his *Opticks*, not the volume he published ten years later, but an *Opticks* in four books, which used optical phenomena to support the Newtonian natural philosophy based on forces between particles. Hence it seems to me a matter of major significance that during this period – in the years, I repeat, immediately following the *Principia* – Newton also invested an enormous effort in alchemy. I suggested before that one use of the quantitative measure of his alchemical papers was the establishment of a rough chronological index of the effort expanded. He wrote about half of the estimated 1,200,000 words on alchemy during the period of seven or eight years that followed the *Principia*. The mere existence of papers from that time cannot, of course, demonstrate a connection between alchemy and the Newtonian concept of force. To me, at least, the papers offer powerful evidence that Newton regarded his alchemical endeavors as a harmonious part of his total philosophical program.

I do not want my argument to be misunderstood. I am seeking the source of the Newtonian concept of forces of attraction and repulsion between particles of matter, the concept that fundamentally altered the prevailing philosophy of nature and ushered in the intellectual world of modern science. I am offering the argument that alchemy, Newton's involvement in which a vast corpus of papers establishes, offered him a stimulus to consider concepts beyond the bare ontology of the mechanical philosophy. It appears to me that the Newtonian concept of force embodies the enduring influence of alchemy upon his scientific thought. As I mentioned, Professor Cohen takes issue with the argument in his recent *Newtonian Revolution*. He presents an analysis of the *Principia*'s development that confines itself to the science of dynamics and its application to orbital motion and treats the concept of attraction as a conclusion that emerged solely from Newton's consideration of such problems. To the suggestion that alchemy influenced Newton, he replies that Newton repeatedly asserted that his success with gravitational attraction led him to consider the possibility of other forces between particles.[62] I wish to say two things in this respect. First, I do not know the assertions to which Professor Cohen alludes. I think he refers primarily to the statement, very similar to the one I quoted above from the "Conclusio," that Newton inserted in the Preface to the *Principia*. What I find in it is an argument from the analogy of nature, not an autobiographical account of his discovery. Second, it appears to me that the technical problems of dynamics, which were of unavoidable importance to Newton's concept of force, can be separated from the conceptual issue with which I have concerned myself in this chapter. Indeed, I believe we have empirical evidence that they were separated in the seventeenth century. Next to Newton, there was no one alive better able to appreciate the technical problems of dynamics than Huygens, Leibniz, and Bernoulli. Each of them studied the *Principia* and appreciated the full extent of its achievement. Even with the book open before him, not one of the three ever admitted the possibility of attractions at a distance. It is my contention that Newton's readiness to consider the possibility derived from the influence of alchemy.

I am not discussing technical dynamics, in which Newton made enormous strides that are obviously related to his concept of force. I am talking rather about a conceptual innovation – an innovation, that is, in relation to the prevailing mechanical philosophy of nature. There are, I insist, strong arguments, summarized in this chapter, for tracing it in part to the influences of alchemy.

J. E. McGuire has recently advanced quite a different argument against the case for alchemy. In a number of articles, McGuire has traced the influence of the Cambridge Platonists on Newton. Why call upon alchemy, he asks, when we have Cambridge Platonism to supply a similar influence?[63] There are also two things I would say in reply to McGuire. First, I see no necessary opposition between us. I do not argue that alchemy exercised the sole influence on Newton. I take McGuire's articles to have demonstrated that Cambridge Platonism, in which one can find a concept of active principles, also influenced Newton. I see no reason why two influences could not operate in the same direction. I say, secondly, that whatever the influence of Cambridge Platonism, the alchemical papers remain. Indeed it is necessary to remark in this respect that for every page in Newton's papers of direct reference to More and Cudworth there are well over a hundred on alchemy. I cannot make those papers disappear.

To say as much is in no way to suggest that Newtonian science – and hence derivatively all of modern science – is a covert form of alchemy. I emphatically reject any attempt to distort my argument in that direction. Hence I must distinguish my position from Castillejo's. No doubt it oversimplifies his book to speak of an equation of Newtonian science with alchemy; but unless I completely misunderstand the work, that statement of his position is far more true than false. With Castillejo's conviction that we need to integrate Newton's alchemical activity into the rest of his intellectual life I am in obvious agreement; beyond that I cannot go. His argument appears to me to neglect the most important aspects of Newton's scientific endeavor – his mathematics, his quantitative science of dynamics, his experimental investigation of light – and to ignore as well the implications of its aftermath – the enormous growth of modern science, three centuries of experimental confirmation, and two centuries of practical confirmation through the successes (and even the disasters) of scientific technology.

Far from equating Newtonian science with alchemy, I emphasize the extent to which Newton altered what he received. His success in practicing alchemy on alchemy itself may be the ultimate measure of its influence on him. If he derived his concept of force partly from the alchemical active principle, he also transformed it in fundamental ways. Above all, he quantified it, so that it could fit smoothly into the structure of his quantitative dynamics. There is no sense in which I deny the relevance of the technical problems internal to dynamics, which

Professor Cohen analyzes so well. Newton may have found an idea of attraction in Sendivogius, but we cannot imagine Sendivogius writing the *Principia*. To that extent Newton transformed what he received.[64]

Hence Newton could see the final result of his work as the perfection of the mechanical philosophy rather than its denial. Physical nature remained for him what it had been for mechanical philosophers: particles of matter in notion. With the quantified concept of force, he called natural philosophy back from its preoccupation with imagining invisible mechanisms and gave decisive demonstration of the power exact mathematical description wields. Perhaps we can best say, using Professor Cohen's approach, that the Newtonian concept of force transformed natural philosophy into modern science. With only modest surprise, I note how close I see myself to Professor Hall for all our surface disagreements. For me also, Newton represents reason; his success in weaving a single fabric from a multiplicity of strands constitutes in my eyes one of the supreme exercises reason has known. We differ, if I understand it correctly, on my readiness to admit that a different standard of rationality in the seventeenth century may have encouraged Newton to open himself to the influence of a tradition that appears to us almost as the antithesis of reason.

Hence also I need to close by pointing as well to the final act in the drama. Newton did in the end turn away from alchemy. Every time I think seriously about Newton and alchemy this final act assumes greater significance. Alchemy formed an integral part of the intense intellectual activity of the early 1690s. The essay "Praxis," composed in the summer of 1693, suggests that the breakdown of that year also had an alchemical dimension. Newton's interest in alchemy did not end suddenly at that moment; there were, for example, dated experimental notes that extended to 1696. Nevertheless, his intense involvement in the art did come to an end about then. A few scraps on alchemy can be dated to his early London years, but only a few. His library contained only three alchemical books published after 1700, two of them by William Y-Worth, presented to him by the author in 1702.[65] Alchemy was the one intellectual pursuit of Newton's Cambridge years that did not follow him to London. Am I wrong then in placing alchemy within the precincts of Newtonian rationality if in the end he turned away from it? "Praxis," with its claim of successful multiplication, does seem to have moved beyond the realm of reason, but 1693 was an extraordinary year for Newton when everything ran over the edge. If that extravagant dream – or nightmare – ended in disillusionment, I suggest that the end of Newton's active involvement in alchemy marked his realization that he had in fact achieved a different success. With his quantified concept of force, he had extracted the essence of the art. Alchemy itself told him to reject the dross as dead and lifeless

matter. The seed had found a fertile matrix where it has flourished ever
since.

Notes

1 *The Correspondence of Isaac Newton*, ed. H. W. Turnbull et al., 7 vols.
(Cambridge, 1959–77), III, 192–3, 195, 215, 216, 217–19.
2 Conduitt's memorandum of 31 August 1726; King's College, Cambridge,
Keynes MS. 130.10, fol. 3ᵛ.
3 David Brewster, *Memoirs of the Life, Writings, and Discoveries of Sir
Isaac Newton*, 2 vols. (Edinburgh, 1855), 11, 371–6.
4 Lord Keynes, "Newton the Man," in Royal Society, *Newton Tercentenary
Celebrations* (Cambridge, 1947), pp. 27–34.
5 B. J. T. Dobbs, *The Foundations of Newton's Alchemy: The Hunting of the
Greene Lyon* (Cambridge, 1975), and "Newton's Copy of *Secrets Reveal'd*
and the Regimen of the Work," *Ambix*, 26 (1979), pp. 145–69. Professor
Dobbs is presently completing a second book that will extend her study into
the alchemical manuscripts that belonged to a later period in Newton's life.
As far as I know, Karin Figala's dissertation has, regrettably, not been
published; see Karin Figala, "Die 'Kompositionshierarchie' der Materie:
Newton's quantitative Theorie und Interpretation der qualitativen
Alchemie," unpublished Habilitationsschrift, Technischen Universität,
Munich. She presented a brief view of her work in "Newton as Alchemist,"
History of Science, 15 (1977), pp. 102–37, an essay-review of Professor
Dobbs's book.
6 Brewster, II, 375.
7 I. Bernard Cohen, *The Newtonian Revolution* (Cambridge, 1980). The
specific passage to which I refer is on p. 10.
8 I quote, with Professor Hall's generous permission, from a private letter to
me about my recent biography of Newton.
9 David Castillejo, *The Expanding Force in Newton's Cosmos* (Madrid, 1981).
10 Bodleian Library, Oxford, MS. Don.b.15.
11 Accounts in the notebook in the Fitzwilliam Museum, Cambridge.
12 Manuscript in the Countway Medical Library, Harvard University, item 3,
fol. 10ᵛ.
13 Jewish National and University Library, Yahuda MS. 259, no. 9.
14 Dobbs, *Foundations*, p. 88.
15 Keynes MS. 30. See my analysis of its content: "Isaac Newton's Index
Chemicus," *Ambix*, 22 (1975), pp. 174–85.
16 John Harrison, *The Library of Isaac Newton* (Cambridge, 1978).
17 Keynes MSS. 51 and 52.
18 Keynes MS. 55 and a manuscript in the Yale Medical Library. Other
examples are Keynes MSS. 22, 24, 31, 33, 39, 50, 58 (part only), 62, 65, and
66.
19 The sheaf is Keynes MS. 67; the notes are Keynes MS. 62.
20 Keynes MS. 33.
21 Newton recorded the visit in two largely identical memoranda: Keynes MS.
26 (published in *Correspondence*, IV, 196–8) and MS. 1075–3 in the Joseph
Halle Schaffner Collection, University of Chicago Library.
22 Professor Dobbs so identified him, convincingly: *Foundations*, p. 112.
23 Early experiments are recorded in Cambridge University Library, Add. MS.
3975, pp. 81–4. Later ones, frequently dated and extending from 1678 to

1696, are found in Add. MSS. 3973 and 3975, pp. 101–58, 267–83. The first examination of Newton's records of his chemical experiments, which remains an indispensable guide to them, is A. R. Hall and Marie Boas [Hall], "Newton's Chemical Experiments," *Archives internationales d'histoire des sciences*, 11 (1958), pp. 113–52.

24 See, for example, Add. MSS. 3973, fols. 5^v–6, 13–13^v, and 3975, p. 143.

25 Dobbs, *Foundations*, pp. 146–63.

26 Add. MSS. 3975, p. 121; 3973, fol. 17; 3975, p. 149.

27 Countway MS., item 4.

28 Ibid., item 3, fol. 7.

29 While similar passages abound, the "Index chemicus" (Keynes MS. 30) is especially rich in them.

30 Keynes MS. 49, fol. 1.

31 Keynes MSS. 38 and 56. Keynes MS. 57, which has no title, is a similar compilation.

32 They are found in Keynes MSS. 40 and 41; Babson College Library, Babson MS. 417; and Dibner Collection, Smithsonian Institution Libraries, Burndy MS. 17.

33 Dobbs, *Foundations*, pp. 251–5.

34 Figala, "Newton as Alchemist," p. 107; D. T. Whiteside, "From his Claw the Greene Lyon," *Isis*, 68 (1977), p. 118.

35 Burndy MS. 10 and Keynes MS. 28.

36 Babson MS. 420.

37 The first two, under different names, in Keynes MSS. 21 and 53.

38 *Correspondence*, III, 265–7.

39 Babson MS. 420, p. 18^a. In the final draft of this passage (p. 17), Newton toned it down somewhat.

40 Add. MS. 3996, fols. 88–135.

41 Add. MS. 4003. Published in A. R. Hall and M. B. Hall, *Unpublished Scientific Papers of Isaac Newton* (Cambridge, 1962), pp. 90–121; English trans., pp. 121–56.

42 Keynes MS. 132.

43 *Correspondence*, I, 362–86.

44 Ibid., I, 365–6.

45 Ibid., I, 368–70.

46 Ibid., II, 288–95.

47 Published in Hall and Hall, *Unpublished Papers*, pp. 214–20; English trans., pp. 220–8.

48 *Principia*, Motte-Cajori trans. (Berkeley, 1934), p. 325. See ed. 1, p. 353, for two important concluding sentences Newton omitted from the second and subsequent editions.

49 Hall and Hall, *Unpublished Papers*, p. 333.

50 *Principia*, pp. 164, 192.

51 *Opticks*, based on 4th ed. (New York, 1952), p. 376.

52 Ibid., p. 401.

53 In an "anonymous" review of John Freind's chemical lectures, *Acta eruditorum* (September 1710), p. 412.

54 Burndy MS. 16, fol. 6^v.

55 Ibid., fols. 5–5^v.

56 I cite from Newton's notes: Keynes MSS. 40, fols. 20, 19^v; 41, fol. 15^v; Babson MS. 417, p. 35.

57 Keynes MS. 19, fols. 1, 3.

58 Add. MSS. 3973, fol. 42; 3975, pp. 281, 104–5; 3973, fols. 13, 21; 3975, pp. 108–9.
59 Hall and Hall, *Unpublished Papers*, p. 333.
60 Ibid., pp. 333–5.
61 *Correspondence*, III, 209–12.
62 See the reference above, note 7, and a fuller discussion in an article published after the paper on which this chapter is based was presented: I. B. Cohen, "The *Principia*, Universal Gravitation, and the 'Newtonian Style,' in Relation to the Newtonian Revolution in Science," in *Contemporary Newtonian Research*, ed. Zev Bechler (Dordrecht, 1982), pp. 67–74.
63 J. E. McGuire, "Neoplatonism and Active Principles: Newton and the *Corpus Hermeticum*," in Robert S. Westman and J. E. McGuire, *Hermeticism and the Scientific Revolution* (Los Angeles, 1977).
64 My inability to write this paragraph without Professor Cohen's concept of transformation must be significant.
65 Harrison, *Library*, items 1138, 1302, 1644.

11

Witchcraft and popular mentality in Lorraine, 1580–1630

ROBIN BRIGGS

Detailed records of early criminal trials are scarce, and one of the most extensive collections to survive for the turn of the sixteenth and seventeenth centuries is that of the ancient duchy of Lorraine, now housed in the Archives Départementales of the Meurthe-et-Moselle at Nancy. Among these documents are well over two hundred complete dossiers for those tried on charges of witchcraft, nearly all of them for the half-century from 1580 to 1630. Although this probably represents only something between 5 and 10 percent of Lorraine's witchcraft prosecutions (for the names of many hundreds of others convicted can be recovered from less complete records), it constitutes an admirable working sample; the present analysis is based on close examination of some seventy trials and a general impression of the remainder. This material is of a kind not normally found in England or France, and only sporadically elsewhere in Europe. It includes full witness depositions, commonly from fifteen to twenty-five witnesses; the interrogation of the accused on the basis of these testimonies; the confrontation of the witnesses and the accused; and normally one or more sessions of interrogation under torture. The nature of the records is very important because they give us an unadulterated view of the first stage of accusations, without any serious likelihood of editing by the lawyers and judges. It is the earlier stages of the trials, rather than the confessions under torture, which enable one to build up a picture of the popular attitudes that had prompted the accusations. The confessions that were eventually extracted from the vast majority of the defendants also have their interest, however; the records generally allow one to distinguish between those admissions made spontaneously and those that resulted from promptings by the judges. This is important, for example, in the case of the sabbat, where the Lorraine material

337

offers us a direct way into popular, as distinct from learned, views about these diabolical festivities.[1]

Lorraine has traditionally been portrayed as the scene of intense witchcraft persecution, and its judges, from the demonologist Nicolas Rémy, *procureur général* of the duchy, down, have acquired an evil name. As so often in the history of witchcraft, there is an element of exaggeration in this. What can be fairly said is that once a suspect reached the courts, his or her chances were poor; the conviction rate generally approached 90 percent. On the other hand, if one takes the reasonable estimate of around 3,000 trials for the period under consideration, this is around 60 a year in a duchy with a population of at least 400,000. As a per capita rate it is not markedly different from the peak rates achieved in Elizabethan Essex, although the proportion of executions was far greater.[2] The accused were a highly selected group, and there are very few clear examples of people who were pulled in because of a casual denunciation made under torture – the chain of accusation that became infamous in some German cities.[3] The attitude of Rémy and other judges may have encouraged people to use the courts, and it was normal to interrogate those who confessed about their accomplices, but this was done with some caution, and there is no real sign that suspects were manufactured by such means. The typical accused had a long local reputation, twenty years being commonplace. He or she was charged with a range of acts of *maléfice*, causing actual harm to neighbors and their animals, stretching many years back. Suspicious noises and nocturnal comings and goings were sometimes mentioned, but village belief was firmly based on the actual damage caused to community and individuals.

A contrast is often drawn between this local belief, founded on specific acts of *maléfice*, and the learned tradition that emphasized the diabolical pact and the sabbat, with witchcraft becoming the most extreme form of heresy. Technically this distinction can certainly be made in Lorraine: The judges sought to obtain confessions to the pact above all, and these were sufficient for a capital sentence even if unaccompanied by admissions of actual evil doing. Such a bald statement would, however, be misleading. The local commentators – Rémy and the legal writer Claude Bourgeois – were far from disregarding the importance of *maléfice*.[4] Judges continued to press for admissions of this even after they had secured the basic confession. Furthermore, the accused always began their confessions with an account, often in pathetic circumstantial detail, of how they had been tempted by the devil in a moment of distress or weakness and had succumbed. The pact was clearly a part of popular belief; perhaps the accused may have regarded it as less of a social sin than harming their neighbors through active witchcraft, since several of them denied any such acts, despite being beaten and brutalized by the devil. At the least, the pact might allow the displacement of guilt for the harm done to neighbors onto the devil, who had allegedly compelled the performance of such evil. Confessions to attendance at the sabbat, however, often had to be elicited by direct questioning, even though most of them reveal a standard popular image, of a rather unimaginative kind, which must again reflect widely held folk beliefs.

There are other reasons why it would be hard to maintain any real division between elite and popular conceptions of witchcraft in Lorraine. The great majority of cases were tried in local courts, some of whose judges were illiterate: The central tribunal of the *échevins* of Nancy reviewed the proceedings, but did not exercise a direct appellate jurisdiction. Rémy's own limitations are interesting here: Despite the classical references with which he interspersed his material, the interest of his *Demonolatry* lies exclusively in the discussion of practical details. His view is really more characteristic of the popular than of the learned tradition, as in the confused passage in which he fails to resolve the question whether it can be right to force a witch into healing her victims. The book is direct and notably accurate when describing actual trials, only to lapse into verbose futility when it moves to general issues. It is, however, remarkably free from any hysterical or paranoid fears of a grandiose international conspiracy of witches, for Rémy viewed the "vile rabble of sorcery" with a certain contempt and was serenely confident in his own invulnerability as a judge. It was in line with such attitudes that he remarked: "For witches make it their chief business to be asked to perform cures so that they may reap some profit, or at least gratitude; since they are for the most part beggars, who support life on the alms they receive."[5]

This last comment will remind many of the analysis of English witchcraft by Keith Thomas and Alan Macfarlane, with its stress on the refusal of charity and subsequent inversion of guilty feelings by the witch's supposed victim.[6] As an explanation of the internal logic of the accusations this remains the biggest single step yet made toward understanding the reality of European witchcraft persecutions, and it can be extensively confirmed by reference to the evidence for Lorraine. While a single example proves nothing, it will at least give the flavor of the material. In 1584 Catherine la Blanche, a widow in her sixties, was on trial. One of the twenty-five witnesses, Cleron Baltaire, said that five years before, when she and her husband had been fattening a bull, Catherine

> vint à sa porte mendier, comme elle faisoit souvent. Elle deposante luy dit Catherine, allez pourchasser et demander vos aulmosnes aultre part, car je ne vous veulx plus rien donner, à rayson que j'ay des enfans pupilz et pauvres en-

fans de feu le frère de mon marit qui sont sur mes bras et
qu'il nous fault nourrir. Pour l'honneur de dieu il vault
mieux de les nourrir que vous et pour ce allez vous en.

Although there was apparently no threat or other reaction from Catherine, Cleron nevertheless blamed her for the subsequent death of the bull.[7] Cases that come so close to the English model do, however, pose some awkward problems. If accusations in a thoroughly Catholic and rather traditional area like this follow an almost identical pattern, what happens to those very plausible general explanations in terms of Protestantism and rapid socioeconomic change?

Some possible answers do suggest themselves and can be developed to illuminate wider aspects of the topic. First, it is easy to overdo the distinction between Protestantism and Catholicism, both at village and elite levels. The faith of the urban elites in Catholic Europe was showing a distinct tendency toward emphasizing individual responsibility, which the whole pastoral effort of the Counter-Reformation was to encourage, while the villagers rather illogically yet sensibly combined magical beliefs in the efficacy of the sacraments with a habit of judging individuals by their actual behavior. The whole business of the diabolical pact was presented as a matter of individual fallibility, even if it was claimed that the devil was too powerful to escape once the fatal step was taken. Accused and judges not infrequently concurred in seeing the trials as a way of reconciling the sinner with God; confession, repentance, and expiation at the stake were saving souls.

Apart from the psychological pressure built up by the legal proceedings themselves, numerous accused witches were probably aware that they had borne their neighbors genuine ill will and may have come to accept responsibility for the ensuing misfortunes. Others remained unconvinced and sometimes tried to revoke their confessions, alleging that to confirm them would risk damning their souls by dying with falsehoods against their name. When Barbelline Goudot was tried in 1604, she revoked her confession on the grounds that "ayant demandé à son père confesseur familliairement sy ayant confessé chose non véritable elle en recevroit peine en l'autre monde lequel luy dit qu'il ne falloit dire que la vérité, qu'il fut la cause qu'elle avoit renyé le tout." She then confessed again, to the relief of her judges, who urged her to further admissions "d'aultant que le crime est sy oculte que le Juge n'en peult sainement juger qu'après la pure et simple confession de celuy ou celle qui en est coupable."[8]

In terms of ideas of personal responsibility, then, there is little to differentiate Protestant and Catholic positions in practice. More surprisingly, what is absent from these records is any evidence of ecclesiastical countermagic in operation, apart from pilgrimages to shrines and the burning of the occasional candle. The *curés* are curiously missing from most trials; they never seem to testify and are involved only indirectly. At the trial of Jeannon Poirson, who was *renvoyée jusqu'au rappel* (the nearest one could get to an acquittal) in 1602, it was alleged that the late *curé* of Leintrey had seen her dancing strangely in the fields, and then said "qu'il n'avoit jamais voulu croire qu'il fut des sorcières mais qu'à ceste heure là il le croyoit."[9] Against this expression of relative scepticism one can set the cases of a *curé* who sought magical remedies from the suspect, and another who diagnosed witchcraft from objects found in victims' bedding.[10] It was probably crucial that the *curés* did not take a more active part in instigating the persecution of witches; had they done so, there would have been far more trials than seem actually to have taken place. The position of the *curé* as a local *notable* and a natural arbiter of disputes would have made him the ideal orchestrator of a persecution. Perhaps his role as confessor to his flock was crucial in inhibiting him, since any accusation might well suggest that he was breaking the secrecy of the confessional.

A second respect in which the situation may be closer than expected to that in England concerns social and economic changes. It is certainly true that peasant society in Lorraine was not disrupted by the development of a full market economy of the kind that was emerging in England. On the other hand, divisions between rich and poor did widen sharply in Lorraine, and most notably, according to the magisterial thesis by Guy Cabourdin, in the period 1580–1630. Substantial amounts of land were transferred from peasant ownership to that of the prosperous few, communal rights were eroded, and peasant indebtedness rose very rapidly.[11] While there are many reasons to be suspicious of the "strain-gauge" explanation of increasing witchcraft tensions, the trials do contain a good deal of circumstantial evidence that would link them to antagonisms between rich and poor. Around 1583 Jean Diez of la Bolle told George Colas that although he was now rich he would become poor, while Jean himself would acquire property; when Jean Diez came to trial in 1592 Colas's widow claimed that the threat had been fulfilled. Despite hard work and a frugal life style they had been reduced to extreme poverty.[12]

Another witch from the same group of trials, Zabel de Sambois, had been unwise enough to get into dispute with the *maire*, Dieudonné Galand, who believed that she had caused him various misfortunes. The *curé* persuaded her to a formal reconciliation and seeking of pardon from the *maire*, on the grounds "que les pauvres doibvent plier pour les riches"; at her trial, however, she objected that the accusations against her were false, "le tout par envie et malveillance et qu'on faict toujours ainsi contre les pauvres gens, et que sy on scavoit tout le fait dudit maire Galand, qu'on ne teindroit pas beaucoup plus de compte de luy qu'on faict d'elle."[13] In 1602 Babelon Henri alleged that "à cause qu'elle est pauvre l'on ne tenoit grande conte au sabat et y avoit bien peu de credit, mais que les riches y ont toujours plus de credit, et sont les plus avant à la besogne.''[14] To emphasize their predominance, the rich sat higher and had more meat. A similar picture was given the following year by Catherine Charpentier, who added that the rich

> disoient, avoir encore des bledz assez en provision, fussent en volonté, et proposoient, de gresler et gaster les bledz et biens de la terre. Que jamais quant à elle, elle n'y voulut consentir, par la crainte qu'elle avoit d'avoir besoing, cognoissant, comme elle faisoit, la pauvreté de son marit, aussy, elle a esté par plusieurs fois battue, dudit son Mre. Persin, qui enclinoit à la volonté des autres.[15]

This theme of social division at the sabbat could be illustrated from several other confessions, and despite its imaginary context there is every reason to suppose that it expresses social strains that were all too real.

Having emphasized likenesses, the third point is one of dissimilarity. Although the psychological spur for accusations was basically identical – a dispute, in which the accuser was quite often seen in an unfavorable light, followed by a misfortune – the range of disputes seems to have been much wider than in the occasional English trials we can follow in comparable detail. Fewer of them turn on the refusal of recognized neighborly services or consideration; although these last are naturally common, they are not really predominant. It does seem plausible to suppose that, as the development of the poor-relief system would suggest, obligations to poorer neighbors had become a source of acute tension in England. In Lorraine the stress was perhaps distributed more widely, and it would be difficult to show that witnesses were commonly of a higher social or economic standing than the accused. Muchembled's suggestion, based on a handful of instances from the Cambrésis, that members of the powerful minority were asserting their social control over their inferiors, would be extremely hard to justify from the mass of Lorraine trials, although, as one would expect, a handful do hint at such antagonisms.[16] In truth, the kinds of tensions revealed are those that must always have been part of village life, as were the misfortunes. The accused sometimes pointed out that it was as reasonable to blame chance or the will of God as to name witchcraft as the cause when animals or children were stricken by sudden or unknown illness. Another subtle difference from the English case concerns the ''inverted guilt'' pattern; this was very commonly present, but far from being a rule. Judges and witnesses alike plainly assumed that bewitchment would follow a quarrel, and a witness who did not recount such an episode as a prelude to misfortune was likely to be specifically asked

if there had been any dispute with the accused. Ill will was not unmotivated, but there is no clear implication that the offenses or the
aggression should have come from the victim.

The accused cannot have been as surprised as they sometimes
claimed to be when they came before the judges. In the great majority
of cases, not only did the witnesses allege a reputation stretching back
many years, the evidence revealed that one or more public accusations
had been made against the supposed witch. The fact that no reparation
had been sought was a major presumption against the suspect, yet there
were powerful motives for taking a chance in letting such insults pass,
for an attempt to obtain an apology or damages could often turn into
a trial on the normal pattern. Every village seems to have contained
individuals whom their neighbors believed to be witches. How did such
identifications take place, and at what point did a formal prosecution
result? At least three quarters of the accused were women; most of
these were at least into their late forties and many much older. The
great majority were poor, their property commonly insufficient even
to meet the modest costs of the trial. Some were beggars, although
Rémy certainly exaggerated here. Other categories found quite commonly were individuals who made themselves obnoxious by their quar-
relsomeness; those who were of dubious sexual morality; and village
herdsmen and women who were often involved in treating the illnesses
of animals and who shade into the category of magical healers, often
themselves prosecuted as maleficent witches. Above all, however, ill
repute was inherited; parents, siblings, or other relatives already accused were a mortal danger. An extreme case was that of Hellenix le
Reytre at Blamont in 1606, whom the judges pressed for details about
her family. It became clear why they did so when she admitted that
her brother had been executed thirty-seven years earlier, while of her
four sisters three had also been executed and the fourth accused.[17] In
many other cases it was claimed that relatives had been suspected,
even if never tried.

Identification might also take place through the white witches or
devins who specialized in countermagical healing. Much of their skill
lay in persuading the client to articulate his own suspicions, but there
could plainly be a random element in the operation. This emerged
alongside the theme of inherited witchcraft when Mengette Estienne
of Le Paire d'Avould was accused and offered the explanation that the
family reputation originated when

> ung jour sa mère allant querir du feu chez ung de leur
> voisin, là où il y avoit ung qui estoit dans ung bain ayans
> mal en ung jambe, et ne pouvant estre gueri il feit aller au
> devin laquel devin dict Que ce pourroit avoir faict quel-
> conque de ses voisines, sur ce ladicte bruict fut donné à sa

mère parce qu'elle avoit esté querir du feu encore qu'elle
n'en eust jamais esté suspitionée.[18]

In other cases knowledge of a visit to the *devin* seems to have induced
the suspect to appear and offer some kind of healing, which would
confirm his or her reputation even if it worked. When Mengeon Laus-
son and his wife Mengeotte were tried in 1620, it emerged that he had
prevented her from undertaking a pilgrimage for a neighbor who had
lost her milk, on the grounds that she was already suspected of causing
similar harm to another woman, and to act as requested would confirm
this belief.[19] Suspects usually knew of the graver suspicions against
them and had to decide what attitude to adopt; although the situation
was horribly dangerous for them, it did at least give them a certain
negative power over their potential accusers. The prime mover in the
accusation against Georgette Herteman of Brouvelieures in 1615 was
the blacksmith Nicolas Mongeot, who believed she had bewitched his
wife; she told others that "elle auroit bien pu fournir quelque chose
pour guerir sa femme, mais puis qu'il s'estoit porté sy terrible, elle la
laisseroit là."[20]

The villagers were equally conscious of the dangers in crossing those
reputed to be witches; many testimonies emphasize how they were
feared and humored. According to the local *tabellion*, Fleuratte Maurice of Docelles was so feared "que personne du village ne fait banquet
de nopces ou autre sans luy envoier quelque present de chair ou autre
vivres."[21] It is striking that in many such cases these individuals were
apparently tolerated for many years before a formal accusation was
brought; although suspected of this appalling antisocial heresy, they
were apparently treated as just one more danger of everyday life, rather
than arousing any immediate or panic-stricken reaction. Many must
have died without coming to trial at all, given the length of the reputations of those who did. It is almost impossible to understand why at
a certain point formal steps were taken, for nothing seems to mark off
those *maléfices* that acted as catalysts from those dating back many
years. We are almost certainly dealing with a situation in which there
was great reluctance to prosecute one's neighbors, in view of the continuing ill will that might result and of the costs that might be incurred
if one came forward as a "partie formelle" to bring the charge. The
witch might be removed, but his or her kin still had to be reckoned
with.

The troubles of Nicolas Mongeot, mentioned above, did not end with
the execution of Georgette Herteman; he appeared again as a witness
at the trial of her husband, Nicolas Herteman, to tell how his wife had
relapsed after Nicolas reproached him "qu'il estoit cause de la mort
de sadite femme, et en quoy on luy avoit faict grand tort, mais que
cela ne dormoit encore et n'estoit oblié." Nicolas Herteman was re-

leased after withstanding the thumbscrews and the rack, leaving one to imagine the future relations between these neighboring families.[22] Another witch mentioned earlier, Hellenix le Reytre, deterred a potential accuser by declaring loudly "qu'elle avoit desja heu cinq proces et les avoit tous gagné. Qu'elle seroit encore bien aysé d'en avoir ung autre pour y faire conformer quelques personnes jusques à leurs chemises."[23] This was an exceptionally aggressive reaction, but several other suspects put the same message across in slightly more veiled terms. Such confrontations emphasize the extent to which witchcraft was a double-edged factor within the complex relationships of village society, allowing a certain status to some of its more rebarbative members.

One way around the dangers of accusing these potentially vindictive neighbors was to seek a direct intervention by the ducal prosecutor or other competent authority. This was difficult to accomplish secretly, however, and involved dealing with a relatively elevated and often distant personage. Much commoner was reliance on the accusations made by the convicted against their accomplices, those with whom they had supposedly gathered at the sabbat. As participants in the world of village gossip, the condemned naturally directed most such nominations at well-known local suspects. Little chains of prosecutions would result, although not all such charges automatically produced further trials without there being any obvious reason for this. Once a trial was under way, rumor and tension would commonly spread through the surrounding villages, with talk of taking all the witches. Numerous testimonies expose the fear and agitation of those who knew themselves threatened; they would sometimes make their relief rather too obvious when they heard that they had not been named. They often talked of flight, but few had the courage to cut loose from their local ties and modest property in this way. It is clear from one exceptional case that good repute and the support of one's neighbors could offer some protection. In 1592 Mathieu Blaise of Saint Margarée was separately accused by three convicted witches, but thirty-seven witnesses produced no serious charge against him, while many testified to his good character and generosity to others. Even Nicolas Rémy was compelled to order his immediate release. Yet Mathieu, whose nickname "le gros" was evidently a reference to his corpulence, did have something of a reputation. One favorable witness told how, talking outside the church of the nearby village of Combrimont a decade earlier, a man had come up to him and said

> que l'on parloit bien des sorciers et sorcières et que sy on
> brusloit Mathieu Blaise il y auroit bien de la gresse. Ce
> qu'ouy par luy deposant, luy dict que sy ledit Mathieu estoit

present et qu'il l'eust ouy, qu'il eust bien reparti en sa re-
verence, lequel devint tout rouge et s'en alla incontinent.[24]
Despite such incidents, and the failed prosecution, the reputation stuck,
so that during a fresh batch of trials in 1603 we find Mathieu being
named again by several of those who confessed.[25]

Once gained, it seems, a reputation for witchcraft was almost impossible to lose. For those who neither antagonized their neighbors
excessively nor engaged in dubious kinds of healing, there were two
main ways in which this kind of reputation was acquired. The first
arose when a sickness was diagnosed as unnatural, either by the *devin*
or by some more orthodox specialist such as the local surgeon, leading
to the idea of bewitchment and inducing the victim to identify a plausible suspect with a grievance against him. The second was through
the general awareness of family background, as expressed for example
in the investigations of prospective marriage partners by members of
the families concerned. As the witchcraft persecutions continued, this
latter mode of generating suspicions must have become more and more
dangerous, so that an increasing number of those accused owed their
reputations to the misfortunes of their relatives.

In theory such identifications might have continued to multiply until
a very high proportion of the population was under suspicion. If a
number of trials is any guide, this does not seem to have happened;
the peak was probably reached in the late sixteenth century, the numbers dropping slightly thereafter until the cataclysm of the Thirty Years
War brought an end to virtually all features of normal life, witchcraft
among them, in the 1630s. It seems likely that some kind of control
mechanism was at work, but its exact nature remains elusive, for this
is just where the documents, by their own character, are least helpful.
It is in fact far easier to understand witchcraft beliefs and persecution
synchronically than diachronically. The kinds of disputes and misfortunes that were used in evidence must have been common to all village
societies. The use of countermagical techniques cannot have been a
complete answer for European peasants, any more than it was for the
Azande in precolonial days; if one's child or cow died anyway or still
worse, continued to languish, one would look for some more positive
action.[26] To employ the witchcraft explanation in such cases was normally a way of seeking practical relief, which might be provided by
extracting a show of goodwill and efforts to cure from the suspect, but
with the dangerous side effect of building up evil reputations.

The natural sanction against those who became too obnoxious, or
failed to cure their supposed victim, was beating or even lynching. To
explain the rise of persecution through the courts one needs to dovetail
popular belief and practices with a number of parallel developments.
These include the extension of the criminal law and the system of public

prosecutors, the spread of demonological theory by the printed book and pamphlet, the general tendency of the social elites and the churches to seek more direct enforcement of social controls, and the rapid socioeconomic changes in rural society in the later sixteenth century. The one thing that does seem plain is that no monocausal explanation is likely to be correct. The related problem of the reasons why persecution through the criminal law ceased cannot be illuminated by the experience of Lorraine, where the devastation of war was followed by a lengthy French occupation, bringing with it the more sceptical attitudes already developed by French lawyers and judges.[27]

Another area of great difficulty is the relationship between popular beliefs about such matters as the pact and the sabbat and the elaborated cumulative accounts given by the demonologists. My own belief is that the confessions were based very largely on an indigenous popular tradition, with relatively little contamination from elite demonology. The occasional vivid description of the sabbat is in the characteristic style of the village storyteller, manipulating elements common to folk belief in many parts of the world. Such stories must certainly have been told at the *veillées*, the winter evening gatherings often known as *poisles* in Lorraine, from the local word for the kitchen in which they customarily took place. The *poisle* appears quite often in the trials, as the scene or cause of disputes, since invitations and friendly behavior were important signs of neighborly feelings. These meetings were an important agency for the maintenance and development of folklore; they were also one of the occasions (alongside visits to mill, forge, and well) for gatherings at which communal action might be discussed or initiated. European folklore generally mixes only small doses of fantasy with primarily realistic elements, so it is not surprising to find that in the accounts of the sabbat given by Lorraine witches there are only a few veiled references to sexual license or to any of the more vivid rituals found in other sources. The exiguous feasting and dancing described are little more than the transposition of the features of a village festival into a different context. Most of the active witchcraft took the form of beating water to arouse hailstorms; these were often said to have been turned aside by the timely ringing of church bells. The distribution of diabolical powder, often referred to in the trials, was generally a personal transaction between devil and witch and was rarely mentioned in connection with the sabbat. Like so much else in the theory, there was no obvious necessity for the powder at all; witches were often represented as having injured their victims without any physical agency being involved.

Such inconsistency is perhaps the most consistent characteristic of Lorraine witchcraft beliefs, which repeatedly demonstrate the adaptability of these popular traditions. They allowed villagers to articulate

their hostility toward members of their society who broke communal
norms too often, to isolate them amid a web of suspicion, and to drive
them into dangerous threats against the potential accusers who surrounded them. Such a mechanism may well have had considerable
effects on the social behavior of individuals; when it was taken up by
the legal system, it resulted in a grim toll of victims. In this, as in so
much else, the witches of Lorraine shared their experiences with those
of many other regions of Europe. There are many reasons to study
them today, and one would certainly be to demonstrate how a rather
commonplace, and indeed commonsense, belief in occult power could
exist through every level of an early modern society.

Notes

1 These documents were employed by E. Delcambre in *Le Concept de la
sorcellerie dans le duché de Lorraine au XVIe et au XVIIe siècle*, 3 vols.
(Nancy, 1948–51), a scholarly but curiously limited work that ignores social
factors and generally eschews analysis.
2 For Essex, see A. Macfarlane, *Witchcraft in Tudor and Stuart England*
(London, 1970).
3 The process is admirably discussed in H. C. E. Midelfort, *Witch Hunting in
Southwestern Germany, 1562–1684* (Stanford, 1972). These chains of
accusation differed from the much smaller ones in Lorraine (discussed later)
in that they commonly extended to many persons never previously
suspected.
4 N. Rémy, *Daemonolatreiae libri tres* (Lyons, 1595), esp. bk. I, chaps. ii–iii,
bk. II, chaps. vii–viii; C. Bourgeois, *Pratique civile et criminelle pour les
justices inférieures du duché de Lorraine* (Nancy, 1614), fol. 43.
5 Rémy, bk. I, chap. xiv; bk. I, chap. ii; bk. III, chap. v.
6 Macfarlane, *Witchcraft*; K. V. Thomas, *Religion and the Decline of Magic*
(London, 1971), pp. 435–583.
7 Archives Départementales, Meurthe-et-Moselle, B 4495. Cited hereafter as
A. D. M.-et-M.
8 Ibid., B 3327, no. 5.
9 Ibid., B 3323, no. 10.
10 Ibid., B 8667, no. 8; B 4126, nos. 2 and 3.
11 G. Cabourdin, *Terre et hommes en Lorraine, 1550–1635,* 2 vols. (Nancy,
1977).
12 A.D. M.-et-M., B 8667, no. 5.
13 Ibid., B 8667, no. 6.
14 Ibid., B 8691, no. 5.
15 Ibid., B 8691, no. 13.
16 R. Muchembled, "Sorcières du Cambrésis," in *Prophètes et sorciers dans
les Pays-Bas (XVIe–XVIIIe siècles)* ed. R. Muchembled, M.-S. Dupont-
Bouchat, and W. Frijhoff (Paris, 1978), pp 210–14.
17 A.D. M.-et-M., B 3335, no. 2.
18 Ibid., B 8667, no. 4.
19 Ibid., B 3804, no. 3.
20 Ibid., B 3789, no. 2.
21 Ibid., B 3789, no. 1.

22 Ibid., B 3792, no. 1.
23 Ibid., B 3335, no. 2.
24 Ibid., B 8667, no. 3.
25 Ibid., B 8691.
26 For the Azande, see the classic work by E. E. Evans-Pritchard, *Witchcraft, Oracles and Magic Among the Azande* (Oxford, 1937).
27 For sceptical attitudes in France, see R. Mandrou, *Magistrats et sorciers en France au XVIIᵉ siècle* (Paris, 1968), and A. Soman, "Les Procès de sorcellerie au parlement de Paris (1565–1640)," in *Annales E.S.C.*, 32 (1977), pp. 790–814.

12

The scientific status of demonology

STUART CLARK

> We use the word "supernatural" when speaking of some native belief, because that is what it would mean for us, but far from increasing our understanding of it, we are likely by the use of this word to misunderstand it. We have the concept of natural law, and the word "supernatural" conveys to us something outside the ordinary operation of cause and effect, but it may not at all have that sense for primitive man. For instance, many peoples are convinced that deaths are caused by witchcraft. To speak of witchcraft being for these peoples a supernatural agency hardly reflects their own view of the matter, since from their point of view nothing could be more natural.[1]

In a treatise on witchcraft first published in Trier in 1589 a German bishop explained that all apparently occult operations that were not in fact miracles could be ascribed in principle to physical causes. For whether or not any particular instance was actually demonic in inspiration, "magic" was simply the art of producing wonderful natural effects outside the usual course of things and above the common understanding of men. It followed that "if this part of philosophy was practised in the schools in the manner of the other ordinary sciences . . . it would lose the name of 'magic' and would be assigned to physics and natural science [*et Physicae naturalique scientiae asscriberetur*]." Likewise, in a set of theses on magical operations and witchcraft published a year later in Helmstädt, a natural philosopher and physician began by arguing that "magical actions and motions are reducible to considerations of physics [*Ad Physicam considerationem reducuntur motus et actiones magicae*]." We might be tempted to read into such statements intimations of that scepticism which (it is said) ultimately undermined the learned belief in the reality of demonic effects, espe-

351

cially those associated with witchcraft, by accounting for them just as adequately in natural scientific terms. But the bishop was in fact Peter Binsfeld, and the notable contribution of his *Tractatus de confessionibus maleficorum et sagarum* to classic demonology, as well as its association with vigorous witch hunting, make it inconceivable that he could have meant to convey any general form of doubt.[2] The more obscure proposer of theses, Martin Biermann, although anxious to refute some of the extreme demonological opinions of Bodin, was no less traditional in his belief in the possibility of limited demonic activity in the world and in the reality of pacts between demons and both magicians and witches.[3]

It seems that insofar as they depend on an assumed disjunction between the "occult" and the "scientific," our expectations about belief and disbelief in such texts may be misleading. Understanding what sort of scepticism was most threatening to orthodox demonology depends on grasping its central intellectual defenses. But since these appear to *include* the use of natural scientific explanations, we need to look again at our assumptions about what it made sense for demonologists to accept as an account of the natural world and its processes. There is still a tendency to think that the flourishing of the debate about demonism and witchcraft somehow contradicted the general cultural, and especially scientific, achievements of the sixteenth and seventeenth centuries. If, however, this debate was not isolated from, or even antagonistic to, other aspects of Renaissance thought, including its science, then the contradiction becomes artificial. It is this wider issue of rationality, as well as the question of what was meant by arguments such as those of Binsfeld and Biermann, that involve us in reconsidering the status of demonology as an attempt to offer an ordered construction of natural reality.

A beginning might be made with those individual scientists who concerned themselves with demonology without any sense of incongruity or of the compromising of their criteria of rational inquiry: from Agostino Nifo, Giovanni d'Anania, and Andrea Cesalpino in sixteenth-century Italy to Henry More, Joseph Glanvill, and Robert Boyle in later seventeenth-century England. Others not primarily concerned with natural philosophy nevertheless combined it with demonology without intellectual embarrassment: for example, Jean Bodin, Lambert Daneau, and the Dutchman Andrea Gerhard (Hyperius). In perhaps the largest group there were the many physicians who made special studies of demonic pathology: the Italian Giovanni Battista Codronchi, the Germans Wilhelm Schreiber and Johann Wier, the Swiss Thomas Erastus, the Englishman John Cotta, and the many French doctors involved in cases of possession, among them Jacques Fontaine, Michel Marescot, and Pierre Yvelin.[4]

Intellectual biography would, however, only drive us back to issues. Some of these were, of course, merely practical. Arguments about the etiology and treatment of the various conditions associated with melancholia provided a general context for many medical incursions into demonology.[5] In the further case of the investigation of demoniacs it has even been suggested that exorcists, possibly displaying an empiricism beyond that of their medical colleagues, carried out what amounted to controlled experiments in order to test for the marks of true possession.[6] Other issues brought theorizing about demons, along with narratives of witchcraft, indirectly into scientific debate, as in the arguments over incorporeal substance in Restoration England. If, for instance, we can now see that Glanvill's demonology was inseparable from his experimental philosophy, it is because behind both lay the perception of a threat to Anglican theology posed by the Sadducism of scientific "materialists" and others.[7] Glanvill thought that the study of spirits could be recommended to the Royal Society without contradicting its standards of inquiry. Nevertheless, in this context the spirits entered scientific investigation, as another natural philosopher and demonologist, George Sinclair, remarked, primarily as "one of the *Outworks of Religion.*"[8] The resulting blend of the newest scientific ideals with the oldest witchcraft beliefs was achieved at a key moment in both their histories. Yet the understandable interest shown in this example should not obscure the real novelty involved. What had changed was not the idea that the devil could be retained in a perfectly natural account of the world; it was the view of nature presupposed by this enterprise.

This can be illustrated if we consider a further set of issues, certainly not unrelated to theological questions (or indeed to Baconian elements in the activities of the Royal Society), but generated directly by what was regarded as the central ontological characteristic of demonic phenomena: the fact that they were extraordinary. The principal themes of sixteenth- and seventeenth-century demonology were the qualities and powers of demonic agents and the effects produced by their activity in the world. These were not merely moral effects: They were either real, physical operations, or they appeared to be, for demons were consummate deceivers. Yet neither were they commonplace. At the very least they were, as Glanvill himself put it, "somewhat varying from the common *Road* of *Nature.*"[9] In fact, for the most part they were prodigious in character and, therefore, often confused with other apparently aberrant phenomena. The key questions faced by demonologists were thus of a causal and criterial kind: What was the exact causal status of demonic effects? What laws did they obey or disobey?

What were the criteria for distinguishing between their true and illusory aspects? Along what point on the axis from miracles through natural wonders to ordinary natural contingencies were they to be placed? Tackling such questions involved making distinctions that were critical for any explanation of phenomena, whether demonic or not – distinctions between what was possible and impossible, or really and falsely perceived, and between both supernature and nature, and nature and artifice. It had to be decided what were the boundary conditions governing miracles, prodigies, marvels, and "prestiges"; how to define and use categories such as "magic" and "occult"; and how to relate the explanatory languages of theology and natural philosophy. However bizarre the resulting discussions may sometimes seem, they were genuine attempts to establish criteria of intelligibility for the understanding of a very wide range of what were taken to be puzzling events, that is, events which were said to have "no certain cause in nature."

This concentration on the interpretation of essentially perverse phenomena is not easily related to any narrowly conceived "scientific revolution" in the same period.[10] But this does not mean that it was peculiar to demonologists. What helped to give the debate about demonism and witchcraft such a general currency toward the end of the sixteenth century was the extent to which its interest in the eccentric in nature was a shared intellectual preoccupation. In his remarkable study, *La Nature et les prodiges: l'insolite au XVIe siècle, en France*, Jean Céard has indicated both the range of the literature dealing with monsters, prodigies, and marvels (as well as with the more general features of "variety" and "vicissitude"), and the fundamental character of the conceptual problems it raised in the overlapping territories of philosophy, theology, and science. More recently the specific case of the monstrous has been canvassed as an important individual indicator of changes in explanatory models in early modern France and England.[11] Demonologists often considered an identical teratology – for example, the monsters generated by incubus or succubus devils – and they usually located demonic prodigies semiologically within a broadly apocalyptic account of God's intentions. On the other hand, their stress on demonic manipulation of the natural world was rather oblique to the theme of nature's own generosity or fecundity in producing forms, which emerges strongly from the literature of the "unusual." The important point, however, is not that they may have given different answers to those engaged in the wider enterprise, but that they confronted the same epistemological puzzles. Wherever and to what extent the devil and witches were actually situated in the causation of irregular events are less significant than the broader identity of purpose. It is in this sense that Céard's work enables us to think of demonology as continuous and not discontinuous with Renaissance natural philosophy.[12]

Moreover, the nature of this link does seem to have been recognized from within the "great tradition" of early modern scientific thought. Francis Bacon's proposal (in his *De augmentis scientiarum*) for a natural history of "pretergenerations" – "the Heteroclites or Irregulars of nature" – has often been cited in the context of prodigy literature, but the general relevance of Bacon's project for demonology is thought to have been negligible. In both its theoretical stance and its actual influence on the early program of the Royal Society, this proposal certainly made the marvelous a central rather than a peripheral category of investigation. Bacon's argument was partly technological – that rarities in nature would lead men to rarities in art – but it was also epistemological; hence, the repetition of the suggestion in Book 2 of the *Novum organum*, at the heart of what we have of his actual logic of inquiry. Singularities and aberrations in nature were not merely correctives to the partiality of generalizations built on commonplace examples; as deviations from the norm they were especially revealing of nature's ordinary forms and processes. This makes the example on which Bacon chose to concentrate in the *De augmentis scientiarum* all the more striking:

> Neither am I of opinion in this history of marvels, that superstitious narratives of sorceries, witchcrafts, charms, dreams, divinations, and the like, where there is an assurance and clear evidence of the fact, should be altogether excluded. For it is not yet known in what cases, and how far, effects attributed to superstition participate of natural causes; and therefore howsoever the use and practice of such arts is to be condemned, yet from the speculation and consideration of them (if they be diligently unravelled) a useful light may be gained, not only for the true judgment of the offences of persons charged with such practices, but likewise for the further disclosing of the secrets of nature.[13]

It would not be totally implausible to transpose even Bacon's point about the technological potential of knowledge of "erring" nature into a demonological context and to ask, for instance, whether the treatment of demoniacs was regarded as offering particularly decisive tests of the efficacy of medical (as well as exorcistic) practices. However, it is the fact that he thought of witchcraft narratives in connection with the epistemological benefits of this knowledge that is so suggestive. For in effect this not only made demonism and witchcraft fit subjects for natural philosophy, but elevated them to the rank of Baconian "prerogative instances," that is, areas of empirical inquiry especially privileged by their unusual capacity to disclose natural processes. This idea surely helps us to understand the role of European demonology in the wider setting. Its appeal in the scientific context was undoubtedly its ability, together with that of prodigy literature in general, to tackle one of the most intractable subject matters known to the period. Adapting Bacon's argument somewhat, we might say it was able to confront empirical and, more so, conceptual issues that, though fundamental to all systematic investigation, were laid bare in an especially illuminating manner by the very waywardness of the phenomena dealt with and the struggle to understand them. In this broader sense demonology was one of the "prerogative instances" of early modern science.

What matters here, again, is not that Bacon should eventually have arrived at the same interpretation of these phenomena as the demonologists. His principle that extraordinary events were worth more attention than ordinary ones had a formal truth, whether it was decided that they were all natural or all demonic. However, if, as we have seen, this was not in fact the nature of the choice that had to be made, then the real intellectual distance between a figure like Bacon and the world of demonology may not in any case be as great as it appears. In the *De augmentis* and the *Novum organum*, Bacon talked as though it was a personified nature itself which erred, not a nature acted on by demonic forces. In the *Sylva sylvarum* he also suggested that it was popular credulity which was responsible for the attribution of purely natural operations to some sort of efficacy in witchcraft. An example was the way the hallucinogenic effects of the "opiate and soporiferous" qualities of magical ointments were mistaken for the (supposedly real) transvections and metamorphoses that appeared in witches' confessions.[14] Above all, Bacon insisted that the only phenomena which were nonnatural were true miracles. It is not surprising that these views have been associated with outright naturalism and, therefore, with philosophical indifference to the problems raised by witchcraft beliefs. Yet all of them can be found in the writings of the demonologists, and the second and third might even be said to be presuppositions of their inquiry. The relative importance of demonically and nondemonically caused events remains the only really contentious issue, and here even Bacon allowed for the first when he remarked that "the experiments of witchcraft are no clear proofs [i.e., of the power of the imagination on other bodies]; for that they may be by a tacit operation of malign spirits."[15] Once again we are faced with the artificiality of bringing the modern notion that there is a difference of kind between the "scientific" and the "occult" to the investigation of what were simply differences of degree between varying conceptions of nature.

That the literature of demonology had any meaning at all in this wider context has been obscured by two misapprehensions about the intentions of its authors. Because the sensational aspects of witchcraft belief – the demonic pact, the sabbat, the reality of *maleficium*, and so on – have caught the modern attention, this has suggested, first of all, that the original texts concentrated narrowly and moralistically on the description of these particular crimes and the appropriate judicial and penal response. Of course, these topics were important, and some – notably the alleged transvection of witches to sabbats and their transmutation into animals – raised just those issues that demanded serious epistemological consideration. But the intention was to examine any phenomenon of sufficiently dubious credentials to warrant the suspicion that it was demonically caused. This led demonolgists way beyond the range of topics and attitudes that have been traditionally associated with witchcraft beliefs. Martin Del Rio defined *magia* as "an art or technique which by using the power in creation rather than a supernatural power produces various things of a marvellous and unusual kind, the reason for which escapes the senses and ordinary comprehension." Within literally a few pages we find him tackling the validity of whole sciences such as natural magic, astrology, mathematics, and alchemy, as well as such questions as whether there is any physical efficacy in the innate qualities of magical practitioners, or in the imagination, or in the use of ritual touching, looking, speaking, breathing, and kissing, and whether characters, sigils, arithmetical and musical notation, words, charms, and amulets have any intrinsic powers.[16]

What is striking in his *Disquisitionum magicarum* and in other demonologies of similar scale, such as Francisco Torreblanca's *Daemonologia* and Giovanni Gastaldi's *De potestate angelica*, is the enormous variety of the subjects examined for their standing in reality and knowledge as well as in morals. At the end of his second volume Gastaldi, having already considered natural and other forms of magic, the traditional topics of witchcraft theory, the arts and prodigies of Antichrist, the healing power of the kings of France, the question of bodily transmutation, and the power of demons over magicians, sorcerers, and evil doers, adds a "Disputatio unica" in which he asks of particular wonders whether they are "natural" or "superstitious." These include the movements of the tides, the possibility of speaking statues, the effects of words and music on animal behavior, the power of fascination, the extraction of solid objects from the human body, and the proper cure for tarantism. Even modest monographs tried to cover the same borderland between the naturally marvelous and the magically specious. Thus, if we turn from Pierre de Lancre's best-known work on the witch trials in Labourd, the *Tableau de l'inconstance des mauvais anges et demons*, to one of his other demonological writings, *L'Incredulité et mescreance du sortilege plainement convaincue*, we find another typical range of topics: the reality of sorcery, fascination, whether touching itself can harm or heal, divination, and how to distinguish between good and evil apparitions.[17]

The repetition of this pattern in many other texts rules out the view that it was random or haphazard; yet witchcraft itself was clearly not the only point of departure. Conversely, such topics and many of the same strategies of argument occur in accounts of curious natural and human behaviors that are not ostensibly demonological at all; for instance, in André du Laurens's treatise on the royal touch, where the idea that this form of ritual healing might be demonic has to be overcome,[18] or in more general surveys of the marvelous such as Claude Rapine (Caelestinus), *De his quae mundo mirabiliter eveniunt*; Scipion Dupleix, *La Curiosité naturelle*; and Gaspar Schott, *Physica curiosa*.[19] Demonology was not, then, anchored only to the question of witchcraft and witch trials. It meshed with other discussions with which it shared common intentions, whether or not its conclusions were the same. This enables us to see more easily how demonology could have been a genuine vehicle for what may be called a scientific debate – a debate concerning the exact status of a variety of extremely questionable phenomena. Indeed, it was this guiding issue that, despite the apparently disparate choice of themes, gave demonology real unity of purpose.

The second misapprehension has more seriously affected our understanding of the intentions behind this literature because it has prevented us from seeing the literature as a contribution to a debate at all, or at least to one of any complexity. This is the idea stemming from such early commentators as G. L. Burr and H. C. Lea, that (again on the issue of the reality of witchcraft) demonology could be divided into *either* belief *or* scepticism, with the assumption that belief was a cut-and-dried affair committing a writer to accepting the whole structure of what was alleged.[20] In fact, what is striking is how few examples there are at each end of the spectrum ranging from total acceptance of all demonic claims – where we find only Bodin and perhaps Rémy (in some passages from his *Daemonolatreiae*) – to total rejection – where we find only Reginald Scot and his English followers. This leaves a vast middle ground occupied by hundreds of texts where genuine attempts are made to discriminate between what is to be accepted and what rejected, where authors are familiar with a number of sceptical positions,[21] and where scepticism as well as belief is evident in their own views as demonologists. Repeatedly we are warned that the subject is controversial and obscure and that, faced with the question of the reality of demonic magic, no rational man would insist that it was all illusory or all true. This is the position adopted by Del Rio, Philipp Ludwig Elich, Francesco Maria Guazzo, Benito Pereira, James VI and I, John Cotta, Noël Taillepied (in the allied field of apparitions), and many others.[22] The example of Henri Boguet's *Discours des sorciers*,

often singled out as an especially dogmatic work, shows just how carefully witchcraft confessions might be tested against assumptions about
real and spurious causal efficacy. What governed his attitude was not
any blanket credulity, but, as Lucien Febvre recognized, the application of standards of what was both possible *and impossible* for human
and demonic agents to effect.[23]

Demonologists did not simply pile up the positive evidence for the
guilt of demonic witchcraft. They tried to separate phenomena correctly attributed to demonic agency from phenomena incorrectly so
attributed, and to both they applied a second set of criteria dealing with
truth and illusion. They therefore had at their disposal four categories
of explanation, or four explanatory languages, dealing, respectively,
with real demonic effects, illusory demonic effects, real nondemonic
effects, and illusory nondemonic effects. And they were well aware,
without this compromising their general acceptance of demonic
agency, of the category errors that could occur when (say) confessions
contained nonetheless impossible feats, when the illusions of the devil
were mistaken for reality, when unfamiliar but quite undemonic natural
contingencies or startling technological achievements were blamed by
the uninformed on demonism, or (above all) when hallucinatory experiences stemming from ordinary diseases or narcotic substances
were attributed to witchcraft. This is clear, for instance, in Pierre Le
Loyer's *Quatres Livres des spectres ou apparitions*, where in the context of a defense of the reality of demonism against the arguments of
"naturalists," a variety of almost Pyrrhonist objections are marshaled
against accepting either the evidence of the senses or the promptings
of reason in cases of apparently aberrant phenomena.[24] Likewise, François Perrault's *Demonologie*, after typical emphasis on the dangers of
both outright scepticism *and* outright credulity, consigns reputedly demonic effects such as *ignis fatuum* and *ephialtes* to the category of the
purely natural.[25] We shall find the same features in discussions of natural magical instances in demonological contexts. The fact that a range
of explanations was open to the great majority of writers enabled them
to probe the conceptual puzzles of their subject matter to an extent
that would have been impossible if, as is often assumed, their options
had been limited to supporting or criticizing witchcraft trials.

This can be illustrated in more detail if we take the central topic of
demonic power and consider the implications of the ways its effects
could be explained. For despite their anxiety to warn readers of the
threat of demonism and witchcraft in the world – and this is, of course,
the tonality that we have tended to recognize most readily – demonologists were also, without exception, committed to exposing the limitations, weaknesses, and deceptions of the devil. In both a theologi-
cally and evangelically critical sense they were attempting to demystify

and deflate demonic pretensions: theological, because of the paramount
need (in the age of Reformation claims and counterclaims) to distinguish between the genuinely and the quasi miraculous; evangelical, because of an audience thought to be prone to believe anything about
demonism and to overreact with "superstitious" countermeasures. It
was always granted that demons had not lost their physical powers
after their fall from grace and that their cumulative experience since
the Creation, their subtle, airy, and refined quality, and their capacity
for enormous speed, strength, and agility enabled them to achieve real
effects beyond human ability. Nevertheless, it was also invariably insisted that such effects were within the boundaries of secondary or
natural causation. They were either forms of local motion or alterations
wrought by the application of actives on passives, even if both types
of operation were (say) enormously accelerated. Explanations of this
are found everywhere in demonology; here they are summarized by
John Cotta:

> Though the divel indeed, as a Spirit, may do, and doth many
> things above and beyond the course of some particular na-
> tures: yet doth hee not, nor is able to rule or commaund
> over generall Nature, or infringe or alter her inviolable de-
> crees in the perpetuall and never-interrupted order of all
> generations; neither is he generally Master of universall Na-
> ture, but Nature Master and Commaunder of him. For Na-
> ture is nothing els but the ordinary power of God in al things
> created, among which the Divell being a creature, is con-
> tained, and therefore subject to that universall power.[26]

Satan might, of course, interfere with the initial specific conditions of
natural events, but he could not dispense with the general laws governing their occurrence.[27]

This situation was not changed, only complicated, by the fact that
where his power to produce real effects gave out, his ingenuity in cam-
ouflaging weaknesses by illusory phenomena took over. He could corrupt sensory perception, charm the internal faculties with "ecstasies"
or "frenzies," use his extraordinary powers over local motion to displace one object with another so quickly that transmutation appeared
to occur, present illusory objects to the senses by influencing the air
or wrapping fantastic shapes around real bodies, and, finally, delude
all the third parties involved so that no testimony damaging to his
reputation as an agent was available. The devil was, therefore, severely
limited in what he could really effect (for, as Boguet pointed out, even
his delusions were species of natural action), but there was nothing
that he might not *appear* to effect.[28] Demonologists consequently went
to considerable lengths to expose such *glaucomata* or "lying wonders"
in order to reveal the ontological and epistemological as well as the

moral duplicity involved. The debate focused on the most spectacular claims – that witches could attend sabbats in noncorporeal form, that demonic sexuality could result in generation, and, above all, that humans could be changed into animals – for in these cases a manifest demonic incompetence to create the real effects that were claimed without breaking natural laws led to complicated strategies of deception on his part, none more involved than the last. Discussions of the possibility of lycanthropy in fact contain some of the most interesting examples of demonologists trying, in what I have suggested was a scientific way, to explain a particularly refractory set of claims.

In Jean de Nynauld's *De la Lycanthropy*, for example, we find the gamut of explanatory languages. He writes to disabuse the ignorant on a subject that surmounts the expectations of the senses but that nevertheless has its causes. Bound by the "divinely instituted course of nature," the devil cannot create fresh forms or change the essential character of existing forms. He can therefore only simulate transmutation of witches into wolves by troubling their imaginations, taking advantage of physiologically induced dream experiences, adding demonic efficacy to the ordinary strength of hallucinogenic unguents, and superimposing the required shapes and properties on their bodies in order to deceive any spectators. Thus while real transmutation cannot occur either nondemonically or demonically, there are real effects resulting from natural conditions and substances that lead to all the required sensory experiences, and that, because they are natural, the devil can manipulate. It might seem tempting to recruit Nynauld as a "sceptic." Yet he does not doubt the existence of witches or their use of potions made from slain infants. What he does is analyze all such phenomena on naturalistic lines in order to reveal the causal relationships between the chemical composition of the narcotic elements in such potions, the sensation of being "transmuted," and the psychosomatic effects of folly and credulity. Similarly, he argues that while no unguent can physically effect transvection to sabbats, this is not always an illusion either, since the devil can achieve it by means of local motion. None of this sets Nynauld apart from a supposed "believer" like Boguet, who accounted in exactly the same terms for the phenomena mistakenly thought to result from real lycanthropy and attendance at the sabbat in spirit only.[29]

This is only the briefest summary of a debate that appears in virtually every text. Although some of its features have attracted attention before, its implications for the scientific status of demonology have, I think, been neglected.[30] At the very least, we cannot go on ascribing to the category of the "supernatural" discussions whose purpose was to establish precisely what was supernatural and what was not. Demonism was said to be part of the realm of the natural, for it lacked just those powers to overrule the laws of nature that constituted truly miraculous agency. It must be stressed, therefore, that demonic intervention did not turn natural into supernatural causation. It is the case that its effects were sometimes labeled "nonnatural" or declared to be not attributable to natural causes. But in context this rarely meant more than either their going beyond what might normally have been expected from the ordinary "flow" of causes and effects, or their unfamiliarity or impossibility in relation to the nature known to and practiced upon by men or (less often) their reflection of the devil's desire to break the restraints he was under.[31] The distinguishing criterion of demonic, and indeed all forms of magic, was not that it was supernatural but that it was *unusual*. Even Nicolas Rémy's contradictory statements might be reconciled along these lines. While appearing to follow Bodin in his view that demonism was irreconcilable with any standard of what was natural, he nevertheless qualified this with several comparisons with what were merely the normal limitations and processes.[32] The danger in this situation of preempting meanings by thinking of the "supernatural" only in its modern sense is well shown by the case of John Cotta, who, after using the term several times in his *The Triall of Witch-Craft*, explained that

> although . . . the Divell as a Spirit doth many things, which in respect of our nature are supernaturall, yet in respect of the power of Nature in universall, they are but naturall unto himselfe and other Spirits, who also are a kinde of creature contained within the generall nature of things created: Opposite therefore, contrary, against or above the generall power of Nature, hee can do nothing.

Cotta's tract is of particular importance in this context because it is dominated by his awareness of the epistemological issue of how one could speak of acquiring "naturall knowledge" – by sense experience, reasoning, or conjecture – of such difficult and inaccessible phenomena. Yet William Perkins had also argued that demonic effects only seemed wonderful because they transcended both the "ordinarie bounds and precincts of nature" and the capacities of men, "especially such as are ignorant of Satans habilitie, and the hidden causes in nature, whereby things are brought to passe."[33]

Others reflected this relativism in preferring to use such terms as "quasi-natural"[34] or "hyperphysical."[35] And Del Rio captured it exactly when he proposed the category of the "preternatural" to describe prodigious effects that seemed miraculous only because they were "natural" in a wider than familiar sense.[36] But whatever terms were used, demonic effects were in principle part of natural processes, and in this sense demonology was from the outset a natural science: that is, a study of a natural order in which demonic actions and effects were presupposed. In fact, despite its reputation for intellectual confusion, demonology derived considerable coherence from a notion that there were limits to nature. As Perkins explained: "What strange workes and wonders may be truely effected by the power of nature, (though they be not ordinarily brought to passe in the course of nature) those the devill can do, and so farre forth as the power of nature will permit, he is able to worke true wonders."[37] This was also, necessarily, the standard in terms of which aspects of witchcraft beliefs could be rejected as illusory. The unity of Boguet's treatise and of his views about the inadmissibility of many demonic phenomena was a function of precisely this criterion. And the same intention in James VI and I's *Daemonologie* to link an account of what was possible in magic, sorcery, and witchcraft with the question "by what naturall causes they may be" drew a special commendation from Bacon.[38] The general application of this principle did not mean that demonologists always ended up locating the boundaries of nature in the same place. It was the fact that there was such uncertainty on this issue at the end of the sixteenth century that made demonology both a debate within itself and a contribution to a wider controversy among philosophers, theologians, and scientists. What is significant is the very adoption of the criterion itself. Beyond nature lay only miracles, which no one claimed devils could perform. The question we have to ask, therefore, is not the one prompted by rationalism (Why were intelligent men able to accept so much that was supernatural?), but simply the one prompted by the history of science (What concept of nature did they share?). And as Kuhn and others have shown, this is not something that can be settled in advance.

For these reasons P. H. Kocher was surely mistaken when he suggested that bringing Satan into nature was a prelude to exiling him from scientific inquiry altogether, and that in the English context it was in effect the first step toward the penetration of demonology by that rationalism which produced the radical scepticism of Reginald Scot. This was to prejudge just what was meant by "scientific" in sixteenth-century science. The reason why so many physicians, including Nynauld and, for that matter, a "sceptic" like Johann Wier himself, felt no incongruity in examining the demonic as well as the ordinary causes of lycanthropy and other aspects of witchcraft was because they were *both* natural forms of causation. Guazzo cited Codronchi, Cesalpino, Valesius, and Fernel in support of the view that a sickness could be both natural and instigated by the devil; to this list might be added Jean Taxil, Jourdain Guibelet, and Giano Matteo Durastante. In these circumstances any choice between one explanation and the other was a matter of emphasis, not of principle.[39]

Demonic effects were not, then, qualitatively different from natural effects, but their causation was obscure and hidden from men. They were, in a word, occult, and this alerts us to another important aspect of the relationship between demonology and science. This is the exactly analogous epistemological stance taken up by demonologists and natural magicians. It has been assumed that the subject of natural magic entered demonological discussions in only two guises. It could be totally assimilated to demonism and then cited in order to further blacken the moral reputation of all forms of magic. Here the literature of witchcraft simply added a further layer of denunciations to a very old tradition of Christian hostility to the magical arts.[40] More significantly, it existed as a threatening source of potentially corrosive scepticism because it could explain mysterious natural effects in a way that usurped the accounts given by demonologists. The suggestion is that, like the other sciences of the "occult" tradition, natural magic had greater explanatory power than Aristotelian natural philosophy in this area.[41] There is, of course, evidence for both these stances, but they were not the only ones, and they may not have been the most typical.[42] In the light of what has been said about the naturalism inherent in quite orthodox demonology, the distinction involved in the second may turn out to be rather overdrawn, at least before 1677 when John Webster made it the foundation of his *The Displaying of Supposed Witchcraft*. Most writers wished to downgrade demonic effects by insisting on their ultimately natural (or more strictly, preternatural) character, while at the same time recognizing their occult appearance to the layman. This suggests a much more positive role for the idea of natural magic in their arguments, one which, far from undermining their belief in demonism, actually enabled them to sustain it.

This is, in fact, just what we find. Natural and demonic magic were at opposite ends of the moral spectrum, but they were epistemologically indistinguishable. The devil was therefore portrayed as a supremely gifted natural magician, the ultimate natural scientist. Paolo Grillandi said that he knew "more of natural things and the secrets of nature than all the men in the world put together," including those of "the elements, metals, stones, herbs, plants, reptiles, birds, fish and the movements of the heavens." King James agreed that he was "farre cunningner [*sic*] then man in the knowledge of all the occult proprieties of nature." In Rémy's view, demons had "a perfect knowledge of the secret and hidden properties of natural things." To Perkins, the devil had "great understanding, knowledge, and capacitie in all naturall things, of what sort, qualitie, and condition soever, whether they be causes or effects, whether of a simple or mixt nature."[43] Such characterizations suggest that even the merely commonplace dismissal of natural magic as satanic was more than a chapter in the history of a reputation. When Benito Pereira explained that it was actually learned from incredibly well-informed demons, this tells us as much about assumptions concerning what devils could know as about any suspicion of the "occult."[44] Moreover, the repeatedly expressed idea that the devil was the most expert natural philosopher put the demonologist in much the same intellectual predicament as the natural magician, or indeed the Aristotelian, when he discussed occult (as opposed to manifest) qualities: that of coming to terms with effects which could be experienced but whose causes might be unknowable. A remark of Perkins puts the epistemological challenge posed by the devil rather effectively:

> Whereas in nature there be some properties, causes, and effects, which man never imagined to be; others, that men did once know, but are now forgot; some which men knewe not, but might know; and thousands which can hardly, or not at all be known: all these are most familiar unto him, because in themselvs they be no wonders, but only misteries and secrets, the vertue and effect whereof he hath sometime observed since his creation.[45]

In these circumstances the fact that demonologists often used the possibility of a natural magic to buttress some of their own central arguments becomes much less surprising than it seems at first. To begin with, there were occasions when writers who in no way doubted the general reality of witchcraft phenomena cited instances from natural magic to suggest that, nevertheless, there were many occult effects in nature which were wrongly confused with demonism simply because their causes were unknown or uncertain. We can see an example in the *De sagarum natura et potestate* of Wilhelm Schreiber (Scribonius), famous for his defense of the water ordeal in witch trials. Schreiber expressed plenty of the ordinary alarmism about witches and their guilt, but he took up a typical position between ascribing too little and too much to them, extremes which (he said) only a proper knowledge of natural philosophy could avoid. By this he meant knowledge both of the ability of unaided nature to generate its own marvels (here he used the play imagery – *lusus naturae* – common in the prodigy literature and in Bacon), and of the capacity of a mimetic and licit natural magic to repeat such marvels artificially. The latter he described traditionally as the most perfect philosophy in its knowledge of the mysteries and secrets of nature and as practiced by the Persian and Egyptian magi and by Moses, Solomon, and Daniel.[46]

A second case arose when demonologists, accepting without question that demonism and witchcraft had *some* sort of efficacy, wished to expose the claim that it lay in the actual means used, where this was (say) a ritual incantation or conjuration or some spurious physical means. This could be done by citing the natural but hidden causal links involved, recognizable only in terms of a knowledge of naturally magical effects. An example here would be De Lancre's attempt to discredit the idea that touching itself had an inherent efficacy. He argued that apparently supportive instances drawn from the unusual behavior of animals, plants, or metals – the torpedo fish, the *echeneis* or remora – or from natural magnetism could be explained in terms of various secret but perfectly natural properties and "antipathies." There were some such effects of which the causes were so hidden that they would never be known, and here men ought to be content with doubt and not strive, in the manner of "naturalists," for explanations at any risk to plausibility. But in other cases the reader might be referred to the works of the natural magicians, to Levinus Lemnius for the bleeding of corpses in the presence of the murderer, and to Jerome Fracastor for the *echeneis*.[47]

Third and most commonly, demonologists cited the science of the occult characteristics of natural things when they wished to reduce the status of demonic operations from the apparently miraculous to the merely wonderful. And this was in fact the context for Peter Binsfeld's remark that magic was just an esoteric form of physics. Because ordinary men were unaware of all nature's secrets, they attributed to the realm of the miraculous demonic effects that originated in natural powers, however elevated. And to this same distinction between popular superstition and learned science could be traced the reputation of natural magic, which appeared equally strange but was really only "a certain hidden and more secret part of Natural Philosophy teaching how to effect things worthy of the highest admiration . . . by the mutual application of natural actives and passives." Examining marvels from this source, such as the salamander, the volcano, and the magnet, would, Binsfeld thought, put the devil's works into proper focus.[48]

Fourth and finally, any remaining strangeness in the character of real demonic effects could be dissipated by the suggestion that they were in fact no more difficult to accept than the parallel claims made by natural magicians for what Boguet called "Nature . . . assisted and helped forward by Art." The speed to which demons accelerated ordinary processes like generation by corruption might (he admitted) invite scepticism. But if alchemists were to be believed, they too could "by a turn of the hand create gold, although in the process of Nature this takes a thousand years." Nor was there any reason to doubt that Satan could make a man appear like a wolf, for "naturalists" such as Albertus Magnus, Cardan, and Della Porta had shown how it was possible to effect similar "prestigitations." Somewhat similarly, Sébastien Michaelis compared demonic effects with the marvels described by Mercurius Trismegistus in his *Asclepius* to show that "there are many effects . . . against and above" the ordinary causation of things. For Rémy the yardstick offered by natural magic was what it revealed of nature itself rather than of art. When he came to consider the question of the reality of the objects supposedly ejected from the bodies of demoniacs, he cited the natural explanations for this being a true phenomenon given by Lemnius and Ambrose Paré (in his *Des Monstres et prodiges*), with the following comment: "If then Nature, without transgressing the limits which she has imposed upon herself can by her own working either generate or admit such objects, what must we think that the Demons will do."[49]

Naturally these arguments were often blended together. Elements of the second and third can be found in Lambert Daneau's dialogue, *De veneficis*, where the apparent (but spurious) efficacy of the forms of words and symbols used in witchcraft is explained away in terms of the natural means (like poisons) interpolated by demons. These are often very strange but never miraculous; instead, they are comparable with technical achievements like the flying wooden dove of Archytas. This reference to one of the classic marvels of the magical tradition (it is also discussed by Agrippa, Campanella, Dee, and Fludd) would not have been lost on Daneau's readers.[50] The idea of natural magic did not therefore always weaken demonology by implying some challenge to theories of demonic agency; on the contrary, it could provide important strengthening points of reference whenever there was a need to contrast or equate this agency with something comparably natural yet occult. Many repeated the standard indictment that the historical natural magic of the Persians and Egyptians had degenerated in time and was now indistinguishable from diabolism. Some, like Pereira and De Lancre, cautioned about the publication of natural magical works on the grounds that free access to such secrets was dangerous. But there was a sense in which the sort of scientific inquiry represented by them – that is, the concept itself of natural magic – remained an intrinsic part of their theories of knowledge. Given the frequency with which it is dealt with in the texts, it may even have been a necessary part of the intellectual structure of demonology.[51] From one direction this may still seem to constitute the debasement of what was undoubtedly a form of science by its association with satanism. The point to be reemphasized is that, considered from a different direction, it illustrates how closely demonological and scientific interests in certain interpretive issues can be identified with each other. Nor must it be forgotten that, conversely, natural magicians were led to a consideration of demonism by the questions raised in their discipline. Della Porta's examination of the powers of the witches' unguent, though excluded from later editions of his *Magiae naturalis*, was widely cited. Georg Pictor's *De illorum daemonum qui sub lunari collimitio versantur* was thought to be sufficiently cognate with the supposititious works of Agrippa for them to be published together in translation in England in 1665. Even Lemnius, who was reputed then and has been since as an outright "sceptic," did not exclude demons from the physical world. In his *De miraculis occultis naturae* they appear among the "accidents" of diseases, insinuating themselves "closely into men's bodies" and mingling with "food, humours, spirits, with the ayre and breath" as well as with violent and destructive tempests. They do not, of course, bulk large in Lemnius's natural philosophy; but neither are they ignored.[52]

This leads on to a final reflection on the entire range of attitudes to demonic magic and witchcraft phenomena in the Renaissance and Reformation period. By establishing that it was (in part) an epistemological debate – a debate about the grounds for ordered knowledge of nature and natural causation – which occupied the middle ground in demonology, we should be in a better position to interpret the views at the extremes. We can see, for instance, why Reginald Scot's radical scepticism stemmed not, as is sometimes suggested, from his espousal of the principles of natural magic, or in particular from the idea that, since miracles had ceased and all created things were left with only their natural capacities, all causation must also be natural. For this only begged the more fundamental question of what *counted* as a natural capacity; and since demonologists themselves endowed devils with such capacities, this was not a sceptical stance that posed any threat.[53] Scot's most telling argument was his reduction (in an Appendix to his *Discoverie of Witchcraft* of 1584) of all demonic agents to a noncorporeal condition, thus removing them from physical nature altogether. When demonologists attacked "naturalism," it was this step which they often had in mind – that is, not merely the commitment to a naturally caused world, but the denial of a devil capable of using such causation for evil ends. It was the fact that the principle of demonic agency's naturalness was not *itself* in doubt which, in other cases of supposedly damaging objections, enabled them to turn sceptical arguments to their own use. At the other extreme we can see that Bodin's reluctance to doubt anything in this area resulted from his view that it was impious to place any advance limits on what was possible in nature. To apply the language of physical events to metaphysical operations was a fundamental category error. Since aspects of magic and witchcraft belonged to this metaphysical reality, there was no criterion for accepting or rejecting them, other than trust. This obliterated the distinction that enabled most other demonologists to make sense of the world. But their case was the case of natural science as a whole. As Jean de Nynauld remarked, Bodin's position made all learning impossible, for "all the means for separating the false from the true would be taken away" if it was admitted that tomorrow the world might (with God's permission) be qualitatively different.[54]

Such issues were not, of course, discussed only at the time of the European "witch craze". Demonologists owed the foundations of their arguments to accounts of broadly the same range of phenomena given by Augustine and Aquinas. The question of what significance was to be given to the marvelous in nature had a very long history indeed. What may be suggested is that the need to reconsider the validity of these phenomena and of the criteria for understanding them was felt especially keenly in the sixteenth and seventeenth centuries, after which consensus was again established. No doubt the witchcraft trials themselves contributed to this. More importantly, the urgency stemmed from the unprecedented intensity of theological controversies concerned with the status and prevalence of miracles, the exact properties of religious objects and forms of words, the possibility of divination in a divinely ordained world, the apocalyptic meaning of prodigies, and so on. It may also be related to the fresh impetus given by disputes about the fundamentals of scientific and philosophical thought to the consideration of problems of epistemology – problems that came to be pursued with special vigor in the various parallel areas of the extraordinary in nature and art. The fact that they were also dealt with in discussions of incubus and succubus devils, flights to the sabbat, and werewolves should not deter us from accepting these, too, as contributions to scientific discourse.

Notes

1 E. E. Evans-Pritchard, *Theories of Primitive Religion* (Oxford, 1965), pp. 109–10.

2 Peter Binsfeld, *Tractatus de confessionibus maleficorum et sagarum*, 2nd ed. (Trier, 1591), pp. 174–6. In this argument Binsfeld follows Francisco Victoria, *Relectiones theologicae* (Lyons, 1587), relectio XII, "De arte magica," pp. 452–3.

3 Martin Biermann (propos.), *De magicis actionibus exetasis succincta* (Helmstädt, 1590), theorem I; cf. theorems XIII and LXXII, sigs. A3^{r-v}, D2^r.

4 Further details of medical interest in demonology, in the context of a supposed "slow progress to an enlightened attitude," are given by Oskar Diethelm, "The Medical Teaching of Demonology in the 17th and 18th Centuries," *Journal of the History of the Behavioural Sciences*, 6 (1970), pp. 3–15.

5 Sydney Anglo, "Melancholia and Witchcraft: The Debate between Wier, Bodin and Scot," and, emphasizing medical viewpoints, Jean Céard, "Folie et démonologie au XVIe siècle," both in *Folie et déraison à la Renaissance*, ed. A. Gerlo (Brussels, 1976), pp. 209–22, 129–43. Evidence of a general affinity of attitudes and methods between demonology and the "new

science" is offered by Irving Kirsch, "Demonology and Science During the Scientific Revolution," *Journal of the History of the Behavioural Sciences*, 16 (1980), pp. 359–68. The same author's "Demonology and the Rise of Science: An Example of the Misperception of Historical Data," *Journal of the History of the Behavioural Sciences*, 14 (1978), pp. 149–57, merely points to coincidences in the timing of new interests in both fields.

6 D. P. Walker, *Unclean Spirits: Possession and Exorcism in France and England in the Late Sixteenth and Early Seventeenth Centuries* (London, 1981), p. 13 and passim; he calls this "an aspect of early modern science that has not yet . . . been investigated." H. C. Erik Midelfort, "Sin, Folly, Madness, Obsession: The Social Distribution of Insanity in Sixteenth-Century Germany," in *Understanding Popular Culture: Europe from the Middle Ages to the Nineteenth Century*, ed. Steven L. Kaplan (forthcoming).

7 Moody E. Prior, "Joseph Glanvill, Witchcraft, and Seventeenth-Century Science," *Modern Philology*, 30 (1930), pp. 167–93; T. H. Jobe, "The Devil in Restoration Science: The Glanvill–Webster Witchcraft Debate," *Isis*, 72 (1981), pp. 343–56.

8 George Sinclair, *Satan's Invisible World Discovered* (Edinburgh, 1685), p. xv.

9 Joseph Glanvill, *Sadducismus triumphatus*, 4th ed. (London, 1726), p. 8.

10 Hence the somewhat artificial linking of the debates of the witch hunt with the classic "revolution" in science in Brian Easlea, *Witch-Hunting, Magic and the New Philosophy: An Introduction to Debates of the Scientific Revolution 1450–1750* (Brighton, 1980), pp. 1–44 and passim.

11 Katharine Park and Lorraine J. Daston, "Unnatural Conceptions: The Study of Monsters in Sixteenth- and Seventeenth-Century France and England," *Past and Present*, no. 92 (1981), pp. 20–54.

12 Jean Céard, *La Nature et les prodiges: l'insolite au XVIᵉ siècle, en France* (Geneva, 1977), passim, esp. pp. 352–64; cf. Lynn Thorndike, *A History of Magic and Experimental Science*, 8 vols. (New York, 1934–58), which, despite its astonishing range, is decidedly unsympathetic to the literature of witchcraft. For Thorndike's distaste for the subject, see V, 69–70; and for a characteristic judgment on a respectable Aristotelian whose demonology involves a "deluded mixture of theology and gross superstition," see his remarks on Cesalpino's *Daemonum investigatio peripatetica* (Florence, 1580), in VI, 325–8.

13 Francis Bacon, *De augmentis scientiarum*, bk. II, chap. 2, in *The Works of Francis Bacon*, ed. J. Spedding, R. L. Ellis, and D. D. Heath, 14 vols. (London, 1857–74), IV, 296; cited hereafter as *Works*. Cf. *Novum organum*, bk. II, aphorisms 28–9, in *Works*, IV, 168–9; *The Advancement of Learning*, bk. II, in *Works*, III, 330–2; *Parasceve ad historiam naturalem et experimentalem*, aphorisms 1–4, in *Works*, IV, 253–7 (all references are to the English trans.). Thomas Sprat, *The History of the Royal Society of London*, 3rd ed. (London, 1722), pp. 214–15, defends the study of "the most unusual and monstrous Forces and Motions of Matter," without mentioning witchcraft. For Boyle and Glanvill on witchcraft and marvels, see Prior, pp. 183–4.

14 Francis Bacon, *Sylva sylvarum: or a Natural History in Ten Centuries*, century X, no. 903; cf. no. 975; in *Works*, II, 642, 664.

15 Ibid., century X, no. 950, in *Works*, II, 658. For an "experiment solitary touching maleficiating," see century IX, no. 888, in *Works*, II, 634.

16 Martin Del Rio, *Disquisitionum magicarum* (Lyons, 1608), bk. I, chaps. 2–5.

17 Giovanni Tommaso Gastaldi, *De potestate angelica sive de potentia motrice, ac mirandis operibus angelorum atque daemonum*, 3 vols. (Rome, 1650–2); cf. Francesco Torreblanca (Villalpandus), *Daemonologia sive de magia naturali, daemoniaca, licita, et illicita, deque aperta et occulta, interventione et invocatione daemonis* (Mainz, 1623), bk. II, pp. 176–403; Pierre de Lancre, *L'Incredulité et mescreance du sortilege plainement convaincue* (Paris, 1622).

18 André du Laurens, *De mirabili strumas sanandi vi solis Galliae Regibus Christianissimus divinitus concessa* (Paris, 1609), chap. 9.

19 I have used the French trans.: Claude Rapine (Caelestinus), *Des Choses merveilleuses en la nature où est traicté des erreurs des sens, des puissances de l'âme, et des influences des cieux*, trans. Jacques Giraud (Lyons, 1557), chap. 8, pp. 113–30 ("On the Operation of Evil Spirits"); Scipion Dupleix, *La Curiosité naturelle rédigée en questions selon l'ordre alphabétique* (Rouen, 1635), pp. 393–4; Gaspar Schott, *Physica curiosa, sive mirabilia naturae et artis* (Würzburg, 1667), bk. I, pp. 1–195. Schott also deals with ghosts, miraculous races, demoniacs, monsters, portents, animal marvels, and meteors.

20 G. L. Burr, "The Literature of Witchcraft," in *George Lincoln Burr*, ed. R. H. Bainton and L. O. Gibbons (New York, 1943), pp. 166–89; H. C. Lea, *Materials Toward a History of Witchcraft*, ed. A. C. Howland, 3 vols. (Philadelphia, 1939; New York, 1957).

21 For a striking example of the presentation of sceptical arguments in a discussion nonetheless committed to the reality of demonism, see Loys le Caron (Charondas), *Questions divers et discours* (Paris, 1579), quest. VIII, ("Si par incantations, parolles ou autres semblables sortileges l'homme peult estre ensorcelé et offensé en ses actions et forces naturelles"), fols. 31ᵛ–43ᵛ; cf. the same author's *Responses du droict français* (Paris, 1579–82), bk. IX, response 43 ("Si les sorciers et sorcières sont dignes de dernier supplice"), pp. 445–50. For the flexibility and variety in theories of witchcraft, see H. C. Erik Midelfort, *Witch Hunting in Southwestern Germany, 1562–1684: The Social and Intellectual Foundations* (Stanford, 1972), pp. 10–29.

22 Del Rio, bk. II, quaest. 6, p. 61; Philipp Ludwig Elich, *Daemonomagia* (Frankfurt, 1607), chap. 5, pp. 60–1; Francesco Maria Guazzo, *Compendium maleficarum*, trans. and ed. Montague Summers (London, 1929), bk. I, chap. 3, p. 7; Benito Pereira, *De magia, de observatione somniorum, et de divinatione astrologia* (Cologne, 1598), bk. I, chap. 1, pp. 4–5; James VI and I, *Daemonologie, in the Forme of a Dialogue* (Edinburgh, 1597), p. 42; John Cotta, *The Triall of Witch-Craft* (London, 1616), dedicatory epistle, sigs. A2–A3ᵛ; Noël Taillepied, *Traité de l'apparition des esprits* (Rouen, 1600), trans. and ed. Montague Summers as *A Treatise of Ghosts* (London, 1933), pp. xvi–xvii, 39–40.

23 Lucien Febvre, "Sorcellerie: sottise ou révolution mentale?" *Annales E.S.C.*, 3 (1948); trans. K. Folca as "Witchcraft: Nonsense or a Mental Revolution?" in *A New Kind of History from the Writings of Febvre*, ed. Peter Burke (London, 1973), pp. 185–92.

24 Pierre Le Loyer, *IIII Livres des spectres, ou apparitions et visions d'esprits, anges et Démons se monstrans sensiblement aux hommes* (Angers, 1586), partly trans. Z. Jones as *A Treatise of Specters or Straunge*

Sights, Visions and Apparitions Appearing Sensibly Unto Men (London, 1605).

25 François Perrault, *Demonologie ou discours en general touchant l'existence puissance impuissance des demons et sorciers* (Geneva, 1656), chaps. 1–3, pp. 1–52.

26 Cotta, chap. 6, p. 34.

27 For standard accounts of demonic power and knowledge and their limitations, see Silvestro da Prierio (Mazzolini), *De strigimagarum daemonumque mirandis* (Rome, 1575), bk. I, chaps. 13–15, pp. 95–126; Otto Casmann, *Angelographia* (Frankfurt, 1597), pt. 2 ("De malis angelis"), chaps. 12–14, 18–20, pp. 428–57, 508–82; Johann Wier, *De praestigiis daemonum et incantationibus ac veneficiis* (Basel, 1568), bk. 1, chaps. 10–18; Torreblanca, bk. II, chaps. 5–10, pp. 191–220; Guazzo, bk. I, chaps. 3–4, 16, pp. 7–11, 57; Gervasio Pizzurini, *Enchiridion exorcisticum* (Lyons, 1668), praeludium, chap. 6, pp. 14–16. For a typical analysis of the devil's powers over local motion, see Leonardo Vairo, *De fascino* (Paris, 1583), bk. II, chap. 13. For a discussion of his interference in the conditions of natural combustion, see Adam Tanner, *De potentia loco motiva angelorum*, quaest. 6, in *Diversi tractatus* ed. Constantine Munich (Cologne, 1629), pp. 90–1.

28 Descriptions of the range of illusion techniques are again in most standard demonologies; e.g., Guazzo, bk. I, chap. 4, p. 9. But see full accounts in Anthoine de Morry, *Discours d'un miracle avenu en la Basse Normandie* (Paris, 1598), pp. 39–56; Andrea Gerhard (Hyperius), "Whether That the Devils Have Bene the Shewers of Magicall Artes," in *Two Commonplaces Taken Out of Andreas Hyperius*, trans. R. V. (London, 1581), pp. 47–81; and André Valladier, "Des Charmes et sortileges, ligatures, philtres d'amour, ecstases diaboliques, horribles, et extraordinaires tentations de Satan . . . ," sermon for Third Sunday in Advent 1612, in his *La Saincte Philosophie de l'ame* (Paris, 1614), pp. 619–41. For the fact that demonic delusions were also naturally caused, see Henri Boguet, *Discours des sorciers*, trans. and ed. Montague Summers and E. A. Ashwin as *An Examen of Witches* (London, 1929), p. xliii.

29 Jean de Nynauld, *De la Lycanthropie, transformation, et extase des sorciers* (Paris, 1615), passim; cf. Boguet, chaps. 17, 47, pp. 46–51, 145–8.

30 P. H. Kocher, *Science and Religion in Elizabethan England* (San Marino, Calif., 1953; New York, 1969), pp. 119–45; Wayne Shumaker, *The Occult Sciences in the Renaissance* (London, 1972), pp. 70–85.

31 This last idea is in Jacob Heerbrand (praeses.), *De magia dissertatio* (Tübingen, 1570), prop. 6, p. 2.

32 Nicolas Rémy, *Daemonolatreiae* (Lyons, 1595), bk. III, chap. 12; bk. I, chap. 6; bk. III, chap. 1; trans. and ed. E. A. Ashwin and Montague Summers as *Demonolatry* (London, 1930), pp. 181–2; cf. pp. xii, 11, 141.

33 Cotta, p. 34; William Perkins, *A Discourse of the Damned Art of Witchcraft* (Cambridge, 1610), epistle; cf. pp. 18–21, 27–8, 159.

34 Paolo Grillandi, *Tractatus de hereticis et sortilegiis* (Frankfurt, 1592), quaest. 7, p. 96.

35 Johann Georg Godelmann, *Tractatus de magis, veneficis et lamiis recte cognoscendis et puniendis* (Frankfurt, 1591), bk. I, chap. 8 ("De curatoribus morborum hyperphysicorum praestigiosis").

36 Del Rio, bk. I, chap. 4, quaest. 3, p. 25; cf. Rémy, bk. II, chap. 5, p. 113.

37 Perkins, p. 23.

38 James VI and I, "To the Reader," cf. p. 42; Bacon, *De augmentis scientiarum*, bk. II, chap. 2, in *Works*, IV, 296. See also Elich, chap. 6, pp. 75ff.

39 Guazzo, bk. II, chap. 8, p. 105; for Taxil and Guibelet, see Céard, "Folie et démonologie," pp. 129–43. Janus Matthaeus Durastantes, *Problemata . . . I, Daemones an sint, et an morborum sint causae* (Venice, 1567), fols. 1–83ᵛ.

40 For a recent survey of this tradition, see Edward Peters, *The Magician, the Witch, and the Law* (Brighton, 1978), passim.

41 H. R. Trevor-Roper, *The European Witch Craze of the Sixteenth and Seventeenth Centuries*, rev. ed. (London, 1978), pp. 58–9; K. V. Thomas, *Religion and the Decline of Magic* (London, 1971), pp. 579, 646; P. W. Elmer, "Medicine, Medical Reform and the Puritan Revolution," unpublished Ph.D. thesis, University of Wales, 1980, pp. 289–302; Jobe, pp. 343–4.

42 For entirely negative accounts of magic, see James VI and I, *Daemonologie*, and Niels Hemmingsen, *Admonitio de superstitionibus magicis vitandis* (Copenhagen, 1575). There is an excellent example of the fully sceptical use of natural magical evidence in Michel Marescot et al., *Discours véritable sur le faict de Marthe Brossier de Romorrantin prétendue démoniaque* (Paris, 1599), pp. 29–30; trans. A. Hartwell as *A True Discourse . . .* (London, 1599), pp. 22–3. The authors do not, however, rule out the possibility of demonic possession in principle. Nor was the argument that only extraordinary effects above the laws of nature could be attributed to the devil (those of the Brossier case not being of this sort) a very telling piece of antidemonology, for the devil was conventionally placed within such laws.

43 Grillandi, quaest. 6, pp. 59, 68–9; these remarks are found in many other texts. James VI and I, p. 44; Rémy, bk. II, chap. 4, p. 107; Perkins, p. 19.

44 Pereira, bk. I, chap. 3, pp. 21–2.

45 Perkins, p. 20; for the epistemological problems posed by occult qualities in the wider context of scientific and philosophical controversy, see Keith Hutchison, "What Happened to Occult Qualities in the Scientific Revolution?" *Isis*, 83 (1982), pp. 233–53.

46 Wilhelm Adolf Schreiber (Scribonius), *De sagarum natura et potestate* (Marburg, 1588), pp. 29–35; for the presence of demons in Schreiber's natural philosophy, see his *Rerum naturalium doctrina* (Basel, 1583).

47 Pierre de Lancre, disc. III ("De l'Attouchement"), fols. 113–77, esp. fols. 124–57. In the same way Boguet referred to Della Porta for the real natural effects of the witches' unguent, and Perkins discussed the well-known natural magical instance of the basilisk or cockatrice, concluding that fascination by breathing or looking alone was either fabulous or the indirect result of natural causes like contagion. There are arguments very similar to De Lancre's in Vairo, bk. II, chap. 10.

48 Binsfeld, pp. 173–8.

49 Boguet, pp. 64, 148–9; Sébastien Michaelis, *Pneumologie, ou discours des esprits*, 2nd ed. (Paris, 1613), trans. W. B. as *Pneumology or Discourse of Spirits* (London, 1613), pp. 5–6; Rémy, bk. III, chap. 1, pp. 139–41.

50 Lambert Daneau, *De veneficis* (Cologne, 1575), pp. 94–5, trans. R. W. as *A Dialogue of Witches* (London, 1575), sigs. H6ᵛ–I6ᵛ. For the dove of Archytas, see Frances Yates, *Giordano Bruno and the Hermetic Tradition* (London, 1964), pp. 147–9, and *The Rosicrucian Enlightenment* (London, 1972), p. 76.

51 A point perhaps insufficiently realized by D. P. Walker, *Spiritual and Demonic Magic from Ficino to Campanella* (London, 1958), pp. 145–85, for even Del Rio thought that to avoid attributing all unusual effects to demonism one had to know of the many things surpassing ordinary scientific inquiry (bk. II, quaest. 5, pp. 60–1).

52 Giovanni della Porta, *Magiae naturalis, sive de miraculis rerum naturalium* (Naples, 1558), bk. II, chap. 26, p. 102; *Henry Cornelius Agrippa: His Fourth Book of Occult Philosophy*, trans. R. Turner (London, 1655), pp. 109–53; Levinus Lemnius, *Occulta naturae miracula* (Antwerp, 1561), fols. 83–87^v, quotations from the English trans., *The Secret Miracles of Nature* (London, 1658), pp. 86–90, 385.

53 In this respect it is instructive to compare Reginald Scot, *The Discoverie of Witchcraft* (London, 1584), bk. I, chap. 7, pp. 14–15, with Le Caron, *Questions divers*, fol. 32^r, where the point is absorbed into conventional demonology.

54 Nynauld, p. 77; the argument is in fact identical to that of Rapine, p. 121. Cf. Jean Bodin, *De la Démonomanie des sorciers* (Paris, 1580), preface and "Refutation des opinions de Jean Wier," fols. 239^v–40^r, 244^r, 247^v, 251^{r-v}.

13

"Reason," "right reason," and "revelation" in mid-seventeenth-century England

LOTTE MULLIGAN

> In the beginning of Time, the great Creator Reason, made the
> Earth to be a Common Treasury to preserve . . . Man.
>   This work to make the Earth a Common Treasury was shewed
> to us by Voice in Trance, and out of Trance, which words were
> these, "Work together, Eate Bread together, Declare this all
> abroad": which Voice was heard three times.[1]

Thus spake Gerrard Winstanley in 1649. How novel was this kind of
dual appeal to reason and revelation? This chapter explores the usages
of the word "reason" (and its cognates) by Winstanley's contemporaries. It follows the prescriptions sketched by J. G. A. Pocock in attempting to "write the history of debates conducted in a culture where
paradigms and other speech structures overlapped and interacted;
where there could be debate, because there was communication, between different 'languages' and language-using groups and individuals."[2]

It is a commonplace that the religious and political controversies of
mid-seventeenth-century England were concerned with the "right
reading" of God's will. The all-important issue for the opponents of
orthodoxy in the ideological war was to establish that their own interpretations of the divine will were right, being based on an unchal-
lengeable source; to undermine the rationale for existing institutions
they claimed for themselves indubitable insights – insights derived from
private illumination of the spirit. While this battle of ideas raged in the
political and religious arenas, a parallel struggle occurred over rival
interpretations of God's determinations in the natural order. The re-
emergence of the writings of Hermes Trismegistus meant that in the
realm of nature, too, knowledge must be based on an illuminist epistemology.

The temptation to link both sets of challengers to orthodoxy has led some historians to seek connections between them. Rattansi's pioneering essay, "Paracelsus and the Puritan Revolution"[3] – an early and successful onslaught on Whig historiography – suggested a framework in which such connections might be studied. His ideas have been elaborated in Charles Webster's *The Great Instauration*[4] and in Christopher Hill's essay, "'Reason' and 'Reasonableness.'"[5] A different (yet related) thesis is offered by Robert Hoopes in *Right Reason in the English Renaissance*.[6]

The first argument of this chapter is that the so-called irrationalism of major hermetic and radical writers of the mid-seventeenth century has been misinterpreted and their unorthodoxy overstressed. Extreme examples of irrationalism like that of the Ranters, the Muggletonians, and Van Helmont certainly existed. But, with few exceptions, influential writers seen by recent historians as belonging to the irrationalist camp should be read as having more in common with their antagonists in the debates than with the extremists. I shall argue that the writings of religious antinomians such as Gerrard Winstanley and William Walwyn, hermetic reformers of the Commonwealth such as John Webster and Samuel Gott, the royalist hermetic-turned-mechanist Walter Charleton, the Anglican hermetic Thomas Vaughan, the Cambridge Platonists Henry More and Nathanael Culverwel, and the Anglican casuist Jeremy Taylor shared a view of "right reason," a view that is not essentially opposed to that of a mechanical philosopher such as Thomas Hobbes. My second argument is that in the usage of the seminal noun phrase "right reason" there was no radical discontinuity between the middle and later seventeenth century; philosophies usually treated as incompatible will be shown here to occupy much common epistemological ground.

Seventeenth-century writers from very different standpoints agreed that reason was a faculty of the mind, God-given both to make sense of the Creation and, through it, to acquire at least a nodding acquaintance with the Creator. They also agreed that revelation provided additional, suprarational knowledge of God and of the natural and moral orders. Where attacks on "reason" occurred, they took the form of a rejection of scholastic, syllogistic *reasoning*. These writers' own approving usages of "reasoning" included the process of logical thinking based on indubitable sense experience or on fundamental logical and moral "principles." They used the term "natural reason" for the *unaided* ratiocinative faculty of the mind to see connections in the realms of natural philosophy, naturalistic ethics, and naturalistic theology. Here the referent of "reason" was not the process, but both the human faculty *and* the conclusions reached by it. In addition (and confusingly) the word "reason" was sometimes substituted for "right reason," a term meaning both the faculty improved by illumination *and* the morally informed conclusions that true Christians (but not natural men) could reach about their duties in the world and about God's purposes in the Creation. The distinction between "mere natural reason" and "right reason" was that noted by Jeremy Taylor and described by Hoopes as "the difference between the 'dry light' of unaided reason, the nonmoral activity of logical disquisition, and the dictates of 'right reason' . . . reason that has been morally purified. Reason is 'right' to the degree that it seeks . . . the knowledge of absolute Truth, that is the Truth of Christianity."[7]

This difference in both the use and mention of "reason" (and its cognates) among disparate groups of writers has led Rattansi, Hoopes, and Hill to distinguish between middle and later seventeenth-century usages. Rattansi, using John Webster and Walter Charleton as examples, sees the reformers and revolutionaries of the 1640s and 1650s as exalting "the knowledge of illumination above that derived from 'carnal reason.'" The changes in Charleton's epistemology, he believes, exemplify the eventual triumph of the mechanical philosophy already espoused by more orthodox writers such as John Wilkins and Seth Ward. The illuminist, fideistic, hermetic strain of those inspired by the revolutionary ideas of the 1640s and 1650s are sharply contrasted by Rattansi with the empirical, rational, mechanical philosophy of the more conservative or latitudinarian temper that triumphed when the revolution had played itself out. He points to "the distinction between the natural magic tradition and the new 'mechanical philosophy' being revealed with great sharpness and clarity."[8] Hoopes's intellectual map of the middle and later seventeenth century distinguishes between the older acceptance of "right reason" with its Christian moral goals, and the later definition of "reason" by the mechanical philosophers as ratiocination independent of moral ends. He uses Jeremy Taylor as a transitional figure who pointed the way to an acceptance of Hobbes's amoral laws of nature by throwing doubt on the universality of "right reason" and who freed God from operating according to humanly defined rationality. Hill sees the changes in seventeenth-century usages as moving from Hooker's unchanging and objective God-given reason, with its connections with eternal truth and virtues, to Hobbes's reason based on human common sense and individual experience, via a period of "unreason"[9] and scepticism about reason's ability to yield either truth or virtue. And he follows Rattansi in linking the fideism of the radical sects with hermetic and Paracelsian traditions.

Each of these modern writers posits radical discontinuities between various seventeenth-century writers in the treatment of "reason." A universally accepted, God-given reason leading to revealed Christian truth and virtue apparently gave way to a period of questioning when the belief in the universal validity of conclusions reached by reason was undermined. By the later years reason had emerged as ratiocination based on sense experience, a *process* that produced knowledge of the morally neutral mechanical laws of nature.

However, an examination of some of the writers used to argue for discontinuity in fact reveals that – throughout the period – they shared an understanding of "right reason" as the pathway to God's eternal truths. Certainly each claimed to be the sober spokesman for the only valid set of conclusions to be drawn from the exercise of right reason. But none denied that it was through right reason that knowledge of God's purposes – both moral and natural – would be achieved. God's eternal laws operating in the world were to be known by reason seasoned with revelation. It is true that seventeenth-century writers were often indiscriminate in how they applied the word "reason," and their conflations have led to modern problems of interpretation. But a more basic difficulty, I believe, has been the concern of historians of seventeenth-century thought to focus on the momentous changes involved in the scientific revolution – a concern that has led them to exaggerate discontinuities with the preceding period and to ignore what was shared between the protagonists of competing natural and moral philosophies. Modern readers have great difficulty in accounting for what are now judged to be dramatic shifts between apparently incompatible world views – shifts that, it seems, occurred over a relatively short time and often in a single work. It is therefore important to give due weight to evidence of continuity, for this makes it easier to understand the nature and degree of these shifts from one to another explanatory model and to grasp how it was possible for seventeenth-century writers to hold at the same time two or more – to us incompatible – models.

Rattansi illustrates his argument – that the revolutionary period of the 1640s and 1650s bred a revival of Paracelsian and Helmontian hermeticism because liberation from "carnal reason" by the illumination of the spirit "had a particular attraction for reformers and revolutionaries" – from the writings of John Webster and Walter Charleton.[10] True, Charleton was no revolutionary, but a royalist and Anglican. Nevertheless, it is argued that the tenor of the times infected him as much as others, such as Webster, who fit the reformer and revolutionary label better. Charleton is useful to Rattansi because later he is alleged to have become aware of the social danger of sectarian and atheistic tendencies released by the revolution. He therefore discarded Van Helmont's mantle and moved toward an acceptance of a socially safer Gassendian mechanical explanation of the natural order.

It is necessary to consider what use Charleton and Webster made of "reason," whether they were concerned with the process of reasoning, and whether they rejected conclusions reached by that process for achieving true knowledge in theology and moral and natural philosophy in favor of direct spiritual illumination. While both rejected (on Baconian grounds)[11] the scholastic syllogistic reasoning of the universities and were part of the context of university reform of the 1650s, it is wrong to suggest that they did so because they rejected outright the use of reason to achieve an understanding of man, God, and the natural and moral order. Webster believed that the spiritual teaching of the Gospel could not be taught as a university discipline because it rested entirely on private illumination. On the other hand, natural knowledge was the proper sphere of human reason. "That what can be discovered of God, and supernatural things, by the power of Reason, and the light of Nature, may be handled as part of natural Philosophy . . . because it is found out by the same means and instruments that other natural Sciences are."[12]

But Webster's "power of reason" was not unaided human reason. For when Adam acquired the language of nature, which enabled him to know its workings – a language sinful man had now forgotten – this knowledge was not learned but *given* him by God; it was "not inventive or acquisitive, but meerly dative from the father of light."[13] Similarly, the insights of the physician by which he recognizes the cause and cure of disease, while requiring human effort of reason and sense in studying anatomy and plant physiology, was nevertheless granted by God.[14] So, while the spiritual nature of God and the mysteries of the Christian religion were closed altogether to reason, all other knowledge was derived from reason illuminated by inspiration.[15]

The case of Walter Charleton is more complex because he transferred his allegiance from some of Van Helmont's hermeticism to a more explicitly mechanical view. Rattansi, Nina Gelbart, and Lindsay Sharp argue for a fundamental shift in Charleton's attitude to the study of nature.[16] But the claim that he rejected reason for illumination as a source of truth in his earliest writings is, I believe, a misinterpretation, and one that results in overstating the drama of his intellectual change. Charleton in fact did not reject *right* reason at any time in his writing career.

In a critical passage in *A Ternary of Paradoxes* (1650), in which he printed translations of some of Van Helmont's work – a passage quoted by Rattansi – he wrote:

> We must quit the dark Lanthorne of Reason, and wholly throw ourselves upon the implicit conduct of Faith. For a deplorable truth it is, that the unconstant, variable, and seductive imposture of Reason, hath been the onely unhappy Cause, to which Religion doth owe all those wide, irreconcileable and numerous rents and schismes . . . made by men

of the greatest Logick . . . every Faction alleaging a ratio-
nall induction, or ground for its peculiar Deflection, from the
unity of Truth.[17]

There are, however, three points to be made. First, Charleton was
explicitly referring here to a knowledge of essences, a metaphysical
knowledge to be gained fully only after death: "The mind, having once
fathomed the extent of her wings, in Metaphysicall speculations, becomes assured, that after her delivery from the Dungeon of Flesh and
Blood, she shall have all her knowledge full . . . in one single act."[18]
So direct knowledge of God and his intellectual essences was to be
gained not by reason but by illuminated faith; but this has nothing to
do with natural knowledge. Second, the section from which the passage
was taken presents a view that Charleton said he now *rejected*: "To
this opinion (I blush not to professe) I had formerly leaned,"[19] and,
after first justifying this belief, he went on:

> These, I say, were the Temptations that first drew me into a
> beliefe, that the Power of Ratiocination seemed too low and
> triviall an Endowment, to make out the Imperiall Preroga-
> tive, of mans being Created in the Image of God: . . . but
> my second thoughts are more wary, and hold it a part of
> prudence, to suspend my positive assent unto this nice Par-
> ticular; as well in respect, this dispute would better beseem
> the Metaphysicall Speculations of the School divine, then
> the grosse and corporeall disquisition of a young Physi-
> cian.[20]

Third, what Charleton had rejected earlier (just as Webster had done)
was the idea of natural reason as ratiocination providing knowledge of
the divine. He was not rejecting, either in his earlier or in his 1650
position, the idea that reason was the proper means for the study of
man and the natural order. He underlined this by distinguishing between the "Metaphysicall Speculations of the divine" and his own
work as a natural philosopher. Like Webster, Charleton subscribed to
the view that faith, reason, and sense had different objects; but all led
to truth. While faith attained truths "above the reach of the other two,"
reason comprehended the dependence of cause and effect, and sense
provided knowledge of qualities. There was no conflict among these
three sources informing our understanding: "All of which Pilots mutually conspire to steer our Mindes . . . towards . . . the main end of
our Creation."[21]

Van Helmont had struggled to free himself from the fetters of reason;
he acknowledged reason as inevitably there in the consideration of
natural and moral matters, but at the same time treated it as an enemy
of the unity of truth and virtue. For him "intellectual Light" – a vision
of truth – was a nonrational process to be gained by prayer, self-ab-

negation, and a passive waiting for the gift of grace.[22] Charleton, by contrast, thought that the difficult and painful efforts of reason and sense were the ways to understand the natural world. That he meant by "reason" not mere natural reason but *right* reason he made clear: "I fix one Eye of Reason on that domestick Security, and internall Serenity, which necessarily redounds from the severe practice of Goodnesse, in this life; and the other of Faith on that infinite Compensation, ordained to reward our pious endeavours, in the next."[23] What was being asserted here was a belief in the proper object of right reason that informed the conduct of a Christian life and included his work as a student of the natural order.

Despite the fact that Charleton chose to translate three of Van Helmont's works, his Prolegomena in *A Ternary* is riddled with propositions contradicting Van Helmont's epistemology. In addition, he was sceptical about some of his science. While subscribing to the magnetic cure of wounds and Van Helmont's work on tartar in wine – because they accorded "with the testimony of experience and were found consonant with Reason"[24] – Charleton's justification of both theories differed from Van Helmont's. Furthermore, he rejected the latter's universal medicine out of hand and, crucially, his acceptance of Van Helmont was always provisional: "If it be thus."[25]

Both Rattansi and Gelbart see Charleton's work during the 1650s as moving from an acceptance of hermeticism to an (albeit idiosyncratic) mechanical view. I have claimed that Charleton did not accept Van Helmont's epistemology or all of his science in 1650. Certainly he did not reject reason for illumination, as Rattansi suggests; he saw himself rather as a Christian physician using right reason in the practice of his medicine. But Gelbart, unlike Rattansi, is careful not to overstate the extent of Charleton's "conversion" between 1650 and 1654, and agrees that he was never either an avowed mystic or a total mechanist.[26] Nevertheless, by concentrating on the subject matter of his pamphlets, rather than on his language and epistemology, she attributes a greater change in his thought than is necessary to account for his statements. Gelbart's criteria for judging that Charleton had rejected hermeticism included his dropping of the microcosm–macrocosm analogy, his denunciation of action at a distance, and his recantation on the efficacy of the magnetic cure. In fact, however, much hermetic language persists in his later books; the microcosm–macrocosm analogy is still there in 1652 and 1657,[27] and he continued to invoke the theory of signatures and the original alphabet of nature.[28] Further, his acceptance of the theory of magnetism to account for the cure of wounds at a distance never relied on the existence of an *anima mundi* but on a mechanical theory of atoms very like Sir Kenelm Digby's. Charleton's rejection of the magnetic cure – on empirical grounds[29] – thus required no wholesale exchange of paradigm. In the same book his account of the lode-stone – an invisible flow of atoms producing mechanically testable effects – invoked a pattern of explanation very like his earlier mechanical account of the sympathetic cure.[30] Nor was Charleton's corpuscular theory as different from Digby's as Gelbart suggests. Although Digby's sympathetic cure was not published until 1657,[31] his earlier friendship with Charleton,[32] and the fact that the same story was used by both as evidence of the cure's efficacy,[33] together suggest that Charleton's theory was not novel; it was more akin to Digby's atomism than to the hermetic explanation based on the *anima mundi*.

Charleton's later works display the essential consistency of the use of the word "reason" throughout his writing life. In 1657 he was ostensibly engaged in proving the immortality of the soul through the use of reason.[34] But here, as in 1650, he continued to argue that faith and reason have different objects and that they are complementary methods of reaching knowledge of God.[35] And this understanding remained even in 1682.[36] For Charleton the concept of right reason remained unchanged.

Rattansi argues for an intellectual affinity between radical and hermetic illuminists by associating Charleton's apparent rejection of Helmontianism with his attack on the religious sects. It has already been argued that no such dramatic rejection took place in the early 1650s, as Charleton did not accept Van Helmont's attack on rationalism. Further, though he clearly feared the contemporary sprouting of heresy due, he said, to the arrogance of the sects who claimed for themselves the ability to "comprehend what God can and determine what he ought to do,"[37] he himself was never tinged with the self-assurance associated with private illumination. Nor was he merely being wise after the event, for he never believed that our understanding of mundane affairs was based on direct revelation. His intellectual development does not show him as a man who shed the chrysalis of a restricting world view to emerge fully fledged as a new philosopher. His writing career serves a more useful historical purpose. It demonstrates rather how his philosophy of knowledge enabled him to bridge the apparent incompatibility between his (admittedly idiosyncratic) hermeticism and his (equally unorthodox) mechanistic world view. Far from rejecting reason for implicit faith and then changing his mind, Charleton clung throughout to a clear perception of the proper ends of both.

Walter Charleton's case demonstrates that it was not necessary for a writer to reject hermetic language and concepts in order to espouse a mechanistic philosophy. But in this he was not special. The literature on social and intellectual reform of the 1640s and 1650s contains many works in which it is difficult to allocate writers to an exclusive philosophical tradition. Hermetic, mechanical, scholastic, and Neoplatonic language may be found even within a single book, which demonstrates that for the writer no essential contradictions were apparent. Samuel Gott's *Nova Solyma* (1648),[38] a Utopian millenarian educational treatise, moved easily within a whole range of terminologies that we are accustomed to regard as incompatible. The science taught in this New Jerusalem was elitist and utilitarian in Baconian fashion. It investigated nature's "hidden spirit and meaning";[39] it searched for the marvelous effects of herbs and fruits and the hidden influence of gems; and, while recognizing that it was beyond man's power to "penetrate beneath the surface of such mysteries," it delegated to adepts in chemistry the task of attempting to resolve them.[40] On the other hand, proof of the Creator's power was adduced through a series of scholastic maneuvers, such as a comparison between his infinity and finite time and matter, and his ability to create something out of nothing.[41] Knowledge of God's perfection and omnipotence was demonstrated in Platonic fashion by mentally removing the imperfections in creation.[42] The universality of religions in the world was explained in mechanical language: "The voice of Nature herself clearly . . . confess[es] a Deity,"[43] while the great work of the Creation displays the design of inscrutable Providence. God the "Great Architect" has demonstrated his plan in the created world.[44] Not only did Gott use the languages of hermeticism, Platonism, Baconianism, scholasticism, and the mechanical philosophy without seemingly finding them contradictory or incompatible, he also believed that knowledge of God and the Creation came from a combination of natural reason and divine inspiration. On the one hand, we are given "a clear and familiar way by the light of God's truth."[45] On the other, Jacob, the exponent of Nova Solyma's virtues, tells his uninitiated friends:

> His truth . . . is a subject far more beyond our natural powers . . . I have certainly gained more advantage . . . by prayer than by book learning; for often those hard knots which I have long anxiously been trying to untie by my studies have suddenly loosed while in the act of prayer . . . I left behind me many wiser than myself still struggling in the unsettled sea of human reason.[46]

This experience, far from being simply a moment of mental clarification, was described as full of "great fear and amazement," during which Jacob's "inner sight being opened . . . he stood . . . arrested by ecstatic musings . . . dazed by excess of heavenly light."[47] Like Charleton, Gott saw the compatibility – indeed, the necessity – of a combination of natural reason and divine inspiration to gain knowledge of God's working in the natural order.

The year in which Charleton published *A Ternary* witnessed a new debate – described by F. B. Burnham as "the beginning of a climactic struggle between the disciples of Hermes Trismegistus and the advocates of the new philosophy."[48] The protagonists were Thomas Vaughan and Henry More. Here too the argument is that the clash of two incompatible epistemologies – illuminism versus rationalism – culminated in a clear victory for the new, rational science. Burnham sees More, as the champion of rational latitudinarianism, triumphing over "the cultural extremes of his day; the Prelatists, Royalists and Scholastics on the conservative wing, and the Sectarians, Anarchists and Hermetics on the left wing."[49] Such a schematized view of the intellectual map of the 1640s and 1650s presents insoluble problems, once it is acknowledged that the opposing camps actually shared much and that there were embarrassing exceptions who do not fit neatly into these politicoreligious philosophical categories. Theodore Hoppen's account of the hermetic beliefs of many important fellows of the Royal Society[50] suggests that there was no swift victory by Latitudinarian rationalists over what was considered to be appropriate matter and method in the investigation of nature. The careers and writings of such men as Gott, Vaughan, Charleton, and Digby certainly make a strict dichotomy difficult to accept.

The sharply defined opposition between Vaughan and More presented by Burnham is tempered by N. L. Brann,[51] who sees both as sharing aspects of Augustinian Platonism carried into the seventeenth century by Cartesian rationalism,[52] where revelation plays a crucial part in knowledge of the Creator and the created universe. His concern, however, is to present More as a supporter of the reality of witches and spirits, which accounts for his quarrels with those he regarded as materialists. While this context for More is appropriate, it does not resolve the problem of his debate with Vaughan, for it is not clear why he should have chosen the latter, a fellow believer in spirits, rather than, say, the materialist Hobbes as his adversary.

My purpose in reexamining this debate is to show that there was no dramatic contrast between the epistemologies of these two protagonists, despite their mutual accusations that the other had abandoned reason. Like Charleton and Webster, Vaughan and More believed in the centrality of right reason for understanding the natural order, and both believed in suprarational means for gaining additional theological knowledge. Again like Charleton and Webster, Vaughan was concerned to offer an alternative to Aristotelian science, and his condemnation of it focused first on syllogistic ratiocination. He made the usual attack on book learning as opposed to the direct study of nature, and his stated aim was to discover God through a study of the creation.[53] He was careful to distinguish between "corrupt" and "right" reason.

He quoted Augustine – "'Deliver us, O Lord, from logic'" – but immediately disclaimed the obvious implication:

> And here I must desire the reader not to mistake me. I do not condemn the use, but the abuse of reason, the many subtleties and reaches of it, which man hath so applied that truth and error are equally disputable. I am one that stands up for a true natural knowledge, grounded – as Nature is – on Christ Jesus, who is the foundation of all things.[54]

Where Vaughan differed from Charleton and Webster was in insisting that natural philosophy and theology were not separate fields to be studied by different methods.[55] While Charleton believed that knowledge of essences was not to be had by natural means, and Webster excluded spiritual knowledge from the study of natural philosophy, Vaughan saw these fields as inseparable from an understanding of God's creation. Knowledge of prime matter and its occult manifestations in particular, and spiritual knowledge in general, were to be revealed through a study of medicine and alchemy. But the natural and the divine were intimately related. Knowledge "ascends by the Light of Nature to the Light of Grace."[56] Unlike Van Helmont, Vaughan believed that reason was not an impediment but an essential part of the study of nature; but he also believed that it required the supplement of illumination for complete understanding. Certainly, to fathom the occult forces in nature required "sudden illustration . . . impossible without a divine assistance."[57] However, despite tantalizing hints that he had achieved positive insights into the occult – for example, his claim "not only to know [prime matter] but after long labours to see it, handle it and taste it"[58] – Vaughan did *not* claim to have received direct, unearned divine illumination himself. "But Reader, be not deceived in me. I am not a man of any such faculties, neither do I expect this blessing in such a great measure in this life."[59] Like Charleton, Vaughan feared what he saw as the irrationalism of the sects, and like Charleton he distinguished his own position from theirs on the grounds that they had only "some empty pretences of the spirit."[60] In contrast, his own beliefs were backed up by the entire rationale of the created universe, underwritten by special, privileged, God-given insights. And these insights were to be earned, not given gratis, as the sects claimed for their illumination. Only an intensive study of nature and the Old and New Testaments would procure full knowledge.[61] God "discovers the laws of nature" to us through our intellectual efforts, not through direct inspiration. *Immediate* illumination as such was not part of natural science.[62]

Henry More began his debate with Vaughan in 1650. His *Observations* were a direct attack on *Anthroposophia theomagia* and *Anima magica abscondita*.[63] But the general purpose behind this onslaught only became clear a year later, when he returned to the attack in *A Second Lash*.[64] Like Charleton *and* Vaughan, he was troubled by the spread of both atheism and enthusiasm and combated them by arguing for the application of reason to religion.[65] Vaughan, too, had been beset by the same concerns and found it important to attack the sects' irrationalism by juxtaposing it with his own theory of knowledge, based on reason aided by revelation. As we shall see, More's position was much closer to that of Vaughan than his invective against Vaughan would suggest.

Despite their hectoring tone, the *Observations* at first read like those of a defender of traditional natural philosophy against the irrational claims of a religious fanatic. More contrasted "preposterous . . . imaginings" with "the light of a purified minde and improved reason,"[66] thus claiming the ground of right reason for his own. The *Observations* are a spirited attack on Vaughan's account of creation, his "fundamentals of science," his high-flown metaphorical language. More presented himself as having more in common with the Aristotelian view of matter and with Cartesian science than with Vaughan's hermeticism.[67]

On closer inspection, however, this self-identification requires modification. In *A Second Lash* More admitted sharing with Vaughan the usefulness of the microcosm–macrocosm analogy.[68] In *Conjectura cabbalistica* (1653) the treatment of Genesis is in many details close to Vaughan's own version of the creation story, with its emphasis on the inward word creating spiritual substances unrelated to matter, and with the creation of the ether "which is liquid as water and yet has the first Principle of Fire which is the first element."[69] But the most striking affinity between their scientific theories is the theory of signatures, which More espoused in one of his self-confessedly naturalistic pieces, in which he stood attired as "a meere naturalist."[70] The signature of plants "is a certaine Key to enter Man into the knowledge and use of the Treasury of Nature. I demand therefore whether it be not a very easie . . . inference . . . that severall herbs are marked with some marke or signe that intimates their virtue, what they are good for."[71]

But More was not primarily concerned to contest the truth of Vaughan's natural philosophy. The brunt of his attack was directed against Vaughan's use of the Bible as a central text for the study of philosophy. "What profane boldnesse is this to distort the high Majesty of the holy Scripture . . . to decide the controversies of the World and of Nature."[72] The Bible is either so obscure about the Creation that men "father their own notions . . . upon the Scriptures," or else it speaks in "the vulgar way" and is useless for philosophy. The misapplications of the Bible to natural philosophy "doe in many well-meaning men eat out the use of their reason," leading them to believe that "these flarings of false light . . . are not from himself but from a Divine Principle . . . And then bidding a dieu to Reason, as having got some Principle above it . . . they treat the casuall figurations of their anxious phansie" as if they were direct messages from God.[73] More was clear that this enterprise of abstracting philosophical principles from the Bible was based on a total rejection of reason: "He that . . . lays aside clear and cautious reason in things that fall under the discussion of Reason upon pretence of hankering after some higher principle . . . casts away one of the most Soveraign Remedies against all melancholic impostures."[74]

It is therefore all the more surprising that, after defending the use of reason in natural philosophy and apparently casting away the Scriptures as a source of philosophical knowledge, More's return to the attack in 1651 shows a very different mood prevailing. The light, hectoring tone of the *Observations* was replaced in *The Second Lash* by an exalted paean to Platonic mysticism, sparked by Vaughan's apparent alignment with just that philosophy:[75]

> How lovely and how magnificent a state is the soul of man
> in whom the life of God in activating her, shoots her along
> with himself through Heaven and Earth, makes her unite
> with . . . the whole world as if she had become God and all
> things. This is the precious clothing and rich ornament of
> the mind, farre above Reason or any other experiment . . .
> This is to be godded with God, and christed with Christ.[76]

More hastened to recognize just how this kind of vision had led to Ranterish immorality and pantheism. But the crucial difference between the two visions of oneness with God rested on More's own reliance on "sound reason and the sober faculties of the soul."[77] So although he could write: "God hath made me . . . Emperor of the World . . . I am inhabitant of Paradise and Heaven upon Earth . . . All Creation is below me," he yet insisted that *his* reason led to conclusions "consistent with the attributes of God, the common notions of Men or the Phenomena of Nature."[78] The quarrel between More and Vaughan no longer resembles a conflict between Burnham's two opposing world views; it looks far more like Brann's squabble between two close relatives struggling to engross for themselves some common family property.

More's closeness to Vaughan is best illustrated at the end of *A Second Lash*. They would hardly have disagreed that the distinguishing hallmark of a true son of God and "member of Christ" was sober morality, nor that "right reason" and God's will required the quenching of sectarian strife. But these eminently rational expressions of the divine will were not left by More as rational inferences from the Scriptures. Rather they were expressed in trancelike terms: "And I had no sooner uttered these words in my mind, but me thought I heard an

Answer from all the Quarters of the Earth from East, West, North and South like the noise of many waters or the voice of Thunder, saying Amen, Hallelujah. This is true."[79] It is hardly surprising that Vaughan could not resist referring to this passage as a "spirituall Ague . . . I believe he is one of the Shakers."[80] After such a finale Vaughan was able to present himself as a sober Christian natural philosopher of the Morian variety: "I am one that stands up for a true Natural Philosophy built, as nature itself is, on Christ Jesus who is the foundation of all things natural and supernatural."[81]

There is another aspect of this controversy that makes it *look* as if Vaughan and More inhabited different mental worlds. While impugning Vaughan's use of the Scriptures as the source of true knowledge of natural philosophy, More allowed himself a proviso: "I will not deny but that some Philosophical Truths may have an happy and useful illustration . . . from passages in Scripture. And their industry is not to be vilified that take any pains therein [as long as there is no] Philosophical abuse thereof."[82] It was not the practice of using the Bible for philosophical ends, it seems, but the rash conclusions which fanatics drew from it that More was castigating. This quarrel was about which biblical exegesis was most in keeping with right reason.

This conclusion is borne out in More's *Conjectura cabbalistica* (1653), which was itself an attempt to see Genesis as a secret key to philosophical, moral, and natural truths – truths to be gained by an entirely nonliteral reading. Searching for the inner mystery rather than the outward history was just what Vaughan had been condemned for. More distinguished his own efforts by disclaiming any divine inspiration for his interpretation:

> Though I call this Interpretation of mine Cabbala yet . . . I received it neither from Man nor Angel. Nor came it to me by divine Inspiration unlesse you will be so wise as to call . . . that Life and Sense that resides in the Rational Spirit . . . inspiration. But such Inspiration . . . is no distractor from, but an accomplisher and enlarger of humane faculties . . . This is the great mystery of Christianity . . . the perfection of the humane nature by participation of the divine . . . [in] our Intellect, Reason and Fancie. But to exclude the use of Reason in search of divine truth is no dictate of the spirit but of headstrong Melancholy and blind Enthusiasme.[83]

His bid was for the ground of inspired, Christian, right reason, a ground he denied to Vaughan. But Vaughan did not seek to "exclude the use of Reason"; he simply claimed that the inspiration necessary for an understanding of the mysteries in the Scriptures did not entail a rejection of right reason. What More made clear, as had Charleton, was the complementary nature of reason and revelation, with their common origin in the nature and will of God. Our ability to use reason, to see connections and agreements between things was "a participation of that divine reason in God . . . whose steady and immovable Reason discovers the connection of all things at once . . . perfected and polished by the holy Spirit."[84]

The similarity between More and Vaughan does not end with the effort of both to extract mystical meanings from Genesis, for this task required a more resembling language than that suggested by More's attack on Vaughan's hyperbolic writing. Having vented his spleen on the latter's "muddy and imaginary" speech, the use of "dry metaphor" and "phantasticall aenigmatic" expressions,[85] More found himself in *Conjectura* writing: "We have thought fit though Aenigmatically, and in a dark Parable, to shadow out . . . the manner of progress to divine perfection, looking upon Man as a Microcosm or a Little World who if he hold out the . . . Progresse of the Spiritual Creation . . . will be figuratively understood."[86] More echoed not only the hermetic analogy but also Vaughan's manner of expression. Thus, "there went up moist vapour from the Earth which . . . concocted by the Spirit of the World . . .became a precious and balmy liquid and fit vehicle of life."[87] This is as far removed from both the approved Aristotelian and the Cartesian views of matter and substance as Vaughan's account, dismissed in *Observations* as "an hideous empty phansie."[88]

More found himself in similar linguistic territory not only with his adversary Vaughan, but in an even more unexpected terrain – that of Gerrard Winstanley at his most mystical. Compare Winstanley's "and that righteous Ruler (God) . . . the tree of Life, begins to walke in the coole of the day, with delight, in the middle of the garden [of Eden] (Mans heart)"[89] with More's "and the Tree of Life was in the midst of this Garden of man's soul."[90] Vaughan had called himself "Eugenius Philalethes" and More, his scourge, became "Alazonomastix Philalethes." His claim to be a fellow member of the Philalethean family and a "Chip from the same Block"[91] seems to have been a joke directed against himself.

Why did More find himself in such unsought-for company? His position was a difficult but consistent one. He wished to take the middle ground between what he saw as the stark atheistic rationalism of the (unspecified) enemies of religion on the one hand, and the rejection of right reason by the current "epidemical disease" of illuminism on the other. His task in *Conjectura* was clear, even if his argument is obscure. In the Dedication to fellow Platonist Ralph Cudworth he justified the search for "the inward and mysterious meaning of the Text" by claiming that biblical literalism had led to an atheistic dismissal of it as "so empty . . . a melancholic concept . . . brought into the world to awe the simpler sort."[92] As a philosopher he thought it safe to ascribe to

Moses a deeper meaning beneath the simple message for simple folk contained in the bare letter. The elitist cabalism he espoused made the Bible a fit text for scholars searching for additional complex but complementary meanings to flesh out the unsophisticated account of creation provided in Genesis.

> For the Truths [of this cabalistic interpretation] themselves, they are such as may well become so holy and worthy a person as Moses, if he would philosophize; they being very precious and choice truths and very highly removed above the conceit of the vulgar and so are the more likely to have been delivered to him or to Adam, first by God for a special Mysterie.[93]

What distinguished his own method of analysis of hidden meanings from the method of those he castigated as full of ''Melancholy and Fancy which they ordinarily call Inspiration''[94] was that *his* conclusions were ''consentatious to Reason.''[95] Further, when fully examined, ''the more irrefutable they will be found, no Hypothesis that was ever yet propounded to man so exquisitely well agreeing with the Phenomena of Nature, the Attributes of God, the Passages of Providence and the rational Faculties of our own minds.''[96] *His* mysticism, therefore, was consonant with reason, and this distinguished it from that of his antagonists, the religious radicals:

> I fear there are no men subject to such mis-interpretation of Scripture as the boldest Religionists and Much-Prophets who are very full of heat and Spirits and have their imagination too often infected with the fumes of those lower parts the full sense and pleasure whereof they prefer before all the subtile delights of Reason and generous Contemplation.[97]

Their mysticism and allegorizing led to such heresies as denying the divinity of Christ, the Second Coming, the afterlife, and the forgiveness of sins, as well as to gross Ranterish immorality. All these consequences followed from their ''unlawful sporting with the Letter.''[98] More needed his mysticism to counter materialist onslaughts, his sophisticated analogizing to satisfy philosophical Christians, and his claim for the rationality of his position to keep the antinomians in their place. And he was able to do this by appealing to the consonance of his own vision with God's ''steady and unmovable Reason,'' communicated to right-thinking Christians through divinely inspired right reason.

In the same year as *Conjectura* More undertook a very different task. In *An Antidote Against Atheisme* he attempted to show, as Charleton had done, how the existence of God could be proved by natural reason. That he saw this task as crucially related to the earlier one is made clear in its Preface, where his antagonists were again explicitly identified as atheists and enthusiasts.[99] But to justify the very different tenor of his earlier book he included an apologia, setting out his tactics in the assault on his enemies:

> But that hee might not be shy of mee, I have conformed my-
> self as neer his own Garbe as I might, without partaking of
> his folly or wickednesse . . . I appeare now in the plaine
> shape of a meere Naturalist, that I might vanquish Athe-
> isme; as I heretofore affectedly symbolized in carelesse
> Mirth and freedome with the Libertines, to circumvent Lib-
> ertinism.[100]

To cope with the enthusiast "I [suffered] myself to be carried into such high Triumphs and Exaltations of Spirit."[101] That this was a different spirit from his enemies' More made plain:

> And I am no more to be esteemed an Enthusiast for such
> passages as these than those wise and circumspect philoso-
> phers Plato and Plotinus, who upon the more then ordinary
> sensible visits of the divine Love and Beauty descending
> into their ravished souls, professed themselves no lesse
> moved . . . Inebriated . . . with the delicious sense of the
> Divine life, that blessed Root and Original of all holy wis-
> dome and virtue.[102]

A respectable philosophical source, and the conjunction of wisdom and virtue in the beatific vision, enabled More to use the language and experience of the enthusiasts without being contaminated by their principles. But it is precisely *because* so much of the language and experience was shared between the protagonists of this mid-seventeenth-century battle that More was able to play all these parts so effectively.

The significance of More's special pleading is heightened when we realize that it occurred precisely in those books in which he sought to adduce rational arguments for the existence of God and to bring to bear naturalistic reasons to combat fanatical illuminism. Already in *An Antidote*, the apparently rationalistic treatise on the existence of God, he exposed himself as one whose visions provided additional validity to his arguments. In *Enthusiasmus triumphatus* (1656) he sought to discredit religious fanatics – those who indulged in mystical interpretations of Scriptures as well as Quakers[103] – by a naturalistic theory of temperament, ascribing their flights of fancy to the effect of a melancholy nature, flatulence, and a hypochondriacal humor.[104] Their revelations resulted from "a ligation of the outward senses, whatever is there represented to the Mind is in the nature of a Dream . . . these Dreams the precipitant and unskillful are forward in conceit to be Representations extraordinary and supernatural, which they call Revelations."[105] Such materialism is worthy of Hobbes himself. But this passage is followed by a very different argument:

> And yet notwithstanding I humbly conceive and hope I may
> doe so without any suspicion of Fanaticism, that there may
> be such a presage in the Spirit of man that is to act in things
> of very high concernment to himself, much more if to the
> publick, as may be a sure guide to him, especially if he con-
> tinue constantly sincere, just and pious. For it is not at all
> improbable but such as act in very publick affairs, in which
> Providence has more than a special hand, that these Agents
> driving on her design may have a very special assistance and
> animation from her . . . but this is Enthusiasm in the better
> sense and therefore not so proper for our Discourse, who
> speak not of that which is true, but of that which is mis-
> taken.[106]

Nowhere did More make it clearer that revelation played a central part
in his epistemology. A *proper* understanding of God's purposes relied
not on unaided, but on inspired, reason. Against the atheists More
defended revelation's legitimacy and quarreled with the fanatics about
the message it conveyed. Fanatics were condemned because their messages were a denial of right reason, while atheists could not attain right
reason in the absence of revelation. The true Christian required both
human reason and genuine divine messages to inform belief and conduct.

Such a synthesis was not exclusive to More. His fellow Platonist
Nathanael Culverwel (writing at the same time) shared his definitions:

> [God's] commands are all rational . . . his Law is the quick-
> ening and wakening of mens reason; his Gospel 'tis the flow-
> ing out of his own reason . . . Spiritual irradiations stamp
> new light, create new reason in the Soul . . . God himself is
> the Eternal spring and head of reason. And that humane
> wisdome is but a created and imperfect copy of his most
> perfect . . . wisdome.[107]

Infusions of the spirit are the means by which human reason may be
transformed from mere natural reason to "new" reason: right reason.

To establish the conventionality of the uses of "right reason" it is
necessary to define the spectrum of its uses and its users. At one end
of the continuum were the usages of the radical sectarians, attacked
by others for their irrationalism and fanatical enthusiasm. It has already
been suggested that More's model of radicalism (from which he pas-
sionately dissociated himself) had as its basis the philosophy of the
Ranters, who "say there is not sinne but that it is onely a conceit."[108]
While choosing their immoralism for the brunt of his attack, he also
had in his sights anyone who claimed to be liberated from the moral
law – the letter of the Old Testament. It is therefore important to stress

that men who shared the antinomianism of the Ranters without subscribing to the antisocial implications of their theology held, with More and the other writers I have discussed here, the one view of right reason.

William Walwyn is usually described by historians as the most rationalistic of the Leveller leaders.[109] A self-styled "antinomian,"[110] his arguments for religious toleration took the form of repeated pleas for "consideration," by which he meant a rational and impartial examination of the issues. "Consideration" was the application of "that uncorrupt rule of reason . . . because it is the truth and nothing but the truth,"[111] and "Reason, experience or the word of God produce not wisdom without consideration . . . without which knowledge and understanding are not true knowledge and understanding."[112] But Walwyn's "consideration" was not what prevailed with him to establish the moral principles of that practical Christianity he preached all his life: "That there is a God: I did never beleeve through any convincing power I have never discerned . . . by any natural argument or reason . . . But it is an unexpressible Power, that in a forcible manner constraines my understanding to acknowledge and beleeve."[113] The foundation of human reason upon which Walwyn's "consideration" was built was unquestioned suprarational illumination.[114]

More dramatically, a similar case may be argued from the writings of Gerrard Winstanley, described by many historians as a pantheistic rationalist.[115] He shared with the Ranters the view that those in whom the spirit worked were absolved from sin, but, like the later Quakers, he drew back from the implication that saints did not obey the moral law.[116] At the start of this chapter we saw that Winstanley referred to God as "the great creator Reason . . . which did make and preserve all things."[117] But Winstanley's "reason" was not human reason. Rather it was "that spirituall power, that guids all mens reasoning in right order and to a right end."[118] Other names for this power were "King of Righteousnesse and Prince of peace." For him, right reason stemmed from a spiritual power which determined the end of human reasoning. To convey the rules that guide behavior, reason was not enough. His visionary digging was inspired by "vision, voice and revelation." Not once, but three times, the voice told Winstanley to "work together, Eate Bread together, Declare this all abroad."[119] While this message was clearly consonant with right reason, it was conveyed by revelation. The fact that More might have regarded this illumination as a false one is beside the point. Like it or not, he shared with intellectual radicals the firm conviction that Christian life and action resulted from reason supported by revelation.

I have argued that writers on the middle ground and radicals shared conventional usages of "right reason." At the other end of the continuum was the orthodox Anglican casuist Jeremy Taylor. I wish now to argue that he, too, believed that the exercise of right reason would lead to a single Christian truth and function as an infallible guide to conduct. He concurred with the others that "meer reason" or ratiocination could err, and certainly that it could produce no reliable conclusions about those truths of religion that were "above" reason. However, because the laws of nature and human reason originated from God, the study of the natural order was also improved by the use of right reason. There was no conflict between the Bible and God's book of nature, because God revealed himself as much in the second as in the first. The use of "right" reason rather than "meer" reason was therefore appropriate for the study of the laws of nature.

The example of Taylor has been used to present a very different argument. Hoopes and Hill both maintain that with Taylor a dramatic shift in attitude to the ubiquity and reliability of right reason occurred, one which allowed the separation of mechanical science from theology, begun by Bacon, to be completed after 1660. Both quote the same passage: "Reason is such a box of quicksilver that it abides no where; it dwells in no settled mansion . . . it looks to me otherwise then to you . . . [it is] as uncertain as the discourses of the people, or the dreams of disturbed fancies."[120] At the beginning of *Ductor Dubitantium* (1660) Taylor put the case of those – clearly the fideists – who argued against the reliability of reason in matters of religion that required, they thought, "new capacities and new illuminations."[121] These men had it that "what is right reason is so uncertain that in the midst of all disputes every man pretends to it, but who hath it, no man can tell."[122] However, Taylor then went on to refute this position. Correct natural reasoning is identical in matters of theology and natural philosophy: "Faith and reason do not divide Theology and Philosophy."[123] In Lockeian manner, matters of faith are to be tested by convincing arguments about the infallible source from which they stem. As in Locke, too, an important distinction followed. Not everything could be known by natural means, for "some things . . . descend upon us immediately from Heaven,"[124] and therefore "*our* right reason, humane reason" cannot fathom everything in the realm of the divine, for God sometimes chooses to act in secret ways that confound our understanding.[125] This makes it possible that conclusions apparently consistent with human reason may be contradicted by the Scriptures, while yet providing a measure of the validity of faith – a kind of veto power. Right reason, or humane reason "is not the affirmative or positive measure of things Divine . . . it is the negative measure."[126] This is Taylor's central contention. Something may be in accordance with natural reason; it may be a natural right "that a man may repel force by force"; but because it is reasonable it is not necessarily right reason. Reason may not verify – it may only falsify – the conclusions of faith.

Accordingly, Taylor went on to argue that the Christian moral law – which he called the law of nature written in our hearts by the finger of God and known by our conscience – cannot be denied by reason, for if a belief were to be falsified by reason it would be rejected by conscience. And while God cannot act *against* reason, he may, like Locke's God, act "above our understanding."[127] God's decrees, like all revealed truths, may not be verified by reason. This leaves Taylor free to propose that human reason is not incontrovertible. On this he was plain. "Every man's reason is not right and every man's reason is not to be trusted,"[128] particularly he had in mind here the reasoning of the Catholics. When he went on to describe reason as "a box of quicksilver," he was describing not right reason but faulty, limited, natural reason. Reason could not aspire to knowledge of spiritual things. Such knowledge requires revelation as an aid to reason.[129]

Thus Taylor's disquisition does not differ essentially from the arguments of the other writers. Taylor was not implicitly removing the moral implications of reason by undermining the reliability of its conclusions about the created universe, as Hoopes and Hill claim. On the contrary, right reason was the language used by God to engrave the moral law of nature in the hearts of men. While it is true that God was free to act outside the constraints of reason, he did not act in contradiction to it. Hoopes argues that as Taylor's God could make and unmake the laws of nature it follows that those laws are bound by no moral imperatives. It is this argument, together with Taylor's emphasis on the unreliability of human reason, that allows Hoopes to present Taylor as a crucial innovator.[130] According to Hoopes, the later seventeenth century saw a shift toward the total separation of ethics and natural philosophy, a shift required for the acceptance of the mechanical world view.

However, Hoopes's analysis may be questioned on three grounds. First, the idea that God is unbound by the laws of nature was no novelty. That argument had traditional roots going back at least to Ockham (if not to Job). Second, we have already seen that Taylor did not deny the validity of conclusions reached by "right" reason, but only those of "unaided" reason. Third, Taylor did not liberate God's laws from ethical constraints. In order to fathom those laws human reason required the supplement of revelation. While reason and conscience provide a negative test or veto power to reject irrational beliefs, it is revelation that provides the positive confirmation of our knowledge of the moral law written in our hearts. We require "right" reason to appreciate the moral law. "But as reason helps our eyes so does revelation inform our reason; and we have no law until by revelation . . . God hath declared . . . a law."[131]

Both Hill and Hoopes use Hobbes in order to make their case for the final rejection of right reason as the means by which the laws of nature are to be understood.[132] Hobbes is seen as the prototype of the mechanical philosopher for whom the universe ran according to laws that are independent of the moral order. Hill quotes Hobbes to make the point that right reason had lost the objective validity it had for Hooker and the proponents of the "old" rationality: "Commonly they that call for right reason to decide any controversy do mean their own."[133] But while it is, of course, true that for Hobbes men arrived at their knowledge of the laws of nature by their reasoning faculties, deducing others from the first imperative of self-preservation, their own estimation of self-interest did not make their reason "right." Only Christians could hope to acknowledge theirs as "right" reason, and only men in a Christian society could hear God's confirmation of that right reason by revelation acting through the agency of the Christian ruler.[134]

I have argued that there were no sharp discontinuities in mid-seventeenth-century conceptions of the operation of reason and revelation. Rather, proponents of widely differing traditions, arguing their cases against each other, competed to establish a monopoly for their own perceptions of the conclusions to be reached by the operation of their right reason. "Right reason! Aye, where is it?"[135] was a cry which expressed the bitterness resulting, not from anguish about the evanescence of the concept, but from the competing claims for private insights into its contents.

The turmoil of the 1640s and 1650s certainly produced anxiety and questioning about previously indubitable moral imperatives. We have seen the concern expressed by some participants in contemporary debates about the dangers from those labeled enthusiasts, atheists, and Catholics. There can be no doubt that each of these writers was critically concerned with the rival arguments put by the proponents of "unreason" to undermine the concept of "right reason." Winstanley the social and religious radical, Vaughan the hermetic, and More the academic Platonist were all haunted by the specter of Ranterish epistemology, which allowed complete liberation from the moral constraints of the Scriptures understood through right reason. They were right to wish to protect themselves from implicit association with men who could write, as Lawrence Clarkson, the Ranter-turned-Muggletonian, did: "You go forth in the strength of Reason's lying imagination, which you call the light within you."[136] But the extremity of such a view only served the purpose of allowing them to stake out their own claim for moderation, and to make their own usages acceptable by invoking the language of "right reason." The flimsiness of the Ranter paper tiger made their own cases appear all the more solid and acceptable. The case for right reason in the mid-seventeenth century remained in the main unshaken.

This chapter has argued that the period from 1640 to 1660 saw no dramatic shift in perception of the role of right reason. Rivals in the ideological battles, presented by historians as holding irreconcilable views of the world, have been shown to share basically similar epistemologies. The shared language of right reason allowed enough flexibility to enable men subscribing to apparently different modes of explanation to communicate with one another. No Kuhnian *Gestalt* switch, it seems, was required. It also allowed others to move with relative ease between these modes. The mechanical universe, it turns out, was not a totally different world from that of its rivals. If there is a problem about irreconcilable world views among seventeenth-century writers, the problem appears to be ours, not theirs.

Notes

1 Gerrard Winstanley, *The True Levellers Standard Advanced* (1649), in *The Works of Gerrard Winstanley*, ed. G. H. Sabine (New York, 1965), pp. 251, 261.

2 J. G. A. Pocock, "The Reconstruction of Discourse: Towards the Historiography of Political Thought," *Modern Language Notes*, 96 (1981), p. 972.

3 P. M. Rattansi, "Paracelsus and the Puritan Revolution," *Ambix*, 5 (1963), pp. 24–32. His argument is elaborated in "Some Evaluations of Reason in Sixteenth- and Seventeenth-Century Natural Philosophy," in *Changing Perspectives in the History of Science*, ed. M. Teich and R. Young (London, 1973), pp. 148–66.

4 Charles Webster, *The Great Instauration: Science, Medicine and Reform, 1626–1660* (London, 1975).

5 Christopher Hill, "'Reason' and 'Reasonableness,'" in Hill, *Change and Continuity in 17th Century England* (London, 1974), pp. 103–26. First published in *British Journal of Sociology, 20* (1969), pp. 235–52.

6 Robert Hoopes, *Right Reason in the English Renaissance* (Cambridge, Mass., 1962).

7 Ibid., p. 167.

8 Rattansi, "Paracelsus," pp. 16, 32.

9 Hill, p. 100.

10 Rattansi, "Paracelsus," p. 26.

11 John Webster, *Academiarum examen* (1653), in *Science and Education in the Seventeenth Century*, ed. Allen G. Debus (London, 1970), pp. 116–17; Walter Charleton, *A Ternary of Paradoxes* (London, 1649), prolegomena, sig. f2.

12 Webster, p. 180.

13 Ibid., pp. 108, 111, 183.

14 Ibid., p. 157.

15 Ibid., pp. 169–70.
16 Rattansi, "Paracelsus," pp. 26, 30. In his later "Some Evaluations of Reason" he argues against "isolating rational and irrational components" of thought and for "regarding it as a unity" (p. 150); Nina Rattner Gelbart, "The Intellectual Development of Walter Charleton," *Ambix*, 18 (1971), pp. 149–68; Lindsay Sharp, "Walter Charleton's Early Life," *Annals of Science*, 30 (1973), pp. 311–40.
17 Charleton, *Ternary*, prolegomena, sig. f2.
18 Ibid., sig. f3.
19 Ibid., sig. f2.
20 Ibid., sig. gl.
21 Ibid., sig. c2.
22 "Afterwards . . . my minde endeavoured to depart, not indeed against, but from the use of Reason: . . but that thing I could not presently obtain, because Reason did continually accompany my Soul against its will . . . Reason did vex the soul with a multiplicity, with a vain complacency of Sciences . . . My minde therefore hath often banished Reason but it hath always privily entered afresh . . . Reason doth demonstrate nothing but a . . . dark knowledge, or a thinking . . . The knowledges of the truth . . . do proceed not from Reason: but from a far different beginning, to wit, the intellectual Light of the Lamp or Candle . . . I stripped myself of all curiousity and appetite of knowing. I betook myself unto rest or poverty of spirit, resigning myself unto the most lovely will of God . . . After two moneths . . . it once again happened to me, that I intellectually understood . . . [This] doth not happen without grace" (John Baptista van Helmont, *Van Helmont's Works*, trans. H. Blunden [1664], pp. 17–18, 22–3).
23 Charleton, *Ternary*, dedication, sig. b2.
24 Ibid., prolegomena, sig. d3.
25 Ibid., sig. e1.
26 Gelbart, p. 150.
27 Walter Charleton, *The Darkness of Atheism Dispelled by the Light of Nature* (London, 1652), p. 157; *The Immortality of the Human Soul Demonstrated by the Light of Nature* (London, 1657), preface.
28 Charleton, *Immortality*, preface.
29 Walter Charleton, *Physiologia Epicuro-Gassendo-Charletoniana* (London, 1654), p. 382.
30 Ibid., pp. 383ff.
31 Sir Kenelm Digby, *A Late Discourse Made in a Solemne Assembly of Nobles and Learned Men at Montpellier in France* (London, 1658). The license for printing was granted in 1657.
32 Sharp, pp. 322–3.
33 Charleton, *Ternary*, prolegomena, sig. d2; *Late Discourse*, pp. 6–12.
34 Charleton, *Immortality*, passim.
35 Ibid., pp. 52, 56–7.
36 Walter Charleton, *The Harmony of Natural and Divine Laws* (London, 1682).
37 Charleton, *Darkness of Atheism*, dedication, sig. a.
38 Samuel Gott, *Nova Solyma the Ideal City: or Jerusalem Regained* (1648).
39 Ibid., I, 169.
40 Ibid., I, 171.
41 Ibid., I, 178ff.
42 Ibid., I, 190.

43 Ibid., I, 196.
44 Ibid., II, 11.
45 Ibid., II, 11.
46 Ibid., I, 222.
47 Ibid., I, 194.
48 Frederic B. Burnham, "The More–Vaughan Controversy: The Revolt Against Philosophical Enthusiasm," *Journal of the History of Ideas*, 35 (1974), p. 33.
49 Ibid., p. 37.
50 K. Theodore Hoppen, "The Nature of the Early Royal Society," *British Journal for the History of Science*, 9 (1976), pp. 1–24, 243–73.
51 Noel L. Brann, "The Conflict between Reason and Magic in Seventeenth-Century England: A Case Study of the More–Vaughan Debate," *Huntington Library Quarterly*, 43 (1980), pp. 103–26.
52 Ibid., p. 103.
53 Thomas Vaughan, *Anima magica abscondita*, in *The Works of Thomas Vaughan*, ed. A. E. Waite (New York, 1968), p. 85.
54 Ibid., p. 86.
55 Thomas Vaughan, *Euphrates*, in *Works*, p. 393.
56 Vaughan, *Anima*, p. 108.
57 Ibid., pp. 110–11.
58 Vaughan, *Euphrates*, p. 397.
59 Vaughan, *Anima*, p. 113.
60 Thomas Vaughan, *Anthroposophia theomagica*, in *Works*, p. 6.
61 Thomas Vaughan, *Magia adamica*, in *Works*, p. 175.
62 Ibid., p. 158.
63 Henry More (Alazonomastix Philalethes), *Observations upon Anthroposophia Theomagica and Anima Magica Abscondita* (London, 1650).
64 Henry More, *A Second Lash of Alazono-Mastix* (Cambridge, 1651).
65 Ibid., p. 11.
66 More, *Observations*, preface.
67 Ibid., pp. 7, 14, 88.
68 More, *Second Lash*, p. 143.
69 Henry More, *Conjectura cabbalistica*, (London, 1653), pp. 24, 147.
70 Henry More, *An Antidote against Atheisme* (London, 1653), p. 65.
71 Ibid.
72 More, *Observations*, pp. 24, 64.
73 Ibid., pp. 67–8.
74 Ibid.
75 More, *Second Lash*, p. 35.
76 Ibid., p. 44.
77 Ibid., p. 47.
78 Ibid., p. 50.
79 Ibid., p. 200.
80 Thomas Vaughan, *The Second Wash: or the Moore Scour'd Once More* (London, 1651), p. 9.
81 Ibid.
82 More, *Observations*, p. 65.
83 More, *Conjectura*, epistle to the reader.
84 Ibid.
85 More, *Observations*, pp. 14, 26, 53.

86 More, *Conjectura*, p. 53.
87 Ibid., p. 36.
88 More, *Observations*, p. 14.
89 Gerrard Winstanley, *Fire in the Bush* (1650), in *Works*, p. 460.
90 More, *Conjectura*, p. 36.
91 More, *Observations*, preface.
92 More, *Conjectura*, epistle dedicatory.
93 Ibid., p. 183.
94 Ibid., p. 88.
95 Ibid., epistle dedicatory.
96 Ibid., p. 183.
97 Ibid., p. 231.
98 Ibid., pp. 244–5.
99 More, *Antidote*, preface, sig. a1.
100 Ibid., sig. b4.
101 Ibid., sig. a3.
102 Ibid., sig. a4.
103 Henry More, *Enthusiasmus triumphatus* (1656), in *A Collection of Several Philosophical Writings of Dr. Henry More* (London, 1662), p. 17.
104 Ibid., p. 12.
105 Ibid., p. 19.
106 Ibid., p. 20.
107 Nathanael Culverwel, *An Elegant and Learned Discourse of the Light of Nature* (London, 1652), p. 120.
108 More, *Second Lash*, p. 46.
109 For an account of the historical evaluations of Walwyn, see Lotte Mulligan, "The Religious Roots of William Walwyn's Radicalism," *Journal of Religious History*, 12 (1982), pp. 162–179.
110 William Walwyn, *A Whisper in the Eare of Mr. Thomas Edwards* (London, 1646), p. 10.
111 William Walwyn, *Some Considerations* (London, 1642), passim.
112 William Walwyn, *A Word in Season* (London, 1646), pp. 1–2.
113 William Walwyn, *A Still and Soft Voice* (London, 1647), p. 12.
114 For the argument concerning the illuminist foundation of Walwyn's philosophy, see Mulligan, passim.
115 The countercase is argued in Lotte Mulligan, John K. Graham, and Judith Richards, "Winstanley: A Case for the Man as He Said He Was," *Journal of Ecclesiastical History*, 27 (1977), pp. 57–75.
116 See Gerrard Winstanley, *Vindication*, in *Works*, pp. 399–403.
117 Winstanley, *True Levellers*, p. 251.
118 Gerrard Winstanley, *Truth Lifting Up Its Head* (1648), in *Works*, p. 105.
119 Winstanley, *True Levellers*, p. 261.
120 Jeremy Taylor, *Ductor Dubitantium or the Rule of Conscience* (London, 1660), I, 231; quoted by Hill, p. 119, and Hoopes, p. 166.
121 Taylor, I, 43.
122 Ibid., I, 43.
123 Ibid., I, 46.
124 Ibid., I, 46.
125 Ibid., I, 47–8.
126 Ibid., I, 48.
127 Ibid., I, 54.
128 Ibid., I, 56.

129 Ibid., I, 58.
130 Hoopes, pp. 165–70.
131 Taylor, I, 468.
132 Hill, p. 119; Hoopes, pp. 171–4.
133 Hill, p. 199. See Thomas Hobbes, *Leviathan* (London, 1651), bk. I, chap. 5; bk. II, chap. 17.
134 *Hobbes*, bk. II, chap. 31.
135 Quoted by Hill, p. 118.
136 Lawrence Clarkson, *The Lost Sheep Found* (London, 1660), p. 38.

INDEX

403